Student Study Guide

to accompany

Anatomy & Physiology

Eighth Edition

Philip Tate

James Kennedy

Rod R. Seeley

 Higher Education

Boston   Burr Ridge, IL   Dubuque, IA   New York   San Francisco   St. Louis
Bangkok   Bogotá   Caracas   Kuala Lumpur   Lisbon   London   Madrid   Mexico City
Milan   Montreal   New Delhi   Santiago   Seoul   Singapore   Sydney   Taipei   Toronto

The McGraw·Hill Companies

Student Study Guide to accompany
ANATOMY AND PHYSIOLOGY, EIGHTH EDITION
PHILIP TATE, JAMES KENNEDY, AND ROD R. SEELEY

Published by McGraw-Hill Higher Education, an imprint of The McGraw-Hill Companies, Inc., 1221 Avenue of the Americas, New York, NY 10020. Copyright © 2008 by The McGraw-Hill Companies, Inc. All rights reserved.

No part of this publication may be reproduced or distributed in any form or by any means, or stored in a database or retrieval system, without the prior written consent of The McGraw-Hill Companies, Inc., including, but not limited to, network or other electronic storage or transmission, or broadcast for distance learning.

3 4 5 6 7 8 9 0 QPD/QPD 0 9 8

ISBN: 978-0-07-326836-1
MHID: 0-07-326836-4

www.mhhe.com

Contents

	Preface	iv
1	The Human Organism	1
2	The Chemical Basis of Life	11
3	Structure and Function of the Cell	26
4	Histology: The Study of Tissues	49
5	Integumentary System	65
6	Skeletal System: Bones and Bone Tissue	76
7	Skeletal System: Gross Anatomy	88
8	Articulations and Movement	119
9	Muscular System: Histology and Physiology	130
10	Muscular System: Gross Anatomy	150
11	Functional Organization of Nervous Tissue	166
12	Spinal Cord and Spinal Nerves	182
13	Brain and Cranial Nerves	193
14	Integration of Nervous System Functions	209
15	The Special Senses	222
16	Autonomic Nervous System	243
17	Functional Organization of the Endocrine System	253
18	Endocrine Glands	260
19	Cardiovascular System: Blood	278
20	Cardiovascular System: The Heart	290
21	Cardiovascular System: Peripheral Circulation and Regulation	306
22	Lymphatic System and Immunity	327
23	Respiratory System	345
24	Digestive System	368
25	Nutrition, Metabolism, and Temperature Regulation	397
26	Urinary System	410
27	Water, Electrolytes, and Acid–Base Balance	425
28	Reproductive System	433
29	Development, Growth, and Aging	457

PREFACE

To the Student

This study guide is designed to accompany *Anatomy and Physiology* by Seeley, Stephens, and Tate. Each chapter in the study guide, and the order of topics within the chapter, corresponds to a chapter in the text. This makes it possible for you to study systematically and makes it easier for you to find or to review information. Read the corresponding material in the text before you use this study guide. It is designed to help you understand and master the subject of anatomy and physiology.

Study Guide Features

Content Learning Activity

This section of the study guide contains a variety of exercises, including matching, completion, ordering, and labeling activities. The content learning activity is not a test; it is a strategy to help you learn. Do not guess! If you learn something incorrectly, it is difficult to relearn it correctly. Use the textbook or your lecture notes for help whenever you are not sure of an answer. The emphasis here is on learning the content. The content questions cover the material in the same sequence as it is presented in the text. Learning the material in this order makes it easier to relate pieces of information to each other and makes it easier to remember the information.

After completing the exercises, check your answers against the answer key. If you answered a question incorrectly, check the text to make sure you now understand the correct answer. Before going on to another section of the study guide, review the section you have just completed to be sure you understand and remember its content. Cover the answers you have written with a piece of paper and mentally answer each question once more as you review. Do not be satisfied until you get at least 90% of the questions correct. Then concentrate on mastering the remaining questions.

Quick Recall

The Quick Recall section asks you to list, name, or briefly describe some aspect of the chapter's content. Although this section can be completed rapidly, do not confine yourself to quickly writing down the answers. As you complete each Quick Recall question, use it to trigger more information in your mind. For example, if a Quick Recall question asks you to name the two major regions of the body, do that, then think of their definition and what their various subparts are, visualize them, and so on. This section should be enjoyable and satisfying because it will demonstrate that you have learned the basic information about the material. Verify your answers using the answer key.

Additional Questions in the Textbook

The questions in the study guide are designed to complement and prepare you for questions found in the textbook. After you have mastered the Content Learning and Quick Recall activities, you are ready to take a practice test. In the Review and Comprehension section at the end of each chapter in the textbook, there is a collection of multiple choice questions that probably are similar to the questions on the exams you will take for a grade. Use these questions as a practice test, but do not try to guess the answers to them because they can be used as a learning tool. If you do not know the answer for sure, admit it to yourself and then find the correct answer using the text. Some of the questions require recall of information. Others may state the information somewhat differently than the way it appeared in either the text or study guide. This is entirely fair because in real life you must be able to recognize the information no matter how it is reworded, and you should even be able to express the information in your own words. Another goal of these questions is to make you think about the relationship between different bits of information or concepts, so some of the questions are more complex than those requiring only recall. Finally, some of these questions ask you to use what you have learned to solve new problems.

After you have answered the multiple choice questions, check the answer key in the text. Make sure you understand why each answer is correct. Check the textbook, ask another student, or talk with your instructor, but make sure you know. The multiple choice questions will show you the areas that you need to concentrate on further. Use them to improve your understanding of anatomy and physiology.

The text also has Predict questions throughout each chapter and Critical Thinking questions at the end of each chapter. These questions provide practice at solving anatomy and physiology problems. They challenge you to apply information to new situations, analyze data and come to conclusions, synthesize solutions, and evaluate problems. Write down your answers to these questions on a separate piece of paper. Writing is a good way to organize your thoughts and most of us can benefit from practice in writing. A good way to determine if you have communicated your thoughts effectively is to have another student read your answers and see if they make sense.

Although the Predict and Critical Thinking questions contain useful information, they are not designed primarily to help you learn specific information. Rather, they emphasize the thought processes necessary to solve problems. If all you do is read the questions and quickly look up the answers, you have defeated the purpose of this type of question. Think about these problems and develop your reasoning skills. Long after you have forgotten a bit of information, these skills will be useful, not only for anatomy and physiology related problems but for many other aspects of your life as well.

We hope that you not only see the benefit of possessing problem-solving skills but will come to appreciated that solving problems is fun!

A Final Thought

Good luck with all the aspects of the anatomy and physiology course you are about to begin. We hope that the study guide makes things a little easier and a little clearer for you. When you have completed the course, we are confident that you will be proud of what you have accomplished. Just remember to enjoy the learning process as you go along.

Philip Tate
James Kennedy
Rodney Seeley

Acknowledgments

The development and production of this study guide involved much more than the work of the authors and we gratefully acknowledge the assistance of the many other individuals involved. We wish to thank our families for their support. Erica Michaels contributed to the development of the design for the study guide. We also wish to acknowledge Melissa Leick at McGraw-Hill and Robin Reed at Carlisle Publishing Services for their assistance. Our thanks to everyone involved with the artwork taken from the textbook, and to D. Michael Dick for his original study guide illustrations. We also wish to recognize the contribution of Sue Pepe to the layout of the study guide. Thanks also to Mickie Bond for her contribution of poetry.

ANATOMY FREAK

I dig. I poke. I pry.
I agonize. I dentify.
All the body parts
That I have grown.

I know each muscle action,
From insertion to contraction,
The brain, the bowel,
The kidney and the bone.

All these parts within me
Used to seem so right.
Some worked on the day shift,
Some mostly at night.

Now my body parts are strangers,
With names I've never heard.
Terms with exotic syllables.
I can't spell a word!

What can stop this torment?
What can ease my plight?
What can make me dream of something
Besides A & P at night?

My organs all implore me
As they grumble and they creak,
NAME ME! NAME ME! NAME ME!

I am an anatomy freak.

      Minyon "Mickie" Bond

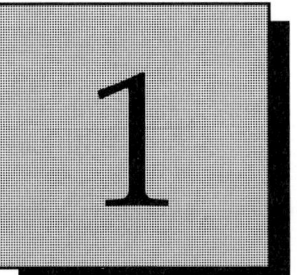

The Human Organism

CONTENT LEARNING ACTIVITY

Anatomy and Physiology

A. Match these terms with the correct statement or definition:

    Anatomical imaging   Gross anatomy
    Anatomy   Histology
    Cytology   Regional anatomy
    Developmental anatomy   Surface anatomy
    Embryology   Systemic anatomy

_____ 1. General term for the scientific discipline that investigates the body's structure.

_____ 2. Study of the structural changes that occur between conception and adulthood.

_____ 3. Subspecialty of developmental anatomy that considers changes from conception to the end of the eighth week of development.

_____ 4. Study of the structural features of cells.

_____ 5. Study of tissues.

_____ 6. Study of structures that can be examined without the aid of a microscope.

_____ 7. Study of the body system by system (a group of structures that have one or more common functions).

_____ 8. Study of the body's organization by areas.

_____ 9. Use of external landmarks, such as bony projections, to locate deeper structures.

_____ 10. Use of x-rays, ultrasound, nuclear magnetic resonance, and other technologies to create pictures of internal structures.

B. Match these terms with the correct statement or definition:

Pathology Systemic physiology
Physiology

_____ 1. Scientific investigation of the processes or functions of living things.

_____ 2. Subdiscipline of physiology that considers the functions of organ systems.

_____ 3. Medical science dealing with all aspects of disease, with an emphasis on the cause and development of abnormal conditions as well as the structural and functional changes resulting from disease.

Structural and Functional Organization

A. Match these terms with the correct statement or definition:

Cell   Organism
Chemical   Organ system
Organ   Tissue
Organelle

_____ 1. Basic unit of all living things.

_____ 2. Structure within a cell that performs one or more specific functions.

_____ 3. Group of similar cells and the materials surrounding them.

_____ 4. Two or more tissue types that perform one or more common functions.

_____ 5. Group of organs classified as a unit because of a common set of functions.

B. Match these terms with the correct statement or definition:

Cardiovascular   Nervous
Digestive   Reproductive
Endocrine   Respiratory
Integumentary   Skeletal
Lymphatic   Urinary
Muscular

_____ 1. Organ system that consists of skin, hair, nails, and sweat glands; protects, regulates temperature, and prevents water loss.

_____ 2. Organ system that consists of bones and cartilage; protects and supports the body, produces blood cells, and stores minerals.

_____ 3. Organ system that consists of muscles attached to the skeleton; allows body movement, maintains posture, and produces body heat.

_____ 4. Organ system that consists of the brain, spinal cord, nerves, and receptors; detects sensation and controls movements.

_____ 5. Organ system that consists of glands that secrete hormones; a major regulatory system.

_____ 6. Organ system that consists of the heart, blood vessels, and blood; transports nutrients, wastes, gases, and hormones.

_____ 7. Organ system that consists of vessels, nodes, and organs; combats disease, maintains tissue fluid balance, and absorbs fats from the digestive tract.

_____ 8. Organ system that consists of the lungs; exchanges gases between the blood and the air and regulates blood pH.

_____ 9. Organ system that consists of the mouth, pharynx, esophagus, stomach, and intestines; breaks down and absorbs nutrients.

_____ 10. Organ system that consists of the kidneys and urinary bladder; removes waste products from the circulatory system and regulates blood pH.

Characteristics of Life

Match these terms with the correct statement or definition:

Development   Morphogenesis
Differentiation   Organization
Growth   Reproduction
Metabolism   Responsiveness

_____ 1. Condition in which the parts of an organism have specific relationships to each other and the parts interact to perform a specific function.

_____ 2. All the chemical reactions taking place in an organism.

_____ 3. Ability to sense changes in the external or internal environment and adjust to those changes.

_____ 4. Ability of cells to increase in size or number.

_____ 5. Changes an organism undergoes through time.

_____ 6. Changes in cell structure and function from generalized to specialized.

_____ 7. Changes in the shape of tissues, organs, and the entire organism.

Homeostasis

A. Match these terms with the correct statement or definition:

Control center	Response
Effector	Set point
Normal range	Stimulus
Receptor	Variable

_____ 1. Condition, such as temperature, that can change.

_____ 2. Ideal, normal value maintained by homeostasis.

_____ 3. Slight increases and decreases of a variable around the set point.

_____ 4. Monitors the value of a variable.

_____ 5. Establishes the set point.

_____ 6. Changes the value of a variable.

_____ 7. Deviation from a set point.

_____ 8. Returns a variable toward the set point; produced by an effector.

B. Match these terms with the correct statement or definition:

Negative feedback
Positive feedback

_____ 1. Maintains homeostasis by reducing or resisting any deviation from an ideal normal value.

_____ 2. Medical therapy is often designed to help this type of feedback.

_____ 3. When a deviation from a normal value occurs, the response is to make the deviation even greater.

_____ 4. Heart rate increases in response to a decrease in blood pressure.

_____ 5. Maintains an elevated blood pressure during exercise.

_____ 6. Causes a decrease in blood pressure as a result of losing blood.

_____ 7. Increases the strength of uterine contractions during delivery.

Body Positions

Match these terms with the correct statement or definition:

Anatomic position   Supine
Prone

_____ 1. Person standing erect with the face directed forward, the upper limbs hanging to the sides, and the palms of the hands facing forward.

_____ 2. Person lying face upward.

Directional Terms

Match these terms with the correct statement or definition:

Anterior
Caudal
Cephalic
Deep
Distal
Dorsal
Inferior
Lateral
Medial
Posterior
Proximal
Superficial
Superior
Ventral

_____ 1. Lower than or toward the tail (two terms).

_____ 2. Higher than or toward the head (two terms).

_____ 3. Toward the front or toward the belly (two terms).

_____ 4. Toward the back (of the body) (two terms).

_____ 5. Farther than another structure from the point of attachment to the trunk.

_____ 6. Away from the midline.

_____ 7. Away from the surface.

Body Parts and Regions

Match these terms with the correct statement or definition:

Arm
Leg
Lower limb
Quadrants
Regions
Thigh
Trunk
Upper limb

_____ 1. Consists of the arm, forearm, wrist, and hand.

_____ 2. Extends from the shoulder to the elbow.

_____ 3. Consists of the thigh, leg, ankle, and foot.

_____ 4. Extends from the hip to the knee.

_____ 5. Extends from the knee to the ankle.

_____ 6. Consists of the thorax, abdomen, and pelvis.

_____ 7. Areas of the abdomen formed by two imaginary lines—one horizontal and one vertical—that intersect at the navel.

_____ 8. Areas of the abdomen formed by four imaginary lines—two horizontal and two vertical.

Planes

A. Match these terms with the correct statement or definition:

Cross (transverse) section
Frontal (coronal) plane
Longitudinal section
Midsagittal plane
Oblique section
Sagittal plane
Transverse plane

_____ 1. Runs vertically through the body and divides the body into right and left portions.

_____ 2. Sagittal plane that divides the body into equal right and left halves.

_____ 3. Runs vertically through the body and divides the body into anterior and posterior parts.

_____ 4. Divides the body into superior and inferior portions and runs parallel to the surface of the ground.

_____ 5. Cut through the long axis of an organ.

_____ 6. Right-angle cut across the long axis of an organ.

B. Match these terms with the correct planes labeled in figure 1.1:

Frontal (coronal) plane
Sagittal (median) plane
Transverse plane

1. _____

2. _____

3. _____

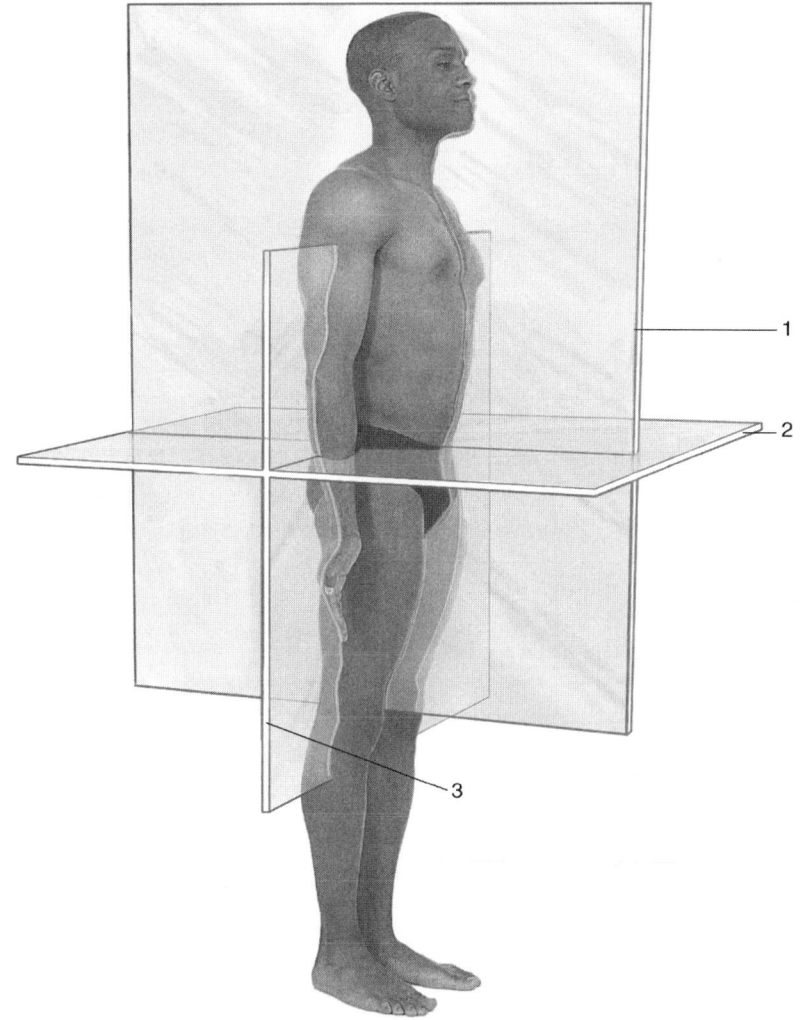

Figure 1.1

C. Match these terms with the correct section in figure 1.2:

Longitudinal section
Oblique section
Cross (transverse) section

1. _____
2. _____
3. _____

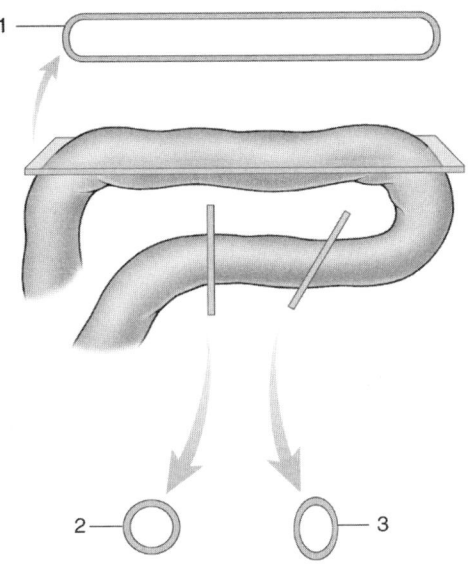

Figure 1.2

Body Cavities

Match these terms with the correct statement or definition:

Abdominal cavity
Abdominopelvic cavity
Mediastinum
Pelvic cavity
Thoracic cavity

_____ 1. Cavity surrounded by the rib cage, bounded inferiorly by the diaphragm.

_____ 2. Medial partition dividing the thoracic cavity; consists of the heart, thymus, trachea, esophagus, and blood vessels.

_____ 3. Cavity bounded primarily by the abdominal muscles.

_____ 4. Small space enclosed by the bones of the pelvis.

_____ 5. Combined abdominal and pelvic cavities.

_____ 6. Cavity containing the lungs and mediastinum.

_____ 7. Cavity containing the stomach, intestines, liver, spleen, pancreas, and kidneys.

_____ 8. Cavity containing the urinary bladder, part of the large intestine, and the internal reproductive organs.

Serous Membranes

A. Match these terms with the correct statement or definition:

    Mesentery   Pleural cavity
    Parietal   Retroperitoneal
    Pericardial cavity   Visceral
    Peritoneal cavity

_____ 1. Portion of a serous membrane in contact with an organ.

_____ 2. Serous membrane–lined cavity that surrounds the heart.

_____ 3. Serous membrane–lined cavity that surrounds the lungs.

_____ 4. Serous membrane–lined cavity within the abdominopelvic cavity.

_____ 5. Two fused layers of serous membrane that anchors some abdominal organs to the body wall.

_____ 6. Location of organs covered only by parietal peritoneum.

B. Match these terms with the correct parts labeled in figure 1.3:

Abdominal cavity
Abdominopelvic cavity
Diaphragm
Mediastinum
Pelvic cavity
Pericardial cavity
Pleural cavity
Thoracic cavity

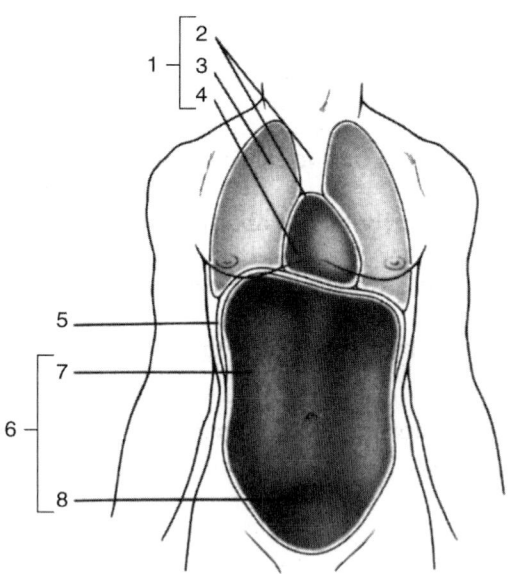

Figure 1.3

1. _____
2. _____
3. _____
4. _____
5. _____
6. _____
7. _____
8. _____

QUICK RECALL

1. From smallest to largest, list the six structural levels of the human body.

2. List the four primary tissue types.

3. Name the three components of many negative-feedback mechanisms.

4. Describe the anatomical position.

5. List the three major planes used to cut the human body and the three sections used to cut an organ.

6. List the three trunk cavities of the human body.

7. Name the three serous membrane–lined cavities within the trunk.

8. List four retroperitoneal organs.

ANSWERS TO CHAPTER 1

CONTENT LEARNING ACTIVITY

Anatomy and Physiology
 A. 1. Anatomy; 2. Developmental anatomy;
   3. Embryology; 4. Cytology; 5. Histology;
   6. Gross anatomy; 7. Systemic anatomy;
   8. Regional anatomy; 9. Surface anatomy;
   10. Anatomical imaging
 B. 1. Physiology; 2. Systemic physiology;
   3. Pathology

Structural and Functional Organization
 A. 1. Cell; 2. Organelle; 3. Tissue; 4. Organ;
   5. Organ System
 B. 1. Integumentary; 2. Skeletal; 3. Muscular;
   4. Nervous; 5. Endocrine; 6. Cardiovascular;
   7. Lymphatic; 8. Respiratory; 9. Digestive;
   10. Urinary

Characteristics of Life
 1. Organization; 2. Metabolism;
 3. Responsiveness; 4. Growth;
 5. Development; 6. Differentiation;
 7. Morphogenesis

Homeostasis
 A. 1. Variable; 2. Set point; 3. Normal range;
   4. Receptor; 5. Control center; 6. Effector;
   7. Stimulus; 8. Response
 B. 1. Negative feedback; 2. Negative feedback;
   3. Positive feedback; 4. Negative feedback;
   5. Negative feedback; 6. Positive feedback;
   7. Positive feedback

Body Positions
 1. Anatomical position; 2. Supine

Directional Terms
 1. Inferior/caudal; 2. Superior/cephalic;
 3. Anterior/ventral; 4. Posterior/dorsal;
 5. Distal; 6. Lateral; 7. Deep

Body Parts and Regions
 1. Upper limb; 2. Arm; 3. Lower limb;
 4. Thigh; 5. Leg; 6. Trunk; 7. Quadrants;
 8. Regions

Planes
 A. 1. Sagittal plane; 2. Midsagittal plane;
   3. Frontal (coronal) plane; 4. Transverse
   plane; 5. Longitudinal section;
   6. Cross (transverse) section
 B. 1. Sagittal (median) plane; 2. Transverse
   plane; 3. Frontal (coronal) plane
 C. 1. Longitudinal section; 2. Cross (transverse)
   section; 3. Oblique section

Body Cavities
 1. Thoracic cavity; 2. Mediastinum; 3. Abdominal
 cavity; 4. Pelvic cavity; 5. Abdominopelvic cavity;
 6. Thoracic cavity; 7. Abdominal cavity; 8. Pelvic
 cavity

Serous Membranes
 A. 1. Visceral; 2. Pericardial cavity; 3. Pleural
   cavity; 4. Peritoneal cavity; 5. Mesentery;
   6. Retroperitoneal
 B. 1. Thoracic cavity; 2. Mediastinum; 3. Pleural
   cavity; 4. Pericardial cavity; 5. Diaphragm;
   6. Abdominopelvic cavity; 7. Abdominal
   cavity; 8. Pelvic cavity

QUICK RECALL

1. Chemical, cell, tissue, organ, organ system, organism
2. Epithelial, connective, muscle, and nervous tissues
3. Receptor, control center, effector
4. A person standing erect with the face directed forward, the upper limbs hanging to the sides, and the palms of the hands facing forward
5. Body: sagittal, transverse (horizontal), and frontal (coronal) planes; organ: longitudinal, cross (transverse), and oblique sections
6. Thoracic, abdominal, and pelvic cavities
7. Pericardial, pleural, and peritoneal cavities
8. Kidneys, adrenal glands, pancreas, parts of the intestine, and urinary bladder

2 The Chemical Basis of Life

CONTENT LEARNING ACTIVITY

Matter, Mass, and Weight

Match these terms with the correct statement or definition:

Gram Matter
Kilogram Weight
Mass

_____ 1. Anything that occupies space and has mass.

_____ 2. Amount of matter in an object.

_____ 3. Gravitational force acting on an object of a given mass.

_____ 4. International unit for mass.

_____ 5. 1/1000 the mass of a kilogram.

Elements and Atoms

A. Match these terms with the correct statement or definition:

Atom Neutron
Electron Proton
Element

_____ 1. Simplest type of matter with unique chemical properties.

_____ 2. Smallest particle of an element that has the chemical characteristics of that element.

_____ 3. Subatomic particle with no electric charge.

_____ 4. Subatomic particle with one negative charge.

B. Match these terms with the correct statement or definition:

Electron cloud
Nucleus

_____ 1. Central part of an atom, which contains protons and neutrons.

_____ 2. Visual representation of the region in which any given electron is most likely to be found.

C. Match these terms with the correct parts labeled in figure 2.1:

Atom
Nucleus
Neutron
Region occupied by electrons
Proton

1. _____

2. _____

3. _____

4. _____

5. _____

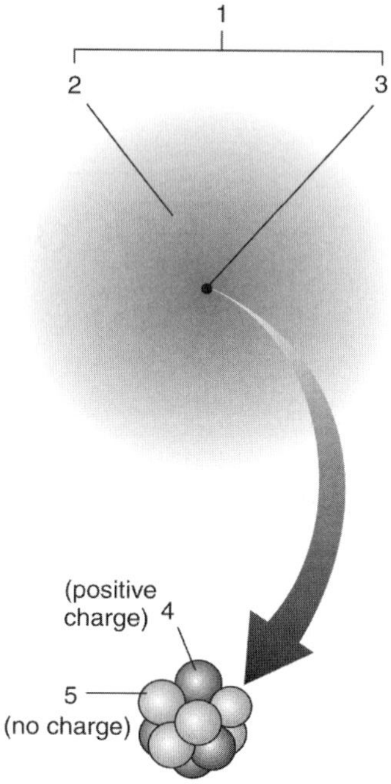

Figure 2.1

D. Match these terms with the correct statement or definition:

Atomic mass
Atomic number
Isotopes
Mass number
Unified atomic mass unit (dalton)

_____ 1. Number of protons in an atom.

_____ 2. Number of protons plus the number of neutrons in an atom.

_____ 3. Two or more forms of the same element that have the same number of protons and electrons but different numbers of neutrons—e.g., hydrogen, deuterium, and tritium.

_____ 4. 1/12 the mass of ^{12}C.

_____ 5. Average mass of the naturally occurring isotopes of an element.

E. Match these terms with the correct statement or definition:

Avogadro's number Mole
Molar mass

_____ 1. Number of atoms in exactly 12 g of ^{12}C.

_____ 2. Amount of a substance that contains Avogadro's number of entities, such as atoms.

_____ 3. Mass of one mole of a substance, expressed in grams.

Electrons and Chemical Bonding

A. Using the terms provided, complete these statements:

Anions   Electrons
Cations   Ions
Covalent bonding   Ionic bonding

Chemical bonding occurs when the outermost (1) are transferred or shared between atoms. If atoms lose or gain electrons, they form charged particles called (2). Positively charged ions are called (3), and negatively charged ions are called (4). Because oppositely charged particles are attracted to each other, cations and anions tend to remain close together; this is called (5). Another kind of chemical bonding, in which atoms share one or more pairs of electrons is called (6).

1. _____
2. _____
3. _____
4. _____
5. _____
6. _____

B. Match these terms with the correct statement or definition:

Double covalent bond   Polar covalent bond
Nonpolar covalent bond   Single covalent bond

_____ 1. Bond in which two atoms share an electron pair.

_____ 2. Bond in which two atoms share four electrons.

_____ 3. Bond in which two atoms share electrons equally.

_____ 4. Bond in which two atoms share electrons unequally.

C. Using the terms provided, complete these statements:

Compound   Molecular mass
Formula   Molecule
Ionic compounds

A(n) _(1)_ is formed when two or more atoms chemically combine to form a structure that behaves as an independent unit. A(n) _(2)_ is a substance formed of two or more different types of atoms that are chemically combined. Although many molecules are compounds, _(3)_ are not molecules because their ions are held together by the force of attraction between opposite charges. The kinds and numbers of atoms (or ions) in a compound can be represented by a(n) _(4)_, which consists of the symbols of the atoms (or ions) plus subscripts denoting the number of each type of atom (or ion). The _(5)_ of a molecule or compound can be determined by adding up the atomic masses of its atoms (or ions).

1. _____
2. _____
3. _____
4. _____
5. _____

D. Match these terms with the correct statement or definition:

Dissociate   Intermolecular forces
Electrolytes   Nonelectrolytes
Hydrogen bond   Solubility

_____ 1. Weak electrostatic attractions between the oppositely charged parts of different molecules, or between ions and molecules.

_____ 2. Formed if the positively charged hydrogen atom of one molecule is attracted to the negatively charged oxygen, nitrogen, or fluorine of another molecule.

_____ 3. Ability of one substance to dissolve in another.

_____ 4. What happens to ions when ionic compounds dissolve in water.

_____ 5. Cations and anions that dissociate in water because they can conduct an electric current.

Chemical Reactions and Energy

A. Using the terms provided, complete these statements:

Anabolism
Catabolism
Chemical reaction
Decomposition
Dehydration
Equilibrium
Hydrolysis
Oxidation–reduction reaction
Oxidation
Metabolism
Products
Reactants
Reduction
Reversible
Synthesis

In a(n) __(1)__, atoms, ions, molecules, or compounds interact either to form or to break chemical bonds. The substances that enter into a chemical reaction are called __(2)__, and the substances that result from a chemical reaction are called __(3)__. When two or more reactants combine to form a new and larger product, the process is called a(n) __(4)__ reaction. Synthesis reactions in which water is a product are called __(5)__ (water out) reactions. All the synthesis reactions that occur within the body are collectively referred to as __(6)__. When a larger reactant is broken down to form two or more smaller products, the process is called a(n) __(7)__ reaction. Decomposition reactions that use water to produce the new molecules are called __(8)__ reactions. All the decomposition reactions that occur within the body are called __(9)__, and all the anabolic and catabolic reactions in the body collectively are called __(10)__. Reactions that can proceed from reactants to products and from products to reactants are called __(11)__ reactions. At __(12)__ the amount of reactants relative to the amount of products remains constant. The loss of an electron by an atom is called __(13)__, whereas the gain of an electron is called __(14)__. Because the complete or partial loss of an electron by one atom is accompanied by the gain of that electron by another atom, this reaction is called a(n) __(15)__.

1. _____
2. _____
3. _____
4. _____
5. _____
6. _____
7. _____
8. _____
9. _____
10. _____
11. _____
12. _____
13. _____
14. _____
15. _____

B. Match these terms with the correct statement or definition:

Energy
Kinetic energy
Potential energy

_____

_____

_____

1. Capacity to do work and to move matter.
2. Stored energy that could do work but is not doing so.
3. Form of energy that actually does work and moves matter.

C. Match these terms with the correct statement or definition:

Chemical
Heat
Mechanical

_____

_____

_____

1. Energy resulting from the position or movement of an object.
2. Potential energy contained in chemical bonds.
3. Energy that flows between objects that are at different temperatures.

Chapter 2

D. Using the terms provided, complete these statements:

Activation energy   Enzymes
Catalyst   Increases
Decreases

The _(1)_ is the minimum energy that reactants must have to start a chemical reaction. A(n) _(2)_ is a substance that increases the rate at which a chemical reaction proceeds without being permanently changed or depleted. Protein molecules in the body that act as catalysts are called _(3)_. An enzyme _(4)_ the rate of a chemical reaction by lowering the activation energy necessary for the reaction to begin. When the temperature decreases, the speed of a chemical reaction _(5)_. Within limits, if the concentration of the reactants increases, the rate of the chemical reaction _(6)_.

1. _____
2. _____
3. _____
4. _____
5. _____
6. _____

Inorganic Chemistry

A. Match these terms with the correct statement or definition:

Hydrogen bonding   Water
Polar covalent bonding

_____ 1. Molecule that accounts for 50%–60% of the weight of a young adult.

_____ 2. Force that holds an oxygen atom and two hydrogen atoms together in a molecule of water.

_____ 3. Force that organizes the water molecules into a lattice that holds the water molecules together.

B. Using the terms provided, complete these statements:

Dissolved   Protects
Hydrolysis and dehydration Specific heat
Ions

Water has many physical and chemical properties that are important for living organisms. Water has a high _(1)_; as a result, water tends to resist large temperature fluctuations. Water _(2)_ body parts from damage resulting from friction. The chemical reactions necessary for life do not take place unless reacting molecules are _(3)_ in water. For instance, ionic compounds, such as sodium chloride, dissociate in water; the resulting _(4)_ can then react with other ions. Water participates directly in _(5)_ reactions.

1. _____
2. _____
3. _____
4. _____
5. _____

C. Match these terms with the correct statement or definition:

Colloid　　　　　Solution
Mixture　　　　Solvent
Solute　　　　　Suspension

_____ 1. Combination of two or more substances physically blended together, but not chemically combined.

_____ 2. Liquid, gas, or solid in which the substances are uniformly distributed, with no clear boundary between the substances.

_____ 3. Substance that dissolves in another substance.

_____ 4. Mixture containing materials that separate from each other unless they are continually, physically blended together.

_____ 5. Mixture in which a dispersed substance (solutelike) is distributed through a dispersing (solventlike) substance—e.g., proteins and water.

Solution Concentrations

Match these terms with the correct statement or definition:

Milliosmole　　Osmole
Osmolality　　　Percent

_____ 1. Method of expressing concentration that states the weight of solute in a given volume of solvent (usually water).

_____ 2. Avogadro's number of particles of a substance in 1 kilogram of water.

_____ 3. Number of osmoles of particles in a solution.

_____ 4. 1/1000 of an osmole.

Acids and Bases

A. Match these terms with the correct statement or definition:

Acids　　Strong
Bases　　Weak

_____ 1. Substances that are proton (H^+) donors.

_____ 2. Substances that accept protons (H^+).

_____ 3. Acids or bases that dissociate almost completely in water.

B. Match these terms with the correct statement or definition:

Acidic solution　　　　　Neutral solution
Alkaline (basic) solution

_____ 1. pH of 7—e.g., pure water.

_____ 2. pH less than 7.

_____ 3. Greater concentration of H^+ than OH^-.

Chapter 2

C. Match these terms with the correct statement or definition:

Buffers   Salts
Conjugate acid–base pair

_____ 1. Molecules consisting of a cation other than H^+ and an anion other than OH^-.

_____ 2. Solution of a conjugate acid–base pair in which the acid component and the base component occur in similar concentration; chemicals that resist changes in the pH of a solution when either acids or bases are added.

_____ 3. Two substances, one that becomes acidic when it gains a hydrogen ion, and the other that is everything left of the acid after a hydrogen ion is lost—e.g., bicarbonate ion and carbonic acid.

Oxygen and Carbon Dioxide

Match these terms with the correct statement or definition:

Carbon dioxide
Oxygen

_____ 1. Composes about 21% of the gas in the atmosphere; essential for most animals.

_____ 2. Used in the final step in a series of reactions that extract energy from food particles.

_____ 3. Produced when organic molecules, such as glucose, are metabolized within the cells of the body.

Carbohydrates

A. Match these terms with the correct statement or definition:

Disaccharides   Monosaccharides
Isomers   Polysaccharides

_____ 1. Simple sugars (e.g., glucose) that are building blocks for other carbohydrates.

_____ 2. Molecules that have the same number and type of atoms but differ in their three-dimensional arrangement—e.g., glucose and fructose.

_____ 3. Sucrose, lactose, maltose, and other double sugars.

_____ 4. Many monosaccharides bound together to form long chains—e.g., glycogen, starch, and cellulose.

B. Match these terms with the correct statement or definition:

Cellulose   Glycogen
Glucose   Starch

_____ 1. Carbohydrate that is found in blood and is a major nutrient for most cells of the body.

_____ 2. Energy-storage molecule in muscle and liver.

_____ 3. Energy-storage molecule in plants.

_____ 4. Structural component of plant cell walls that is not digestible by humans.

Lipids

A. Match these terms with the correct statement or definition:

Carboxyl group Monounsaturated
Fatty acids Polyunsaturated
Glycerol Saturated

_____ 1. These two combine to form triglycerides (triacylglycerols).

_____

_____ 2. Straight chain of carbon atoms with carboxyl group attached.

_____ 3. Oxygen and a hydroxyl group attached to a carbon atom; responsible for the acidic nature of some molecules.

_____ 4. Fatty acid with single covalent bonds between carbon atoms.

_____ 5. Fats with one double covalent bond between carbon atoms.

_____ 6. Fats with two or more double covalent bonds between carbon atoms.

B. Match these terms with the correct statement or definition:

Eicosanoids Phospholipids
Fats Steroids
Fat-soluble vitamins

_____ 1. Molecules that consist of one glycerol and three fatty acids; triglycerides (triacylglycerols).

_____ 2. Molecules with polar and nonpolar ends that are important structural components of the cell membrane.

_____ 3. Group of important chemicals derived from fatty acids—e.g., prostaglandins, thromboxanes, and leukotrienes.

_____ 4. Molecules composed of carbon atoms bound together in four ringlike structures—e.g., cholesterol, bile salts, estrogens, progesterone, and testosterone.

_____ 5. Small, nonpolar molecules essential for many of the normal functions of the body.

Proteins

A. Using the terms provided, complete these statements:

Amino acid   Primary
Denaturation Quaternary
Peptide   Secondary
Polypeptide   Tertiary

The building blocks for proteins are 20 types of (1) molecules. Covalent bonds formed between amino acid molecules during protein synthesis are called (2) bonds. A molecule of many amino acids bonded together by peptide bonds is called a(n) (3). The (4) structure of a protein is determined by the sequence of amino acids bound to each other, whereas the (5) structure of a protein results from the folding or bending of the polypeptide chain caused by the hydrogen bonds between amino acids. Two common shapes that result are helices and pleated sheets. If these hydrogen bonds are broken, (6) of the protein occurs, and the protein is nonfunctional. Folding of helices or pleated sheets by the formation of covalent bonds between sulfur atoms of two amino acids and attraction or repulsion of part of the protein molecule to water produces the (7) structure of proteins. The spatial relationships produced as two or more proteins combine to produce a functional unit is the (8) structure.

1. _____
2. _____
3. _____
4. _____
5. _____
6. _____
7. _____
8. _____

B. Using the terms provided, complete these statements:

Activation energy
Active site
Catalysts
Chemical reactions
Coenzymes
Cofactors
Induced fit model
Lock-and-key model
Protease

Enzymes are protein (1) that increase the rate at which a reaction proceeds without themselves being permanently changed. Enzymes are highly specific because of their three-dimensional shape, which determines the structure of the enzyme's (2). According to the (3), a reaction occurs when the reactants (key) bind to the active site (lock) on the enzyme. The view of enzymes and reactants as rigid structures has been modified by the (4), in which the enzyme is able to change shape slightly and better fit the reactants. Enzymes function by lowering the (5) required to start a reaction. Some enzymes require additional, nonprotein substances called (6) to be functional. These substances can be ions or a complex of organic molecules. Organic cofactors, such as certain vitamins, are called (7). Enzymes are often named by adding *-ase* to the name of the molecules on which they act. For instance, an enzyme that breaks down proteins is called a(n) (8). Enzymes control the rate at which most (9) proceed in living systems.

1. _____
2. _____
3. _____
4. _____
5. _____
6. _____
7. _____
8. _____
9. _____

Nucleic Acids and ATP

Match these terms with the correct statement or definition:

Adenosine triphosphate (ATP)
Deoxyribonucleic acid (DNA)
Nucleotide
Ribonucleic acid (RNA)

_____ 1. Double helix of nucleotides; the genetic material of cells.

_____ 2. Single strand of nucleotides; its organic bases include uracil.

_____ 3. Building block for nucleic acids, consisting of a monosaccharide, a nitrogenous organic base, and a phosphate group.

_____ 4. Energy currency of cells; synthesized from oxidation of glucose.

Chapter 2

Molecular Diagrams

Match these terms with the correct molecular diagram in figure 2.2:

Amino acid
ATP
Fatty acid
Glycerol
Monosaccharide

Nucleotide
Polypeptide
Polysaccharide
Triglyceride

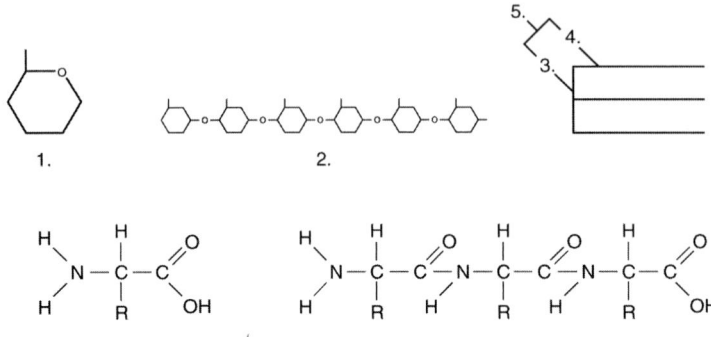
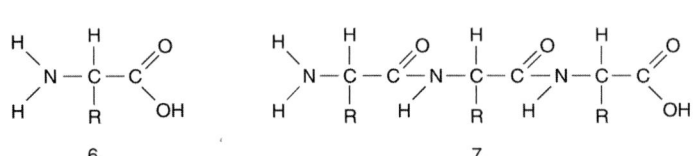

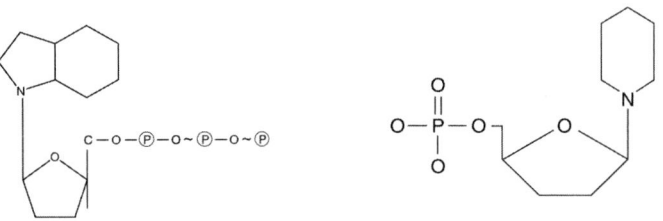

Figure 2.2

1. _____
2. _____
3. _____
4. _____
5. _____
6. _____
7. _____
8. _____
9. _____

QUICK RECALL

1. List three subatomic particles and give their charge.

2. List two types of bonds between atoms. List one intermolecular force.

3. Name four types of chemical reactions.

Chapter 2

4. List three types of energy and give an example that applies to the human organism for each.

5. List three factors that affect the rate of chemical reactions.

6. List four properties of water that make it well suited for living organisms.

7. Name the four types of organic molecules found in living things. For each type of organic molecule, list its building blocks.

8. List three kinds of carbohydrates and give an example for each in the human organism.

9. List three kinds of lipids and give an example where each is found in the human organism.

10. List the organic bases for DNA, RNA, and ATP.

Chapter 2

ANSWERS TO CHAPTER 2
CONTENT LEARNING ACTIVITY

Matter, Mass, and Weight
1. Matter; 2. Mass; 3. Weight; 4. Kilogram;
5. Gram

Elements and Atoms
- A. 1. Element; 2. Atom; 3. Neutron; 4. Electron
- B. 1. Nucleus; 2. Electron cloud
- C. 1. Atom; 2. Region occupied by electrons; 3. Nucleus; 4. Proton; 5. Neutron
- D. 1. Atomic number; 2. Mass number; 3. Isotopes; 4. Unified atomic mass unit (dalton); 5. Atomic mass
- E. 1. Avogadro's number; 2. Mole; 3. Molar mass

Electrons and Chemical Bonding
- A. 1. Electrons; 2. Ions; 3. Cations; 4. Anions; 5. Ionic bonding; 6. Covalent bonding
- B. 1. Single covalent bond; 2. Double covalent bond; 3. Nonpolar covalent bond; 4. Polar covalent bond
- C. 1. Molecule; 2. Compound; 3. Ionic compounds; 4. Formula; 5. Molecular mass
- D. 1. Intermolecular forces; 2. Hydrogen bond; 3. Solubility; 4. Dissociate; 5. Electrolytes

Chemical Reactions and Energy
- A. 1. Chemical reaction; 2. Reactants; 3. Products; 4. Synthesis; 5. Dehydration; 6. Anabolism; 7. Decomposition; 8. Hydrolysis; 9. Catabolism; 10. Metabolism; 11. Reversible; 12. Equilibrium; 13. Oxidation; 14. Reduction; 15. Oxidation–reduction reaction
- B. 1. Energy; 2. Potential energy; 3. Kinetic energy
- C. 1. Mechanical; 2. Chemical; 3. Heat
- D. 1. Activation energy; 2. Catalyst; 3. Enzymes; 4. Increases; 5. Decreases; 6. Increases

Inorganic Chemistry
- A. 1. Water; 2. Polar covalent bonding; 3. Hydrogen bonding
- B. 1. Specific heat; 2. Protects; 3. Dissolved; 4. Ions; 5. Hydrolysis and dehydration
- C. 1. Mixture; 2. Solution; 3. Solute; 4. Suspension; 5. Colloid

Solution Concentrations
1. Percent; 2. Osmole; 3. Osmolality; 4. Milliosmole

Acids and Bases
- A. 1. Acids; 2. Bases; 3. Strong
- B. 1. Neutral solution; 2. Acidic solution; 3. Acidic solution
- C. 1. Salts; 2. Buffers; 3. Conjugate acid–base pair

Oxygen and Carbon Dioxide
1. Oxygen; 2. Oxygen; 3. Carbon dioxide

Carbohydrates
- A. 1. Monosaccharides; 2. Isomers; 3. Disaccharides; 4. Polysaccharides
- B. 1. Glucose; 2. Glycogen; 3. Starch; 4. Cellulose

Lipids
- A. 1. Glycerol and fatty acids; 2. Fatty acids; 3. Carboxyl group; 4. Saturated; 5. Monounsaturated; 6. Polyunsaturated
- B. 1. Fats; 2. Phospholipids; 3. Eicosanoids; 4. Steroids; 5. Fat-soluble vitamins

Proteins
- A. 1. Amino acid; 2. Peptide; 3. Polypeptide; 4. Primary; 5. Secondary; 6. Denaturation; 7. Tertiary; 8. Quaternary
- B. 1. Catalysts; 2. Active site; 3. Lock-and-key model; 4. Induced fit model; 5. Activation energy; 6. Cofactors; 7. Coenzymes; 8. Protease; 9. Chemical reactions

Nucleic Acids and ATP
1. Deoxyribonucleic acid (DNA); 2. Ribonucleic acid (RNA); 3. Nucleotide; 4. Adenosine Triphosphate (ATP)

Molecular Diagrams
1. Monosaccharide; 2. Polysaccharide; 3. Glycerol; 4. Fatty acid; 5. Triglyceride; 6. Amino acid; 7. Polypeptide; 8. ATP; 9. Nucleotide

QUICK RECALL

1. Protons: positive charge; neutrons: no charge; and electrons: negative charge
2. Ionic and covalent bonds; hydrogen bonds
3. Synthesis, decomposition, oxidation–reduction, and reversible reactions
4. Mechanical energy: walking or heartbeat; chemical energy: food molecules; heat energy: metabolism and body temperature
5. Presence of catalysts (enzymes), temperature, and concentration of reactants
6. High specific heat, protects from friction, necessary for chemical reactions, and a good mixing medium
7. Carbohydrates: monosaccharides; fats (triglycerides): glycerol and fatty acids; proteins: amino acids; nucleic acids: nucleotides
8. Monosaccharides: glucose, fructose, galactose, ribose, deoxyribose; disaccharides: sucrose, lactose; maltose; polysaccharides: glycogen
9. Fats: energy storage, padding, insulation; phospholipids: cell membrane; eicosanoids: response of tissue to injuries; steroids: cholesterol, bile salts, and sex hormones; fat-soluble vitamins: vitamins A, D, E, and K
10. DNA: adenine, thymine, guanine, cytosine; RNA: adenine, uracil, guanine, cytosine; ATP: adenine

Structure and Function of the Cell

CONTENT LEARNING ACTIVITY

How We See Cells

Match these terms with the correct statement or definition:

Electron microscope
Light microscope
Scanning electron microscope
Transmission electron microscope

_____ 1. Used to view the general features of cells.

_____ 2. Used to study the fine features of cells; there are two types.

_____ 3. Used to see the cell surface and the surface of internal cell structures.

_____ 4. Used to see "through" parts of cells.

Plasma Membrane

A. Match these terms with the correct statement or definition:

Extracellular
Glycocalyx
Glycolipid
Glycoprotein
Intracellular
Intercellular
Membrane potential

_____ 1. Inside a cell.

_____ 2. Outside a cell.

_____ 3. Between cells.

_____ 4. Charge difference across the plasma membrane.

_____ 5. Carbohydrates combined with lipids.

_____ 6. Carbohydrates combined with proteins.

_____ 7. Collection of glycolipids, glycoproteins, and carbohydrates on the outer surface of the plasma membrane.

B. Match these terms with the correct statement or definition:

    Cholesterol   Hydrophobic
    Hydrophilic   Phospholipid

_____ 1. Predominant lipid of the plasma membrane; forms a double layer of molecules called a lipid bilayer.

_____ 2. Polar ends of phospholipids, which are exposed to water inside and outside the cell.

_____ 3. Helps determine the fluid nature of the plasma membrane.

C. Match these terms with the the correct plasma membrane parts labeled in figure 3.1:

    Carbohydrate chains   Integral protein
    Cholesterol   Nonpolar region of phospholipid
    Cytoskeleton   Peripheral protein
    Glycocalyx   Phospholipid bilayer
    Glycolipid   Polar region of phospholipid
    Glycoprotein

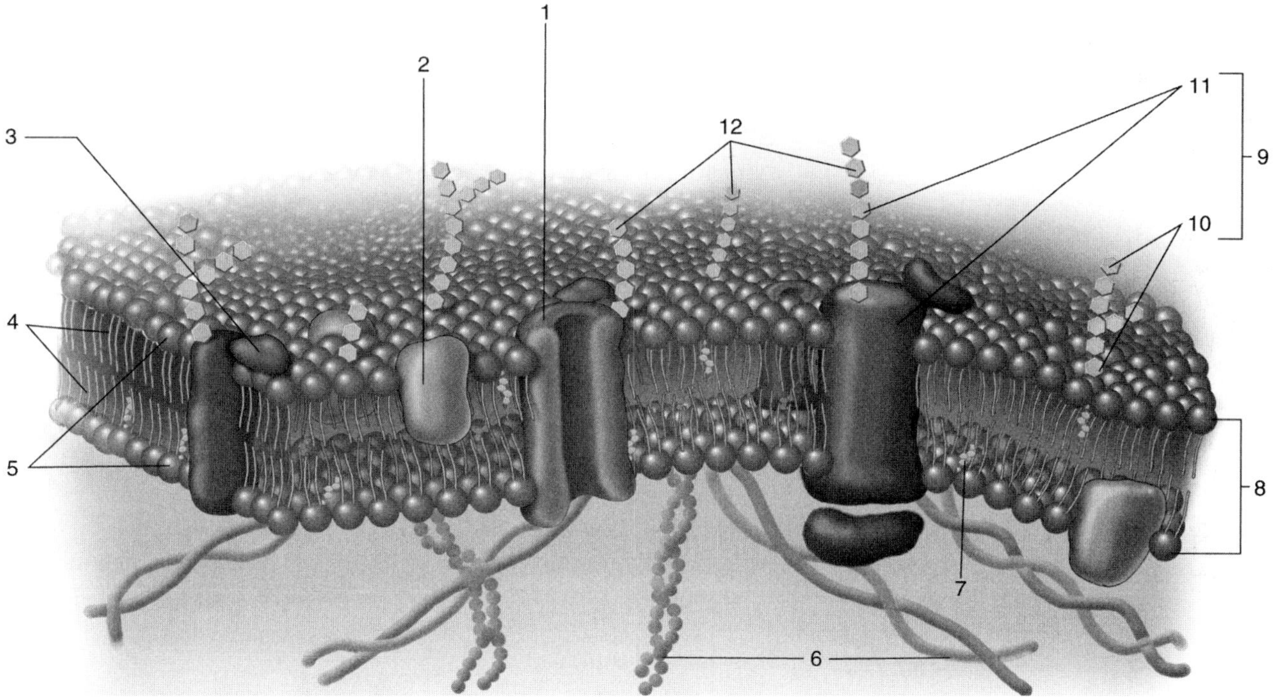

Figure 3.1

1. _____   5. _____   9. _____
2. _____   6. _____   10. _____
3. _____   7. _____   11. _____
4. _____   8. _____   12. _____

Chapter 3

D. Match these terms with the correct statement or definition:

Fluid-mosaic model Peripheral protein
Integral protein Protein
Lipid

_____ 1. Basic structure of the plasma membrane and some of its functions are determined by this type of molecule.

_____ 2. Type of molecule responsible for many of the functions of the plasma membrane.

_____ 3. Proteins and other molecules are suspended in the lipid bilayer; distributes molecules, allows repair of the plasma membrane, and enables plasma membranes to fuse together.

_____ 4. Type of protein that deeply penetrates the lipid bilayer, often from one surface to the next.

E. Match these terms with the correct statement or definition:

Cadherins Marker molecules
Integrins

_____ 1. Cell surface molecules that allow cells to identify each other or other molecules.

_____ 2. Attachment protein that attaches cells to other cells.

_____ 3. Attachment protein that attaches cells to extracellular molecules.

F. Match these terms with the correct statement or definition:

Channel protein Transport protein
Ligand-gated ion channel Voltage-gated ion channel
Nongated ion channel

_____ 1. Integral protein involved with the movement of ions or molecules across the plasma membrane; includes channel proteins, carrier proteins, and ATP-powered pumps.

_____ 2. Integral protein that forms a passageway through which small molecules or ions pass through the plasma membrane.

_____ 3. Ion channel that is always open.

_____ 4. Ion channel that opens in response to small proteins or glycoproteins.

_____ 5. Ion channel that opens when there is a change in charge across the plasma membrane.

G. Match these terms with the correct statement or definition:

Antiport
ATP-powered pump
Carrier protein
Symport
Uniport

_____ 1. Transport protein that changes shape and moves ions or molecules from one side of the plasma membrane to the other; also called a transporter.

_____ 2. Movement of two different ions or molecules in the same direction by a carrier protein.

_____ 3. Movement of two different ions or molecules in the opposite direction by a carrier protein.

_____ 4. Transport protein that changes shape and requires ATP to move ions or molecules across the plasma membrane.

H. Match these terms with the correct statement or definition:

Enzyme
G protein
Receptor molecule

_____ 1. Protein with a receptor site to which a ligand can combine; can be part of a ligand-gated ion channel; functions as part of an intercellular communication system that enables coordination of the activities of cells.

_____ 2. Protein complex on the inner surface of the plasma membrane that is activated when a ligand binds to its receptor; the activated protein stimulates a cell response.

_____ 3. Membrane protein that can catalyze chemical reactions on either the inner or outer surfaces of the plasma membrane.

Movement Through the Plasma Membrane

Match these terms with the correct statement or definition:

Membrane channels
Lipid bilayer
Selectively permeable
Transport protein
Vesicle

_____ 1. Allows only certain substances to pass through the plasma membrane.

_____ 2. Molecules that are soluble in lipid dissolve in this layer; acts as a barrier to most polar substances.

_____ 3. Allow molecules of only a certain size range to pass through; water moves through them rapidly.

_____ 4. Large polar molecules cannot pass through the plasma membrane in significant amounts unless transported by this protein.

_____ 5. Transports large, nonlipid-soluble molecules, small pieces of matter, and even whole cells.

Diffusion

A. Match these terms with the correct statement or definition:

Concentration gradient	Solvent
Diffusion	Viscosity
Solute

_____ 1. Predominant liquid or gas in a solution.

_____ 2. Tendency for solutes to move from an area of higher concentration to an area of lower concentration in a solution.

_____ 3. Product of the constant random motion of all atoms, molecules, or ions in a solution.

_____ 4. Concentration difference between two points.

_____ 5. Measure of how easily a liquid flows.

B. Match these terms with the correct statement or definition:

Decrease
Increase

_____ 1. Change in the rate of diffusion when there is an increase in the concentration gradient or an increase in temperature.

_____ 2. Change in the rate of diffusion when there is a decrease in the size of diffusing molecules.

_____ 3. Change in the rate of diffusion when there is an increase in viscosity.

Osmosis

Match these terms with the correct statement or definition:

Crenation	Isosmotic
Hyperosmotic	Isotonic
Hypertonic	Lysis
Hyposmotic	Osmosis
Hypotonic	Osmotic pressure

_____ 1. Diffusion of water across a selectively permeable membrane; can occur through membrane channels called aquaporins.

_____ 2. Force required to prevent the movement of water by osmosis across a selectively permeable membrane.

_____ 3. Solution with fewer solute particles and a lower osmotic pressure than another solution.

_____ 4. When a cell is placed in this type of solution, the cell, by definition, shrinks.

_____ 5. Cell shrinkage.

_____ 6. Rupture of the cell, which may occur from placing a cell into a hypotonic solution.

Filtration

Using the terms provided, complete these statements:

Concentration difference Lower
Higher Pressure difference
Larger Smaller

1. _____
2. _____
3. _____

Filtration depends on a(n) _(1)_ on either side of a partition. Liquid moves through the partition from the side with the _(2)_ pressure. Holes in the partition prevent particles that are _(3)_ than the holes from moving through the partition.

Mediated Transport Mechanisms

A. Match these terms with the correct statement or definition:

Competition Specificity
Saturation

_____ 1. Carrier protein transports only a single type of molecule.

_____ 2. Result of similar molecules binding to the carrier protein.

_____ 3. Rate of transport is limited by the number of carrier proteins.

B. Match these terms with the correct statement or definition:

Active transport
Facilitated diffusion

_____ 1. Does not require metabolic energy (ATP).

_____ 2. Always moves substances from a higher to a lower concentration.

_____ 3. Can move substances against a concentration gradient.

C. Using the terms provided, complete these statements:

Against Into
Down Out of

1. _____
2. _____
3. _____
4. _____

In secondary active transport, a Na^+–K^+ pump moves Na^+ _(1)_ a cell, establishing a concentration gradient for Na^+. As Na^+ move _(2)_ their concentration gradient, Na^+ and glucose molecules can be transported _(3)_ cells by carrier proteins. The energy provided by Na^+ movement can result in the movement of glucose _(4)_ its concentration gradient.

Chapter 3

Endocytosis and Exocytosis

Match these terms with the correct statement or definition:

Endocytosis Phagocytosis
Exocytosis Pinocytosis

_____ 1. Includes both phagocytosis and pinocytosis; can exhibit specificity because it is receptor-mediated.

_____ 2. Means "cell eating"; the ingestion of solid particles.

_____ 3. Means "cell drinking"; the ingestion of molecules dissolved in liquid.

_____ 4. Vesicle fuses with the plasma membrane and the content of the vesicle is expelled from the cell.

Cytoplasm

A. Match these terms with the correct statement or definition:

Cytoplasmic inclusion Cytosol
Cytoskeleton Organelles

_____ 1. Fluid portion of cytoplasm; a solution containing dissolved ions and molecules and a colloid with suspended molecules, especially proteins.

_____ 2. Supports the cell and holds organelles in place; responsible for cell movements.

_____ 3. Aggregate of chemicals within cells.

_____ 4. Small structures within cells that are specialized for particular functions.

B. Match these terms with the correct statement or definition:

Actin filament Microtubule
Intermediate filament

_____ 1. Composed primarily of tubulin; located in centrioles, spindle fibers, cilia, and flagella.

_____ 2. Provides support in microvilli; involved with cell movement, such as muscle contraction; also called a microfilament.

_____ 3. Provides mechanical strength to cells.

Nucleus

A. Match these terms with the correct statement or definition:

Centromere
Chromatid
Chromatin
Chromosome
Kinetochore
Nuclear envelope
Nucleoplasm
Nucleus

_____ 1. Large, membrane-bound organelle usually located near the center of the cell; contains DNA.

_____ 2. Outer boundary of the nucleus; consists of two membranes with nuclear pores.

_____ 3. Discrete unit consisting of nuclear DNA and histones.

_____ 4. Chromosomes dispersed throughout the nucleus as delicate filaments during most of the life cycle of a cell.

_____ 5. Two parts of a chromosome at the beginning of cell division.

_____ 6. Attachment site between chromatids.

_____ 7. Attachment site for microtubules within centromeres.

B. Match these terms with the correct statement or definition:

Gene
Nucleolar organizer regions
Nucleolus
RNA

_____ 1. Somewhat round, dense region within the nucleus; lacks a membrane.

_____ 2. Molecule through which DNA functions; can leave the nucleus through the nuclear pores.

_____ 3. DNA within the nucleolus from which rRNA is produced.

_____ 4. Sequence of nucleotides in a DNA molecule that specifies the structure of a protein or an RNA molecule.

Ribosomes and Endoplasmic Reticulum

Match these terms with the correct statement or definition:

Cisternae
Ribosome
Rough endoplasmic reticulum
Smooth endoplasmic reticulum

_____ 1. Site where mRNA and tRNA come together and assemble amino acids to form proteins; free or attached to endoplasmic reticulum.

_____ 2. Tubules and flattened sacs with many ribosomes attached.

_____ 3. Large amounts of this structure are in cells that secrete proteins.

_____ 4. Found in the cell when lipid synthesis, detoxification processes, or storage of calcium ions occurs.

Golgi Apparatus and Secretory Vesicles

Match these terms with the correct statement or definition:

Golgi apparatus
Secretory vesicle
Transport vesicle

_____ 1. Modifies, packages, and distributes proteins and lipids manufactured by the rough and smooth endoplasmic reticulum.

_____ 2. Sac that pinches off of the endoplasmic reticulum and fuses with the Golgi apparatus.

_____ 3. Sac that pinches off from the Golgi apparatus; can contain proteins that become part of the plasma membrane or function as enzymes.

_____ 4. Releases its contents to the exterior of the cell by exocytosis.

Lysosomes, Peroxisomes, and Proteasomes

Match these terms with the correct statement or definition:

Autophagia
Catalase
Lysosomes
Peroxisomes
Proteasomes

_____ 1. Vesicles that function as intracellular digestive systems.

_____ 2. Process of digesting cell organelles that are no longer functional.

_____ 3. Membrane-bound vesicles containing enzymes that produce hydrogen peroxide while breaking down fatty acids and amino acids.

_____ 4. Enzyme in peroxisomes that breaks down hydrogen peroxide, which can be toxic to cells.

_____ 5. Large protein complexes with enzymes that break down and recycle proteins.

Mitochondria

Using the terms provided, complete the following statements:

ATP
Cristae
DNA
Electron-transport chain
Matrix
Oxidative metabolism

Mitochondria are the major sites of (1) production within cells. Mitochondria have inner and outer membranes separated by a space. The inner membranes have numerous infoldings called (2) that project like shelves into the interior of the mitochondria. A complex series of mitochondrial enzymes forms two major enzyme systems, which are responsible for (3) and most ATP synthesis. The enzymes of the citric acid (Krebs) cycle are located in the (4), and the enzymes of the (5) are embedded with the inner membrane. The information for making some mitochondrial proteins is contained with (6) located within mitochondria.

1. _____
2. _____
3. _____
4. _____
5. _____
6. _____

Centrioles, Spindle Fibers, Cilia, Flagella, and Microvilli

Match these terms with the correct statement or definition:

Centrioles
Centrosome
Cilia
Flagella
Microvilli
Spindle fibers

_____ 1. Specialized zone of cytoplasm containing two centrioles; the center of microtubule formation.

_____ 2. Small, cylindrical organelles oriented perpendicular to each other.

_____ 3. Attach to chromosomes at the kinetochore; aid chromosome separation in cell division.

_____ 4. Short appendages that are capable of moving; move small particles across the cell surface.

_____ 5. Appendages that are longer than cilia; move sperm cells.

_____ 6. Extensions of the plasma membrane that are smaller than cilia; increase cell surface area.

Cell Diagram

Match these terms with the cell parts labeled on figure 3.2:

Centrioles
Centrosome
Chromatin
Cilia
Cytoskeleton
Golgi apparatus
Lysosome
Microvilli
Mitochondrion
Nuclear envelope
Nuclear pore
Nucleolus
Nucleus
Peroxisome
Phagocytic vesicle
Plasma membrane
Proteasome
Ribosome (free)
Rough endoplasmic reticulum
Secretory vesicles
Smooth endoplasmic reticulum

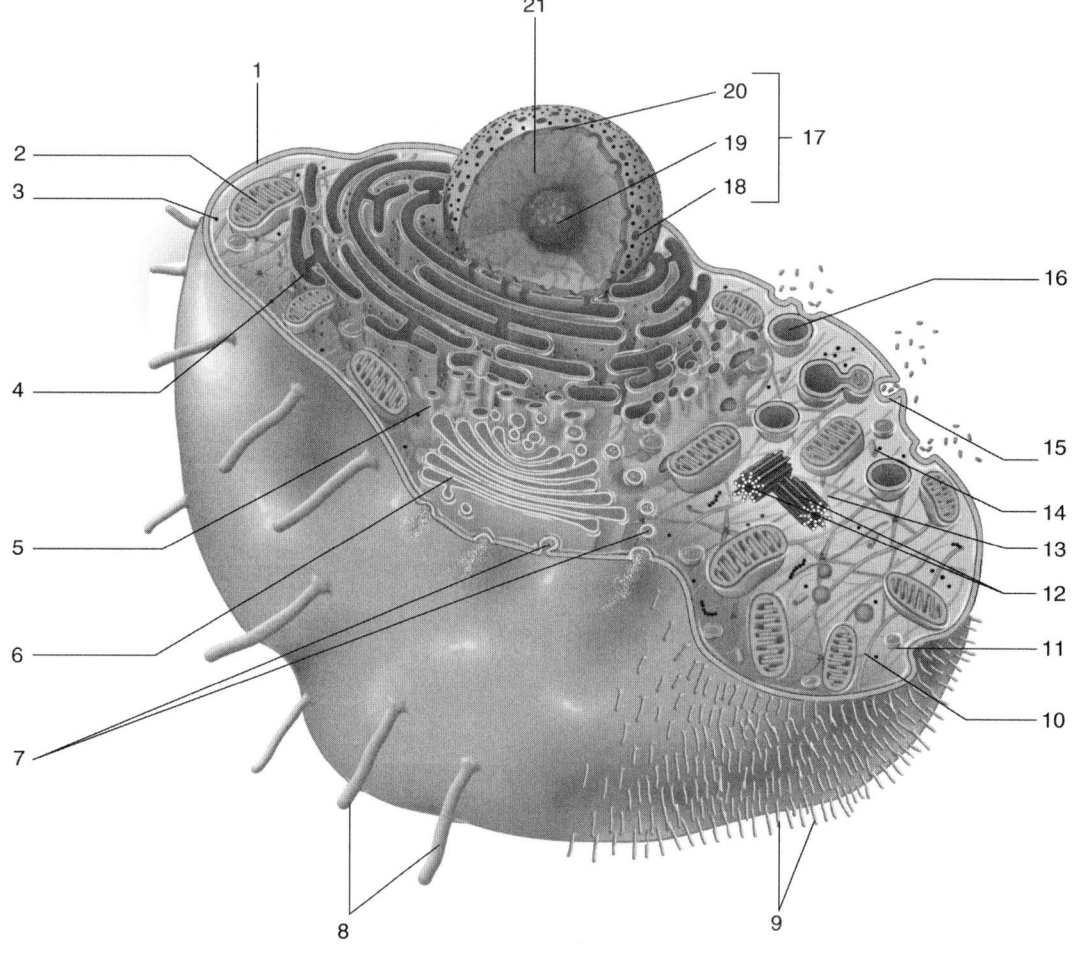

Figure 3.2

1. _____
2. _____
3. _____
4. _____
5. _____
6. _____
7. _____
8. _____
9. _____
10. _____
11. _____
12. _____
13. _____
14. _____
15. _____
16. _____
17. _____
18. _____
19. _____
20. _____
21. _____

Genes and Gene Expression

A. Match these terms with the correct statement or definition:

mRNA Transcription
rRNA Translation
tRNA

_____ 1. Process the occurs when double strands of a DNA segment separate and nucleotides pair with the nucleotides of the DNA to form RNA.

_____ 2. Type of RNA that carries information in groups of three nucleotides, called codons, and each codon codes for a specific amino acid.

_____ 3. Type of RNA with an anticodon; binds to a specific amino acid.

_____ 4. Synthesis of polypeptide chains at the ribosome in response to information contained in mRNA molecules.

B. Match these terms with the correct statement or definition:

Adenine Guanine
Codon Thymine
Cytosine Triplet
Gene Uracil

_____ 1. Sequence of three nucleotides in DNA.

_____ 2. Triplets in DNA required for the synthesis of a specific protein or RNA molecule.

_____ 3. Group of three nucleotides in mRNA; responsible for the genetic code.

_____ 4. Two nucleotides that can pair with adenine.

_____

_____ 5. Nucleotide that pairs with cytosine.

C. Match these terms with the correct statement or definition:

Promoter Terminator
RNA polymerase Transcription factors

_____ 1. Enzyme that synthesizes a complementary RNA molecule from DNA.

_____ 2. Sequence of DNA nucleotides to which RNA polymerase attaches, causing DNA to unwind.

_____ 3. Proteins necessary for RNA polymerase to interact with the promoter.

_____ 4. Sequence of DNA nucleotides that signals RNA polymerase to detach from DNA and release synthesize mRNA.

Chapter 3

D. Match these terms with the correct statement or definition:

Alternative splicing
Exon
Intron
Pre-RNA
Post-transcriptional processing
Start codon
Stop codon

_____ 1. Region of DNA that codes for a portion of a protein.

_____ 2. Region of DNA that does not code for a portion of a protein.

_____ 3. mRNA containing introns.

_____ 4. Removal of introns and splicing together of extrons by spliceosomes; adding a 7-methyl guanosine cap and poly-A tail to mRNA.

_____ 5. Produciton of different mRNA from a single gene.

_____ 6. Sequence of three nucleotides in mRNA that signal the beginning of translation.

E. Match these terms with the correct statement or definition:

Polyribosome
Post-translational processing
Proenzyme
Proprotein

_____ 1. Protein with extra pieces that are removed by enzymes to form a functional protein.

_____ 2. Proproteins that can be modified to form functional enzymes.

_____ 3. Adding polysaccharides to a protein or combining of amino acid chains to form a protein.

_____ 4. Cluster of ribosomes attached to an mRNA; the ribosomes produce identical proteins.

Cell Life Cycle

A. Match these terms with the correct statement or definition:

G_1 phase
G_2 phase
G_0 phase
Interphase
S phase

_____ 1. Time between cell divisions.

_____ 2. Part of the cell cycle in which cells carry out routine metabolic activities.

_____ 3. Part of the cell cycle in which cells prepare for cell division.

_____ 4. Part of the cell cycle in which cells synthesize DNA.

_____ 5. Phase in which cells are "resting" and do not divide unless stimulated; not part of the cell cycle.

B. Match these terms with the correct statement or definition:

DNA ligase   Lagging strand
DNA polymerase   Leading strand

_____ 1. Enzyme that adds nucleotides to a DNA template.

_____ 2. DNA strand that forms as a continuous strand.

_____ 3. DNA strand that forms as short segments called Okazaki fragments.

_____ 4. Enzyme that splices together Okazaki fragments.

Cell Division: Mitosis and Cytokinesis

A. Match these terms with the correct statement or definition:

Cytokinesis
Mitosis

_____ 1. Division of the nucleus into two nuclei.

_____ 2. Division of the cell's cytoplasm to produce two new cells.

B. Match these terms with the correct statement or definition:

Astral fibers   Chromatids
Centromere   Spindle fibers

_____ 1. Identical pieces of DNA

_____ 2. Specialized region that joins two chromatids together.

_____ 3. Microtubules that radiate from the centrioles and end blindly.

_____ 4. Microtubules that project toward the equator and may attach to the centromeres of the chromosomes.

C. Match these terms with the correct statement or definition as they relate to the events in mitosis:

Anaphase   Prophase
Metaphase   Telophase

_____ 1. Chromosomes become visible.

_____ 2. Chromosomes (each with two chromatids) align along the equator with spindle fibers attached to their centromeres.

_____ 3. Separation of chromatids occurs.

_____ 4. Migration of each set of chromosomes toward the centrioles is completed.

D. Match these terms with the phases of mitosis and the cell parts involved in mitosis labeled in figure 3.3:

Anaphase
Astral fiber
Centrioles
Centromere
Chromatid
Chromatin
Chromosomes (identical)
Cleavage furrow
Interphase
Metaphase
Prophase
Spindle fiber
Telophase

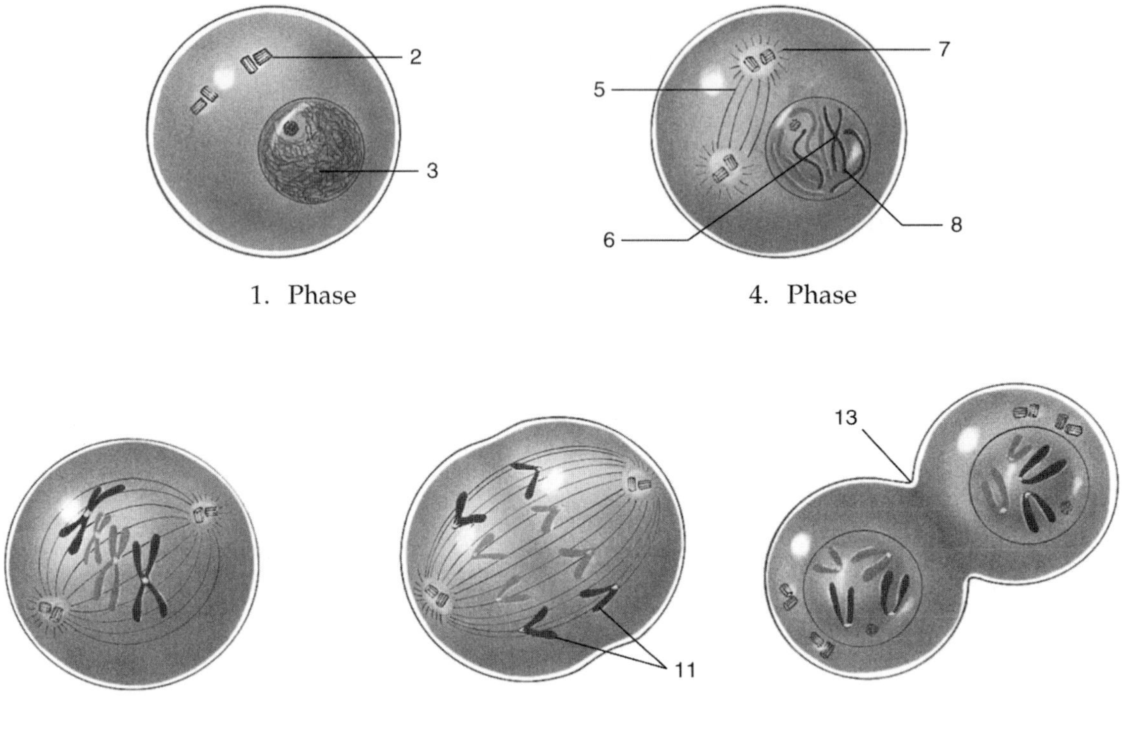

1. Phase
4. Phase
9. Phase
10. Phase
12. Phase

Figure 3.3

1. _____
2. _____
3. _____
4. _____
5. _____
6. _____
7. _____
8. _____
9. _____
10. _____
11. _____
12. _____
13. _____

Genetics

Match these terms with the correct statement or definition:

Genetics
Genomic medicine
Mendelian genetics

_____ 1. Study of heredity—i.e., the characteristics inherited by children from their parents.

_____ 2. Used to determine the risk of inheriting certain genetic diseases.

_____ 3. Genetic approach to the diagnosis and management of diseases.

Mendelian Genetics

Match these terms with the correct statement or definition:

Albinism
Alleles
Dominant allele
Genotype
Heterozygous
Homozygous
Phenotype
Polydactyly

_____ 1. Actual genes a person possesses.

_____ 2. Appearance of an individual, resulting from the expression of the genes he or she possesses.

_____ 3. Alternate forms of genes.

_____ 4. Masks the effects of the recessive allele for a trait.

_____ 5. Two alleles for a trait are identical.

_____ 6. Two alleles for a trait are slightly different.

_____ 7. Condition in which an abnormal, recessive allele causes the production of a defective enzyme, resulting in an inability to synthesize melanin.

_____ 8. Condition in which an abnormal, dominant allele causes extra fingers or toes.

Modern Concepts of Genetics

A. Match these terms with the correct statement or definition:

Diploid
Gametes
Haploid
Somatic cells
23
23 pairs

_____ 1. All the cells of the body except the gametes, or sex cells.

_____ 2. Sperm cells and oocytes.

_____ 3. Number of chromosomes in a somatic cell.

_____ 4. Number of chromosomes in a gamete.

_____ 5. Total number of chromosomes in a human somatic cell.

_____ 6. Total number of chromosomes in a human gamete.

Chapter 3

B. Match these terms with the correct statement or definition:

Autosomal
Genome
Homologous
Karyotype
Locus
Sex chromosomes
XX
XY

_____ 1. All the chromosomes, except the sex chromosomes.

_____ 2. X or Y chromosomes; one pair in a normal somatic cell.

_____ 3. Sex chromosomes in a normal female somatic cell.

_____ 4. Sex chromosomes in a normal male somatic cell.

_____ 5. Display of the chromosomes in a somatic cell during metaphase of mitosis.

_____ 6. Pair of chromosomes, one of which is derived from a person's father and the other is derived form a person's mother.

_____ 7. All the genes in the haploid number of chromosomes from one parent.

_____ 8. Part of a chromosome occupied by a gene.

C. Match these terms with the correct statement or definition:

Codominance
Complete dominance
Incomplete dominance
Multiple alleles
Mutated allele (gene)
Pleiotropy
Polygenic
Sex-linked traits

_____ 1. Many different forms of an allele exist.

_____ 2. Different form of an allele; also called an allelic variant or a polymorphism.

_____ 3. The homozygous dominant and the heterozygote have the same phenotype.

_____ 4. Two alleles at the same locus are expressed so that separate, distinguishable phenotypes occur at the same time.

_____ 5. The dominant gene does not completely mask the effects of the recessive gene in the heterozygote.

_____ 6. Phenotype that results from the interaction of many genes.

_____ 7. Production of multiple, often seemingly unrelated, phenotypic effects by one gene.

_____ 8. Traits affected by genes on sex chromosomes; can be X-linked or Y-linked.

D. Match these terms with the correct statement or definition:

ABO blood group   Hemophilia
Albinism   Marfan syndrome
Beta thalassemia   Phenylketonuria
Height

_____ 1. Example of a recessive trait with multiple alleles.

_____ 2. Example of complete dominance.

_____ 3. Example of codominance.

_____ 4. Example of incomplete dominance.

_____ 5. Example of a polygenic trait.

_____ 6. Example of pleiotropy.

_____ 7. Example of an X-linked recessive trait.

E. Match these terms with the correct statement or definition:

Carrier   One
DNA replication Punnett square
False   True
Meiosis   Two

_____ 1. Process by which haploid gametes are derived from diploid cells.

_____ 2. Used to determine the probability of inheriting dominant and recessive traits.

_____ 3. Heterozygous person with an abnormal recessive gene.

_____ 4. Occurs before meiosis begins.

_____ 5. Number of cell divisions in meiosis.

_____ 6. True or false: As a result of the first meiotic division, each daughter cell receives one replicated member of a homologous pair of chromosomes.

_____ 7. True or false: As a result of the second meiotic division, each daughter cell receives one chromatid from each chromosome.

Chapter 3

F. Match these terms with the correct statement or definition:

Aneuploidy
Cancer
Down syndrome
Genetic disorder
Mutagen
Mutation
Nondisjunction

_____ 1. Failure of structure, function, or both as a result of abnormalities in a person's genetic makeup.

_____ 2. Change in a gene that usually involves a change in the number or kinds of nucleotides in DNA.

_____ 3. Agents, such as chemicals, radiation, or viruses, causing mutations.

_____ 4. During meiosis, the process through which one daughter cell receives both homologous chromosomes of a pair and the other daughter cell receives none of the chromosomes in that pair.

_____ 5. Having a greater than normal or less than normal number of chromosomes.

_____ 6. Aneuploidy in which the individual has three chromosomes 21; also called trisomy 21.

_____ 7. Results from mutations in somatic cells that cause uncontrolled cell division and changes in normal cell functions.

QUICK RECALL

1. List four major functions of cells.

2. List the three major types of molecules in the plasma membrane and give their functions.

3. List the factors that affect the rate and direction of diffusion in a solution.

4. List the types of movement of materials across plasma membranes.

5. List three characteristics of mediated transport mechanisms.

Complete the following chart by writing in the organelle described by the structures and functions listed.

	ORGANELLE	STRUCTURE	FUNCTION
6.	_____	Composed primarily of protein units called tubulin.	Support for the cytoplasm of the cell; involved in cell division; essential component of centrioles, spindle fibers, cilia, and flagella.
7.	_____	Small protein fibrils that form bundles, sheets, or networks.	Provide structure to cytoplasm and mechanical support to microvilli; responsible for muscle's contractile capabilities.
8.	_____	Protein fibrils, intermediate in size between actin filaments and microtubules.	Provide mechanical strength to cells.
9.	_____	Granule in the cytoplasm.	Aggregates of chemicals include hemoglobin and glycogen.
10.	_____	Surrounded by double-layered envelope with pores.	Contains DNA in the form of chromatin (chromosomes), which produces RNA.
11.	_____	One to four rounded, dense, well-defined nuclear bodies.	Production of ribosomal subunits.
12.	_____	Two subunits composed of ribosomal RNA and protein.	Site where mRNA and tRNA come together to assemble amino acids into proteins.

Chapter 3

13.	_____	Broad, flattened sacs and tubules that interconnect, ribosomes attached.	Synthesis of proteins.
14.	_____	Broad, flattened sacs and tubules that interconnect, no ribosomes attached.	Lipid synthesis, detoxification, and calcium ion storage.
15.	_____	Closely packed stacks of curved cisternae composed of smooth endoplasmic reticulum.	Modifies, packages, and distributes proteins and lipids.
16.	_____	Membrane-bound vesicle pinched off from the Golgi apparatus.	Contents released to the exterior of the cell by exocytosis.
17.	_____	Membrane-bound vesicle that contains hydrolytic enzymes.	Breakdown of phagocytized particles; autophagia.
18.	_____	Membrane-bound vesicle that contains a variety of enzymes, such as catalase.	Detoxifies harmful molecules, breaks down hydrogen peroxide.
19.	_____	Large protein complexes with enzymes.	Breaks down and recycles proteins.
20.	_____	Small, spherical, rod-shaped or filamentous organelle; double membrane with infoldings of the inner membrane called cristae.	Most ATP synthesis in the cell.
21.	_____	Nine evenly spaced, longitudinally oriented, parallel units; each unit consists of three parallel microtubules joined together.	Moves to each side of the nucleus during cell division; special microtubules called spindle fibers develop from the surrounding region.
22.	_____	Appendages from the surface of the cell with two centrally located microtubules and nine peripheral pairs of microtubules joined together.	Movement of materials over the surface of the cells.
23.	_____	Cylindrical extensions of the plasma membrane supported by microfilaments.	Increase cell surface area for absorption.

24. List the three phases of the cell life cycle.

25. Name the four phases of mitosis.

26. Compare the number of cell divisions, the number of cells produced, and the number of chromosomes in each cell produced by mitosis and meiosis.

ANSWERS TO CHAPTER 3

CONTENT LEARNING ACTIVITY

How We See Cells
1. Light microscope; 2. Electron microscope; 3. Scanning electron microscope; 4. Transmission electron microscope

Plasma Membrane
A. 1. Intracellular; 2. Extracellular; 3. Intercellular; 4. Membrane potential; 5. Glycolipid; 6. Glycoprotein; 7. Glycocalyx
B. 1. Phospholipid; 2. Hydrophilic; 3. Cholesterol
C. 1. Membrand channel; 2. Integral protein; 3. Peropheral protein; 4. Nonpolar region; 5. Polar region; 6. Cytoskeleton; 7. Cholesterol; 8. Phospholipid bilayer; 9. Glycocalyx; 10. Glycolipid; 11. Glycoprotein; 12. Carbohydrate chains
D. 1. Lipid; 2. Protein; 3. Fluid-mosaic model; 4. Integral protein
E. 1. Marker molecules; 2. Cadherins; 3. Integrins
F. 1. Transport protein; 2. Channel protein; 3. Nongated ion channel; 4. Ligand-gated ion channel; 5. Voltage-gated ion channel
G. 1. Carrier protein; 2. Symport; 3. Antiport; 4. ATP-powered pump
H. 1. Receptor molecule; 2. G protein; 3. Enzyme

Movement Through the Plasma Membrane
1. Selectively permeable; 2. Lipid bilayer; 3. Membrane channels; 4. Carrier proteins; 5. Vesicle

Diffusion
A. 1. Solvent; 2. Diffusion; 3. Diffusion; 4. Concentration gradient; 5. Viscosity
B. 1. Increase; 2. Increase; 3. Decrease

Osmosis
1. Osmosis; 2. Osmotic pressure; 3. Hyposmotic; 4. Hypertonic; 5. Crenation; 6. Lysis

Filtration
1. Pressure difference; 2. Higher; 3. Larger

Mediated Transport Mechanisms
A. 1. Specificity; 2. Competition; 3. Saturation
B. 1. Facilitated diffusion; 2. Facilitated diffusion; 3. Active transport
C. 1. Out of; 2. With; 3. Into; 4. Against

Endocytosis and Exocytosis
1. Endocytosis; 2. Phagocytosis; 3. Pinocytosis; 4. Exocytosis

Cytoplasm
A. 1. Cytosol; 2. Cytoskeleton; 3. Cytoplasmic inclusions; 4. Organelles
B. 1. Microtubule; 2. Actin filament; 3. Intermediate filament

Nucleus
A. 1. Nucleus; 2. Nuclear envelope; 3. Chromosome; 4. Chromatin; 5. Chromatid; 6. Centromere; 7. Kinetochore
B. 1. Nucleolus; 2. RNA; 3. Nucleolar organizer region; 4. Gene

Ribosomes and Endoplasmic Reticulum
1. Ribosome; 2. Rough endoplasmic reticulum; 3. Rough endoplasmic reticulum; 4. Smooth endoplasmic reticulum

Golgi Apparatus and Secretory Vesicles
1. Golgi apparatus; 2. Transport vesicle; 3. Secretory vesicle; 4. Secretory vesicle

Lysosomes, Peroxisomes, and Proteasomes
1. Lysosomes; 2. Autophagia; 3. Peroxisomes; 4. Catalase; 5. Proteasomes

Mitochondria
1. ATP; 2. Cristae; 3. Oxidative metabolism; 4. Matrix; 5. Electron-transport chain; 6. DNA

Centrioles, Spindle Fibers, Cilia, Flagella, and Microvilli
1. Centrosome; 2. Centrioles; 3. Spindle fibers; 4. Cilia; 5. Flagella; 6. Microvilli

Cell Diagram
1. Plasma membrane; 2. Mitochondrion; 3. Ribosome (free); 4. Rough endoplasmic reticulum; 5. Smooth endoplasmic reticulum; 6. Golgi apparatus; 7. Secretory vesicles; 8. Cilia; 9. Microvilli; 10. Cytoskeleton; 11. Peroxisome; 12. Centrioles; 13. Centrosome; 14. Proteasome; 15. Phagocytic vesicle; 16. Lysosome; 17. Nucleus; 18. Nuclear pore; 19. Nucleolus; 20. Nuclear envelope; 21. Chromatin

Genes and Gene Expression
- A. 1. Transcription; 2. mRNA; 3. tRNA; 4. Translation
- B. 1. Triplet; 2. Gene; 3. Codon; 4. Thymine, Uracil; 5. Guanine
- C. 1. RNA polymerase; 2. Promoter; 3. Transcription factors; 4. Terminator
- D. 1. Exon; 2. Intron; 3. Pre-RNA; 4. Post-transcriptional processing; 5. Alternative splicing; 6. Start codon
- E. 1. Proprotein; 2. Proenzyme; 3. Post-translational processing; 4. Polyribosome

Cell Life Cycle
- A. 1. Interphase; 2. G_1 phase; 3. G_2 phase; 4. S phase; 5. G_0
- B. 1. DNA polymerase; 2. Leading strand; 3. Lagging strand; 4. DNA ligase

Cell Division: Mitosis and Cytokinesis
- A. 1. Mitosis; 2. Cytokinesis
- B. 1. Chromatids; 2. Centromere; 3. Astral fibers; 4. Spindle fibers
- C. 1. Prophase; 2. Metaphase; 3. Anaphase; 4. Telophase
- D. 1. Interphase; 2. Centrioles; 3. Chromatin; 4. Prophase; 5. Spindle fiber; 6. Centromere; 7. Astral fiber; 8. Chromatid; 9. Metaphase; 10. Anaphase; 11. Chromosomes (identical); 12. Telophase; 13. Cleavage furrow

Genetics
1. Genetics; 2. Mendelian genetics; 3. Genomic medicine

Mendelian Genetics
1. Genotype; 2. Phenotype; 3. Alleles; 4. Dominant allele; 5. Homozygous; 6. Heterozygous; 7. Albinism; 8. Polydactyly

Modern Concepts of Genetics
- A. 1. Somatic cells; 2. Gametes; 3. Diploid; 4. Haploid; 5. 23 pairs; 6. 23
- B. 1. Autosomal; 2. Sex chromosomes; 3. XX; 4. XY; 5. Karyotype; 6. Homologous; 7. Genome; 8. Locus
- C. 1. Multiple alleles; 2. Mutated allele (gene); 3. Complete dominance; 4. Codominance; 5. Incomplete dominance; 6. Polygenic traits; 7. Pleiotropy; 8. Sex-linked traits
- D. 1. Phenylketonuria; 2. Albinism; 3. ABO blood group; 4. Beta thalassemia; 5. Height; 6. Marfan syndrome; 7. Hemophilia
- E. 1. Meiosis; 2. Punnett square; 3. Carrier; 4. DNA replication; 5. Two; 6. True; 7. True
- F. 1. Genetic disorder; 2. Mutation; 3. Mutagen; 4. Nondisjunction; 5. Aneuploidy; 6. Down syndrome; 7. Carrier

QUICK RECALL

1. Energy use, synthesis of molecules, communication, and reproduction and inheritance
2. Lipids: phospholipids form the lipid bilayer that separates the inside of the cell from the outside; cholesterol: contributes to the fluid nature of the membrane; proteins: markers, attachment sites, membrane channels, carrier proteins, ATP-powered pumps, receptors, and enzymes
3. Magnitude of concentration gradient, temperature of solution, size of diffusing molecules, and viscosity of the solvent
4. Diffusion, osmosis, filtration, mediated transport (i.e., facilitated diffusion, active transport, and secondary active transport), endocytosis (i.e., phagocytosis, pinocytosis, and receptor-mediated endocytosis) and exocytosis
5. Specificity, saturation, and competition
6. Microtubules
7. Actin filaments
8. Intermediate filaments
9. Cytoplasmic inclusion
10. Nucleus
11. Nucleolus
12. Ribosome
13. Rough endoplasmic reticulum
14. Smooth endoplasmic reticulum
15. Golgi apparatus
16. Secretory vesicle
17. Lysosome
18. Peroxisome
19. Proteasome
20. Mitochondria
21. Centriole
22. Cilia
23. Microvilli
24. G_1, G_2, and S phases
25. Prophase, metaphase, anaphase, and telophase
26. Mitosis has one cell division, producing two cells, each of which has the same number of chromosomes as the parent cell. Meiosis has two divisions and produces four cells, each of which has half the number of chromosomes as the parent cell.

4 Histology: The Study of Tissues

CONTENT LEARNING ACTIVITY

Tissues and Histology

Match these terms with the correct statement or definition:

Autopsy
Biopsy
Extracellular matrix
Histology
Tissue level of organization
Tissues

_____ 1. Similar cells and the substances surrounding them.

_____ 2. Level of organization that contains all the tissue types.

_____ 3. Noncellular substances surrounding cells.

_____ 4. Microscopic study of tissues.

_____ 5. Removing tissue samples from patients.

Embryonic Tissue

Match these terms with the correct statement or definition:

Ectoderm
Endoderm
Mesenchyme
Mesoderm
Neural crest cells
Neuroectoderm

_____ 1. Inner germ layer; forms the lining of the digestive tract and its derivatives.

_____ 2. Middle germ layer; forms tissues such as muscle and bone.

_____ 3. Outer germ layer; forms the skin and other structures.

_____ 4. Part of the outer germ layer that becomes the nervous system.

_____ 5. Groups of cells that break away from neuroectoderm; become peripheral nerves, skin pigment, and many tissues of the face.

_____ 6. Embryonic tissue from which connective tissues arise; formed from mesoderm and neural crest cells.

Epithelial Tissue

Using the terms provided, complete these statements:

Basal
Basement membrane
Cell contacts
Diffusion
Extracellular
Free (apical)
Lateral
Mitosis
Surfaces
Tissues

Epithelial cells have very little (1) matrix between them. Epithelium covers (2) of the body and forms glands that are derived developmentally from surfaces. Most epithelial tissues have one (3) surface not attached to other cells, a(n) (4) surface attached to other epithelial cells, and a(n) (5) surface, which is usually attached to a(n) (6). The basement membrane, which is extracellular material secreted on the basal surface, helps attach epithelial cells to underlying (7) and helps guide cell migration during tissue repair. Specialized (8) bind adjacent epithelial cells together. Blood vessels in the underlying connective tissue do not penetrate the basement membrane; all gases and nutrients must reach the epithelium by (9). Because epithelial cells retain the ability to undergo (10), damaged cells can be replaced with new epithelial cells.

1. _____
2. _____
3. _____
4. _____
5. _____
6. _____
7. _____
8. _____
9. _____
10. _____

Functions of Epithelial Tissue

Using the terms provided, complete these statements:

Absorption
Barrier
Permits
Protecting
Secreting

Major functions of epithelial tissue include (1) underlying structures, such as the oral epithelium preventing abrasion of underlying structures. Epithelial tissue acts as a(n) (2), preventing the movement of many substances, such as water and toxins, through the epithelial layer. Conversely, epithelium (3) the movement of many substances, such as oxygen and carbon dioxide, through the epithelial layer. Epithelial tissues function in (4) substances, such as sweat, enzymes, and mucus. The cell membranes of certain epithelial tissues contain carrier proteins that regulate the (5) of materials into the cell.

1. _____
2. _____
3. _____
4. _____
5. _____

Classification of Epithelium

A. Match these types of epithelium with the correct statement or definition:

Keratinized stratified squamous
Nonkeratinized (moist) stratified squamous
Pseudostratified columnar
Simple columnar
Simple cuboidal
Simple squamous
Stratified columnar
Stratified cuboidal
Transitional

_____ 1. Single layer of cube-shaped cells.

_____ 2. Multiple layers of tall, thin cells.

_____ 3. Single layer of flat or scalelike cells that rests on the basement membrane.

_____ 4. Single layer of cells; all cells are attached to the basement membrane, but only some of them reach the free surface.

_____ 5. Multiple layers of cells in which the outermost cell layer is living and covered with a fluid layer.

_____ 6. Multiple layers of cells; outer cells are dead and dry.

_____ 7. Layers of cells that appear cubelike when an organ is relaxed and flattened when the organ is distended by fluid.

B. Match these terms with the types of epithelial tissue in figure 4.1:

Pseudostratified columnar
Simple columnar
Simple squamous
Transitional

1. _____

2. _____

3. _____

4. _____

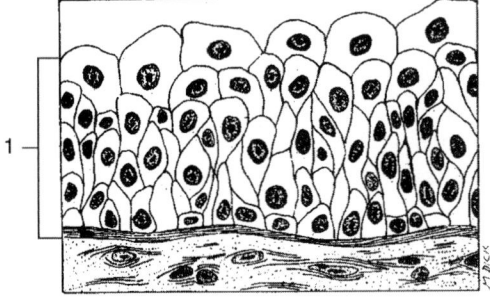
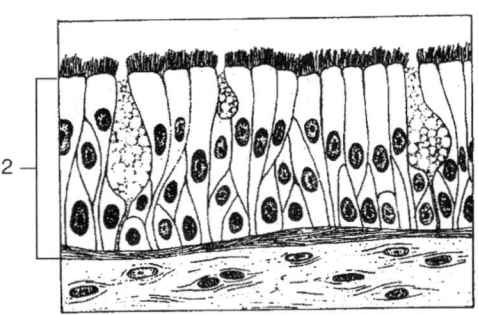
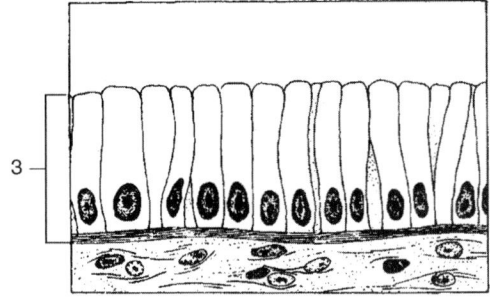
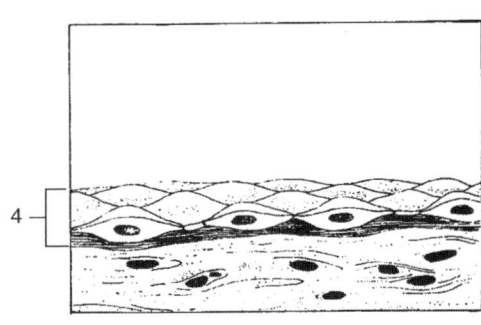

Figure 4.1

Functional Characteristics

A. Match these terms with the correct statement or definition:

Simple epithelium
Stratified epithelium

_____ 1. Found in organs where the principal function is diffusion, filtration, secretion, or absorption.

_____ 2. Found in areas where protection from abrasion is a major function.

_____ 3. Found in areas such as the mouth, skin, throat, anus, and vagina.

B. Match these terms with the correct statement or definition:

Cuboidal or columnar
Squamous

_____ 1. Epithelial cells involved with secretion or absorption.

_____ 2. Epithelial cells involved with diffusion or filtration.

_____ 3. Epithehial cells with greater cytoplasmic volume—e.g., goblet cells.

C. Match these terms with the correct statement or definition:

Ciliated   Microvillar
Folded   Smooth

_____ 1. Cell surface that reduces friction.

_____ 2. Cell surface that greatly increases surface area.

_____ 3. Propels materials along the cell surface.

_____ 4. Cell surface with alternate rigid and flexible sections.

D. Using the terms provided, complete these statements:

Cilia   Hemidesmosomes
Desmosomes   Intercalated disks
Gap junctions   Zonula adherens
Glycoproteins   Zonula occludens

Epithelial cells secrete __(1)__ that attach the cells to the basement membrane and to one another. This relatively weak binding is reinforced by __(2)__, disk-shaped structures with especially adhesive glycoproteins that bind cells to one another. __(3)__, similar to one-half of a desmosome, attach epithelial cells to the basement membrane. Tight junctions consist of the __(4)__, which acts as a weak glue to hold cells together, and the __(5)__, which acts as a permeability barrier. __(6)__ are small contact regions between cells containing protein channels that allow the passage of ions and small molecules between cells as a means of intercellular communication. Gap junctions between ciliated epithelial cells may coordinate movements of __(7)__. Specialized gap junctions between cardiac cells are called __(8)__.

1. _____
2. _____
3. _____
4. _____
5. _____
6. _____
7. _____
8. _____

Glands

A. Match these terms with the correct statement or definition:

Endocrine
Exocrine

_____ 1. Glands with a duct that is lined with epithelium.

_____ 2. Ductless glands that secrete hormones into the bloodstream.

B. Match these terms with the correct statement or definition:

Acinar or alveolar   Straight
Coiled   Tubular
Compound   Unicellular
Simple

_____ 1. Exocrine glands composed of one cell—e.g., goblet cells.

_____ 2. Exocrine glands with ducts that have few branches.

_____ 3. Exocrine glands with ducts that branch repeatedly.

_____ 4. Exocrine glands with ducts that end in tubules.

_____ 5. Exocrine glands with ducts that end in saclike structures.

_____ 6. Exocrine glands with ducts that coil.

_____ 7. Tubular exocrine glands with ducts that have no coiling.

C. Match these terms with the correct statement or definition:

Apocrine   Merocrine
Holocrine

_____ 1. Exocrine glands that secrete products with no loss of actual cellular material—e.g., water-producing sweat glands.

_____ 2. Exocrine glands that discharge fragments of the gland's cells into the secretion—e.g., mammary glands.

_____ 3. Exocrine glands that shed entire cells—e.g., sebaceous glands.

Chapter 4

D. Match these terms with the types of exocrine glands in figure 4.2:

Compound acinar
Compound tubular
Simple acinar
Simple branched acinar
Simple coiled tubular
Simple straight tubular

1. _____
2. _____
3. _____
4. _____
5. _____
6. _____

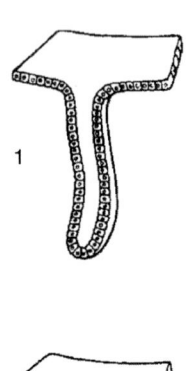
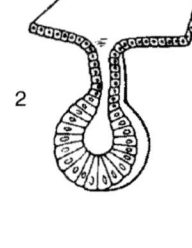
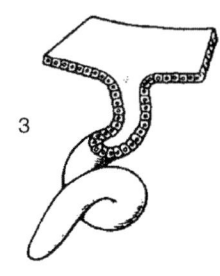
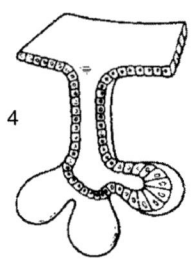
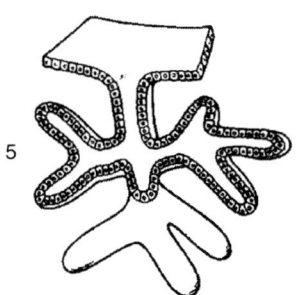
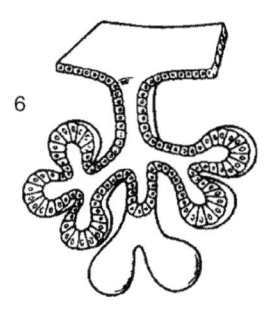

Figure 4.2

Cells of Connective Tissue

A. Match these terms with the correct statement or definition:

Blasts
Clasts
Cytes

_____ 1. Suffix for connective tissue cells that create extracellular matrix.

_____ 2. Suffix for connective tissue cells that maintain extracellular matrix.

_____ 3. Suffix for connective tissue cells that break down the extracellular matrix for remodeling.

B. Match these terms with the correct statement or definition:

Adipocytes
Macrophages
Mast cells
Stem cells

_____ 1. Cells that contain a large, centrally located lipid droplet with a thin layer of cytoplasm around it.

_____ 2. Cells that release histamine and heparin in response to injury; located beneath membranes in loose connective tissue.

_____ 3. Fixed or wandering cells derived from monocytes; phagocytize foreign and injured cells.

_____ 4. Undifferentiated mesenchymal cells that persist in adult connective tissue.

Extracellular Matrix

A. Match these terms with the correct statement or definition:

Collagen fibers
Elastin fibers
Reticular fibers

_____ 1. These protein fibers are the most common protein in the body and are strong and flexible but inelastic.

_____ 2. These protein molecules are very short, thin collagen fibers that branch to form a network.

_____ 3. This protein gives the tissue in which it is found an elastic quality; it contains polypeptide chains that recoil when stretched.

B. Using the terms provided, complete these statements:

Aggregates
Chondronectin
Fibronectin
Glycosaminoglycans
Ground substance
Hyaluronic acid
Monomers
Osteonectin
Proteoglycans
Water

The _(1)_ is the shapeless background against which collagen fibers are seen through the microscope. Molecules in ground substance include _(2)_, a long, unbranched polysaccharide that is very slippery, and _(3)_, which are formed from proteins and polysaccharides. Proteoglycan _(4)_ are composed of 80–100 polysaccharides, called _(5)_, each attached at one end to a common protein core. If the protein cores of proteoglycan monomers attach to hyaluronic acid, proteoglycan _(6)_ are formed. Proteoglycans trap large quantities of _(7)_, which allows them to return to their original shape when compressed or deformed. Several adhesive molecules are found in ground substance, including _(8)_ in cartilage, _(9)_ in bone, and _(10)_ in fibrous connective tissue.

1. _____
2. _____
3. _____
4. _____
5. _____
6. _____
7. _____
8. _____
9. _____
10. _____

Classification of Connective Tissue

A. Match these terms with the correct statement or definition pertaining to embryonic connective tissue:

Mesenchyme   Mucous connective tissue (Wharton's jelly)

_____ 1. All adult connective tissue types develop from this embryonic connective tissue.

_____ 2. Embryonic connective tissue located in the umbilical cord.

B. Match these terms with the correct statement or definition pertaining to adult connective tissue:

Dense   Fibrocytes
Fibroblasts   Loose (areolar)

_____ 1. Fibrous connective tissue in which protein fibers form a lacy network, with numerous fluid-filled spaces.

_____ 2. Fibrous connective tissue in which protein fibers form thick bundles and fill nearly all the extracellular space.

_____ 3. Cells that produce the fibers of connective tissue.

_____ 4. Fibroblasts that are completely surrounded by matrix.

C. Match these terms with the correct statement or definition:

Dense irregular
Dense regular

_____ 1. Dense connective tissue that contains protein fibers predominantly oriented in the same direction; strong in one direction.

_____ 2. Dense connective tissue that contains protein fibers that can be arranged as a meshwork of randomly oriented fibers; strong in many directions.

D. Match these terms with the correct statement or definition:

Dense irregular collagenous   Dense regular collagenous
Dense irregular elastic   Dense regular elastic

_____ 1. Connective tissue in tendons and ligaments; collagen fibers oriented in the same direction.

_____ 2. Connective tissue in special ligaments, such as the nuchal ligament; collagen and elastin fibers oriented in the same direction.

_____ 3. Connective tissue with randomly oriented collagen fibers; located in the skin (dermis) and capsules of the kidneys and spleen.

_____ 4. Connective tissue with randomly oriented fibers of collagen and elastin; located in the walls of elastic arteries.

E. Match these terms with the correct statement or definition:

Adipose tissue   Reticular tissue
Reticular cells

_____ 1. Consists of adipocytes (fat cells), which contain large amounts of lipid and can be either brown or yellow (white) in color.

_____ 2. Forms the framework of the spleen, lymph nodes, bone marrow, and liver.

_____ 3. Cells that produce reticular fibers.

Cartilage

Match these terms with the correct statement or definition:

Chondrocytes   Hyaline cartilage
Elastic cartilage   Lacunae
Fibrocartilage   Perichondrium

_____ 1. Cartilage cells.

_____ 2. Spaces in which cartilage cells are located.

_____ 3. Dense irregular connective tissue on the surface of cartilage.

_____ 4. Very smooth cartilage with a glassy, translucent matrix; located in the rib cage, rings of bronchi and trachea, and joints.

_____ 5. Cartilage with thick bundles of collagen dispersed through the matrix; located in areas that must withstand a great deal of pressure, such as the knee and intervertebral discs.

_____ 6. Cartilage with elastic fibers in addition to collagen and proteoglycans; located in external ears.

Bone

Match these terms with the correct statement or definition:

Cancellous (spongy) bone   Osteocytes
Compact bone   Red marrow
Hydroxyapatite   Yellow marrow
Lacunae

_____ 1. Specialized salt crystal; the mineral (inorganic) portion of bone.

_____ 2. Bone cells.

_____ 3. Spaces occupied by bone cells.

_____ 4. Bone with spaces between trabeculae.

_____ 5. More solid bone with almost no space between lamellae.

_____ 6. Located in bone cavities; contains yellow adipose tissue.

_____ 7. Hemopoietic tissue in bone cavities.

Identification of Connective Tissue

Match these terms with the correct tissue type or structure labeled in figure 4.3:

Adipose
Bone
Cartilage
Chondrocyte
Dense regular connective tissue

Fat droplet
Fibroblast
Lacuna
Osteocyte

1. _____
2. _____
3. _____
4. _____
5. _____
6. _____
7. _____
8. _____
9. _____

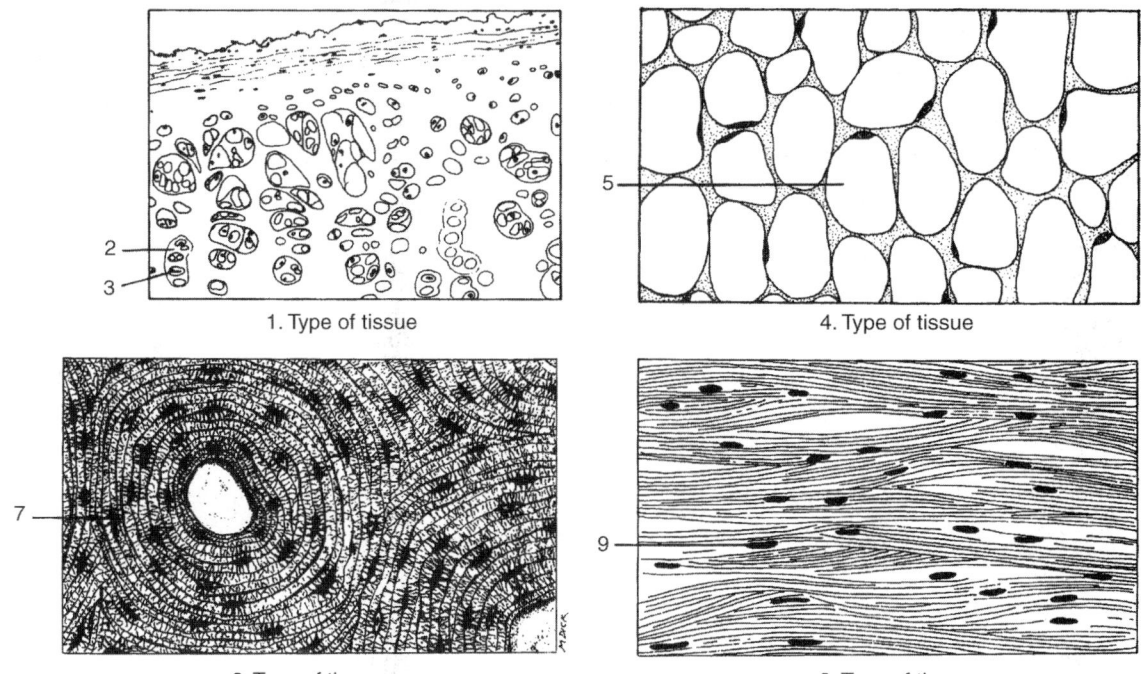

Figure 4.3

Muscle Tissue

A. Match these terms with the correct statement or definition:

Cardiac
Skeletal

Smooth

_____ 1. Striated, voluntary muscle cells.

_____ 2. Striated, involuntary muscle cells.

_____ 3. Nonstriated, involuntary muscle cells.

B. Match these terms with the correct tissue type or structure labeled in figure 4.4:

Cardiac muscle
Intercalated disk

Skeletal muscle
Smooth muscle

1. _____   3. _____

2. _____   4. _____

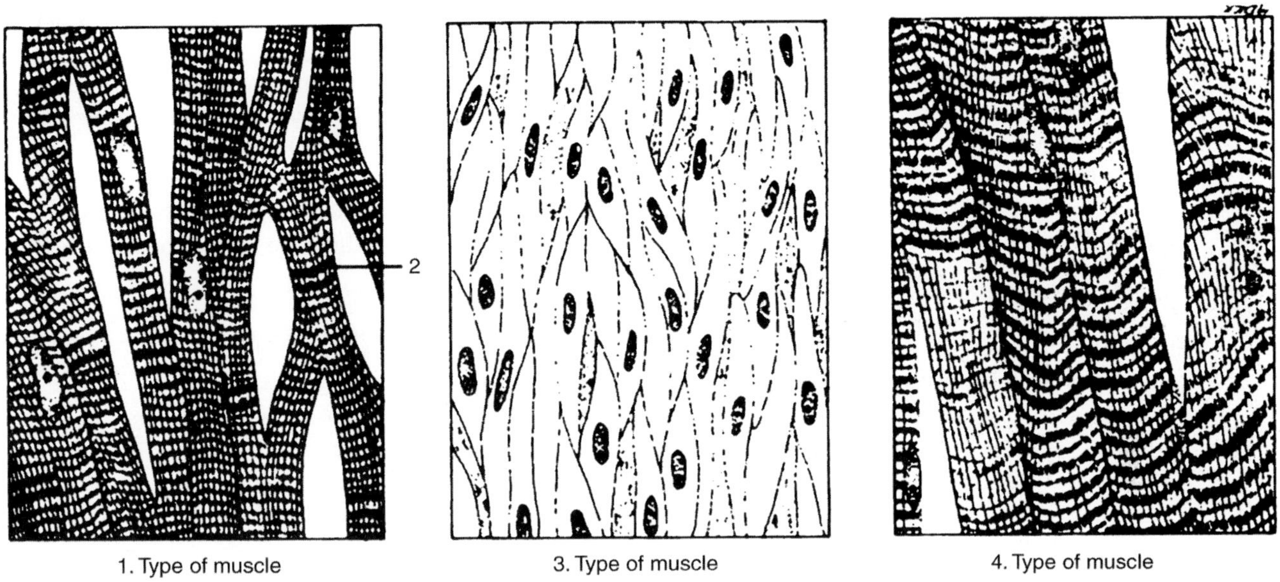

1. Type of muscle

3. Type of muscle

4. Type of muscle

Figure 4.4

Nervous Tissue

Match these terms with the correct statement or definition:

Action potentials
Axon
Bipolar
Cell body

Dendrite
Multipolar
Neuroglia
Unipolar

_____ 1. Electric signals conducted by nerve cells.

_____ 2. Part of a neuron that contains the nucleus.

_____ 3. Nerve cell process that usually receives electrical signals and conducts them toward the cell body.

_____ 4. Nerve cell process that conducts action potentials away from the cell body.

_____ 5. Neuron with several dendrites and one axon.

_____ 6. Neuron with one dendrite and one axon.

_____ 7. Support cells of the brain, spinal cord, and peripheral nerves that nourish, protect, and insulate neurons.

Chapter 4

Membranes

A. Match these terms with the correct statement or definition:

Mucous membranes   Synovial membranes
Serous membranes

_____

1. Consist of epithelial cells and their basement membrane, which rests on a thick layer of loose connective tissue called the lamina propria; line cavities that open to the outside of the body.

_____

2. Consist of simple squamous epithelium (mesothelium), its basement membrane, and a delicate layer of loose connective tissue; line cavities that do not open to the exterior.

_____

3. Modified connective tissue cells associated with the dense connective tissue of the joint capsule; line freely movable joints.

B. Match these terms with the correct statement or definition:

Mucus   Synovial fluid
Serous fluid

_____

1. Fluid that lubricates serous membranes.

_____

2. Viscous substance produced by goblet cells and multicellular glands.

_____

3. Fluid rich in hyaluronic acid; lubricates freely movable joints.

Inflammation

Match these terms with the correct statement or definition:

Blood vessel dilation   Edema
Clotting   Mediators of inflammation
Disturbance of function Pain

_____

1. Chemical substances that are released or activated in tissues and adjacent blood vessels after a person is injured.

_____

2. Expansion of blood vessels, which produces symptoms of redness and heat.

_____

3. Swelling of a tissue because of fluid accumulation; results from increased permeability of blood vessels.

_____

4. Occurs when certain proteins from the blood react; "walls off" the site of the injury from the rest of the body.

_____

5. Result of edema and some mediators stimulating nerves.

_____

6. Result of pain, limitation of movement resulting from edema, and tissue destruction.

Tissue Repair

A. Using the terms provided, complete these statements:

   Labile   Replacement
   Permanent   Scar
   Regeneration   Stable

In (1) , the new cells are of the same type as those that were destroyed, whereas, in (2) , a new type of tissue develops that eventually causes (3) production and the loss of some tissue function. (4) cells continue to divide throughout life. Damage to these cells can be completely repaired by regeneration. (5) cells do not actively replicate after growth ceases, but they do retain the ability to divide if necessary and are capable of regeneration. (6) cells have a very limited ability to replicate, and, if killed, are replaced by a different type of cell.

1. _____
2. _____
3. _____
4. _____
5. _____
6. _____

B. Using the terms provided, complete these statements:

   Fibrin   Scab
   Granulation tissue   Scar
   Phagocytic   Secondary union
   Primary union (intention)   (intention)
   Pus   Wound contraction

If the edges of a wound are close together, such as in a surgical incision, the wound heals by a process called (1) . The wound fills with blood and a clot forms. The clot is filled with a threadlike protein, (2) , and the surface of the clot dries to form a (3) , which seals the wound. Some of the white blood cells that move into the damaged tissue are (4) cells called neutrophils. As these cells ingest bacteria and tissue debris, they may be killed and accumulate as a mixture of dead cells and fluid called (5) . Eventually, the clot is replaced by (6) , which consists of fibroblasts, collagen, and capillaries. A large amount of granulation tissue is converted to a (7) . If the edges of the wound are not close together or if there has been extensive loss of tissue, the process is called (8) . In this process, wound regeneration takes longer, and (9) occurs when fibroblasts in the granulation tissue contract. As a result, disfiguring and debilitating scars may result.

1. _____
2. _____
3. _____
4. _____
5. _____
6. _____
7. _____
8. _____
9. _____

Quick Recall

1. List the three primary germ layers and one derivative of each.

2. List eight kinds of epithelium based on numbers of cell layers and shape of cells.

3. List five functions performed by epithelial cells.

4. List the four types of free cell surfaces of epithelial cells.

5. Name five ways that cells are mechanically bound together.

6. List three types of exocrine glands based on how products leave the cell.

7. Name three types of protein fibers located in connective tissue.

8. List four types of dense connective tissue in the human body.

9. List two types of connective tissue with special properties.

10. Name three types of cartilage in the human body.

11. Name two types of bone in the human body.

12. List three types of muscle cells in the human body.

13. List the three major categories of membranes in the human body.

14. List the five major symptoms of inflammation.

15. List the three categories into which cells can be classified according to their regenerative ability.

ANSWERS TO CHAPTER 4

CONTENT LEARNING ACTIVITY

Tissues and Histology
 1. Tissues; 2. Tissue level of organization; 3. Extracellular matrix; 4. Histology; 5. Biopsy

Embryonic Tissue
 1. Endoderm; 2. Mesoderm; 3. Ectoderm; 4. Neuroectoderm; 5. Neural crest cells; 6. Mesenchyme

Epithelial Tissue
 1. Extracellular; 2. Surfaces; 3. Free (apical); 4. Lateral; 5. Basal; 6. Basement membrane; 7. Tissues; 8. Cell contacts; 9. Diffusion; 10. Mitosis

Functions of Epithelial Tissue
 1. Protecting; 2. Barrier; 3. Permits; 4. Secreting; 5. Absorption

Classification of Epithelium
 A. 1. Simple cuboidal; 2. Stratified columnar; 3. Simple squamous; 4. Pseudostratified columnar; 5. Nonkeratinized (moist) stratified squamous; 6. Keratinized stratified squamous; 7. Transitional
 B. 1. Transitional; 2. Pseudostratified columnar; 3. Simple columnar; 4. Simple squamous

Functional Characteristics
- A. 1. Simple epithelium; 2. Stratified epithelium; 3. Stratified epithelium
- B. 1. Cuboidal or columnar; 2. Squamous; 3. Cuboidal or columnar
- C. 1. Smooth; 2. Microvillar; 3. Ciliated; 4. Folded
- D. 1. Glycoproteins; 2. Desmosomes; 3. Hemidesmosomes; 4. Zonula adherens; 5. Zonula occludens; 6. Gap junctions; 7. Cilia; 8. Intercalated disks

Glands
- A. 1. Exocrine; 2. Endocrine
- B. 1. Unicellular; 2. Simple; 3. Compound; 4. Tubular; 5. Acinar or alveolar; 6. Coiled; 7. Straight
- C. 1. Merocrine; 2. Apocrine; 3. Holocrine
- D. 1. Simple straight tubular; 2. Simple acinar; 3. Simple coiled tubular; 4. Simple branched acinar; 5. Compound tubular; 6. Compound acinar

Cells of Connective Tissue
- A. 1. Blasts; 2. Cytes; 3. Clasts
- B. 1. Adipocytes; 2. Mast cells; 3. Macrophages; 4. Stem cells

Extracellular Matrix
- A. 1. Collagen fibers; 2. Reticular fibers; 3. Elastin fibers
- B. 1. Ground substance; 2. Hyaluronic acid; 3. Proteoglycans; 4. Monomers; 5. Glycosaminoglycans; 6. Aggregates; 7. Water; 8. Chondronectin; 9. Osteonectin; 10. Fibronectin

Classification of Connective Tissue
- A. 1. Mesenchyme; 2. Mucous connective tissue (Wharton's jelly)
- B. 1. Loose (areolar); 2. Dense; 3. Fibroblasts; 4. Fibrocytes
- C. 1. Dense regular; 2. Dense irregular
- D. 1. Dense regular collagenous; 2. Dense regular elastic; 3. Dense irregular collagenous; 4. Dense irregular elastic
- E. 1. Adipose tissue; 2. Reticular tissue; 3. Reticular cells

Cartilage
1. Chondrocytes; 2. Lacunae; 3. Perichondrium; 4. Hyaline cartilage; 5. Fibrocartilage; 6. Elastic cartilage

Bone
1. Hydroxyapatite; 2. Osteocytes; 3. Lacunae; 4. Cancellous (spongy) bone; 5. Compact bone; 6. Yellow marrow; 7. Red marrow

Identification of Connective Tissue
1. Cartilage; 2. Lacuna; 3. Chondrocyte; 4. Adipose; 5. Fat droplet; 6. Bone; 7. Osteocyte; 8. Dense regular connective tissue; 9. Fibroblast

Muscle Tissue
- A. 1. Skeletal; 2. Cardiac; 3. Smooth
- B. 1. Cardiac muscle; 2. Intercalated disk; 3. Smooth muscle; 4. Skeletal muscle

Nervous Tissue
1. Action potentials; 2. Cell body; 3. Dendrite; 4. Axon; 5. Multipolar; 6. Bipolar; 7. Neuroglia

Membranes
- A. 1. Mucous membranes; 2. Serous membranes; 3. Synovial membranes
- B. 1. Serous fluid; 2. Mucus; 3. Synovial fluid

Inflammation
1. Mediators of inflammation; 2. Blood vessel dilation; 3. Edema; 4. Coagulation; 5. Pain; 6. Disturbance of function

Tissue Repair
- A. 1. Regeneration; 2. Replacement; 3. Scar; 4. Labile; 5. Stable; 6. Permanent
- B. 1. Primary union (intention); 2. Fibrin; 3. Scab; 4. Phagocytic; 5. Pus; 6. Granulation tissue; 7. Scar; 8. Secondary union (intention); 9. Wound contraction

Quick Recall

1. Ectoderm: brain, spinal cord, and peripheral nerves; endoderm: lining of digestive system, trachea, bronchi, lungs, liver, and thyroid; mesoderm: bone, cartilage, tendons, muscle, and blood
2. Simple squamous, simple cuboidal, simple columnar, stratified squamous, stratified cuboidal, stratified columnar, pseudostratified, and transitional epithelium
3. Protecting underlying structures, acting as a barrier, permitting the passage of substances, secreting substances, and absorbing substances
4. Smooth, microvillar, ciliated, and folded
5. Glycoproteins, desmosomes, hemidesmosomes, tight junctions, (zonula adherens and zonula occludens), and gap junctions
6. Merocrine, apocrine, and holocrine
7. Collagen, reticular fibers, and elastin
8. Dense regular collagenous, dense irregular collagenous, dense regular elastic, and dense irregular elastic
9. Reticular and adipose
10. Hyaline, fibrocartilage, and elastic cartilage
11. Compact and cancellous
12. Skeletal, cardiac, and smooth muscle
13. Serous, mucous, and synovial membranes
14. Redness, heat, swelling, pain, and disturbance of function
15. Labile, stable, and permanent

5 Integumentary System

CONTENT LEARNING ACTIVITY

Skin

Match these terms with the
correct statement or definition:

Dermis
Epidermis

_____ 1. Connective tissue.

_____ 2. Layer of epithelial tissue.

_____ 3. Most superficial layer of the skin.

Epidermis

A. Match these terms with the
correct statement or definition:

Desquamate   Melanocytes
Keratinization   Merkel cells
Keratinocytes   Strata
Langerhans cells

_____ 1. Cells that produce a protein called keratin; the most abundant epidermal cell.

_____ 2. Cells in the epidermis that are part of the immune system.

_____ 3. Cells in the epidermis specialized to detect light touch and superficial pressure.

_____ 4. To slough or be lost from the surface of the epidermis.

_____ 5. Process that occurs in epidermal cells during their movement from deeper epidermal layers to the surface.

_____ 6. Layers of cells within the epidermis.

B. Match these terms with the correct statement or definition:

Keratohyalin
Lamellar bodies
Stratum basale
Stratum corneum
Stratum granulosum
Stratum lucidum
Stratum spinosum

_____ 1. Deepest portion of the epidermis; a single layer of cells; the site of production of most epidermal cells.

_____ 2. Epidermal layer superficial to the stratum basale, consisting of 8 to 10 layers of many-sided cells.

_____ 3. Derives its name from protein granules contained in the cells and is superficial to the stratum spinosum.

_____ 4. Nonmembrane-bound protein granules located in the cells of the stratum granulosum.

_____ 5. Structures that move to the cell membrane and release their lipid contents into the intercellular space; the lipids are responsible for the permeability characteristics of the epidermis.

_____ 6. Clear, thin zone above the stratum granulosum; absent in most skin.

_____ 7. Most superficial stratum of the epidermis; dead cells with hard protein envelope and filled with keratin, which provides structural strength.

C. Match these terms with the parts labeled in figure 5.1:

Dermis
Epidermis
Stratum basale
Stratum corneum
Stratum granulosum
Stratum lucidum
Stratum spinosum

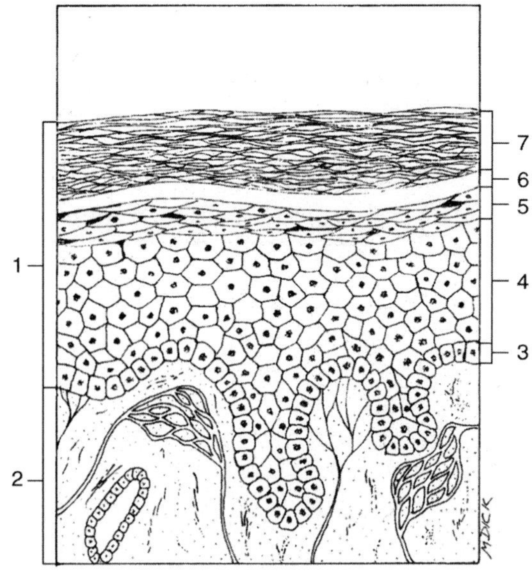

1. _____
2. _____
3. _____
4. _____
5. _____
6. _____
7. _____

Figure 5.1

Thick Skin and Thin Skin

Match these terms with the correct statement or definition:

Callus Thick skin
Corn Thin skin

_____ 1. The papillae of the dermis of this type of skin comprise curving ridges that produce fingerprints and footprints.

_____ 2. In this type of skin, the stratum lucidum is usually absent; hair is found in this type of skin.

_____ 3. Most abundant type of skin.

_____ 4. Thickened area of thin or thick skin resulting from a greatly increased number of layers of stratum corneum.

_____ 5. Cone-shaped structure that develops in thin or thick skin over a bony prominence.

Skin Color

Using the terms provided, complete the following statements:
- Albinism
- Carotene
- Cyanosis
- Melanin
- Melanocytes
- Melanosomes
- Ultraviolet light

_(1)_, a brown to black pigment, is responsible for most skin color. It is produced by _(2)_, irregularly shaped cells with many long processes that extend between the keratinocytes of the stratum basale and stratum spinosum. Melanin is packaged into vesicles called _(3)_, which are released from the cell processes by exocytosis. A single mutation can prevent the manufacture of melanin, resulting in _(4)_. Exposure to _(5)_ increases melanin production. _(6)_ is a yellow pigment found in plants, such as carrots. When large amounts of this pigment are consumed, the excess accumulates in the stratum corneum and fat cells of the dermis and hypodermis, causing the skin to develop a yellowish tint. A decrease in blood oxygen content produces _(7)_, a bluish skin color, whereas an abundant supply of oxygenated blood produces a reddish hue.

1. _____
2. _____
3. _____
4. _____
5. _____
6. _____
7. _____

Dermis

Match these terms with the correct statement or definition:

Cleavage lines Reticular layer
Papillae Striae
Papillary layer

_____ 1. Deep layer of dermis; dense irregular connective tissue that blends into the hypodermis.

Chapter 5

_____ 2. Lines visible through the epidermis produced by rupture of the dermis.

_____ 3. Projections from the dermis toward the epidermis.

Hypodermis

Match these terms with the correct statement or definition:

Hypodermis
Loose connective tissue

_____ 1. Layer of tissue connecting the skin to underlying structures.

_____ 2. Type of tissue found in the hypodermis.

_____ 3. Also called subcutaneous tissue or superficial fascia.

Hair

A. Match these terms with the correct statement or definition:

Lanugo   Vellus hairs
Terminal hairs

_____ 1. Delicate, unpigmented hair that covers a fetus.

_____ 2. Long, coarse, pigmented hairs that replace lanugo on the scalp, eyebrows, and eyelids.

_____ 3. Short, fine, unpigmented hairs that replace lanugo over most of the body.

_____ 4. Type of hair that replaces vellus hairs at puberty.

B. Match these terms with the correct statement or definition:

Cortex   Medulla
Cuticle   Root
Hair bulb   Shaft

_____ 1. Portion of hair protruding above the surface of the skin.

_____ 2. Expanded knob at the base of the hair root.

_____ 3. Central axis of the hair that consists of two or three layers of cells containing soft keratin.

_____ 4. Forms the bulk of the hair; consists of cells containing hard keratin.

_____ 5. Outermost layer of the hair shaft and root, composed of a single overlapping layer of cells containing hard keratin.

C. Match these terms with the correct statement or definition:

Arrector pili Matrix
Dermal root sheath Melanocytes
Epithelial root sheath

_____ 1. Surrounds the epithelial root sheath.

_____ 2. Layers of cells immediately surrounding the root of the hair.

_____ 3. Mass of undifferentiated epithelial cells inside the hair bulb that produces the hair and internal epithelial root sheath.

_____ 4. Produce the pigment responsible for hair color.

_____ 5. Smooth muscle cells that attach to the hair follicle dermal root sheath and the papillary layer of the dermis; cause hair to "stand on end."

D. Match these terms with the correct parts labeled in figure 5.2:

Cortex Hair follicle
Cuticle Hair root
Dermal papilla Hair shaft
Dermal root sheath Internal epithelial root sheath
External epithelial root sheath Matrix
Hair Medulla
Hair bulb

1. _____
2. _____
3. _____
4. _____
5. _____
6. _____
7. _____
8. _____
9. _____
10. _____
11. _____
12. _____
13. _____

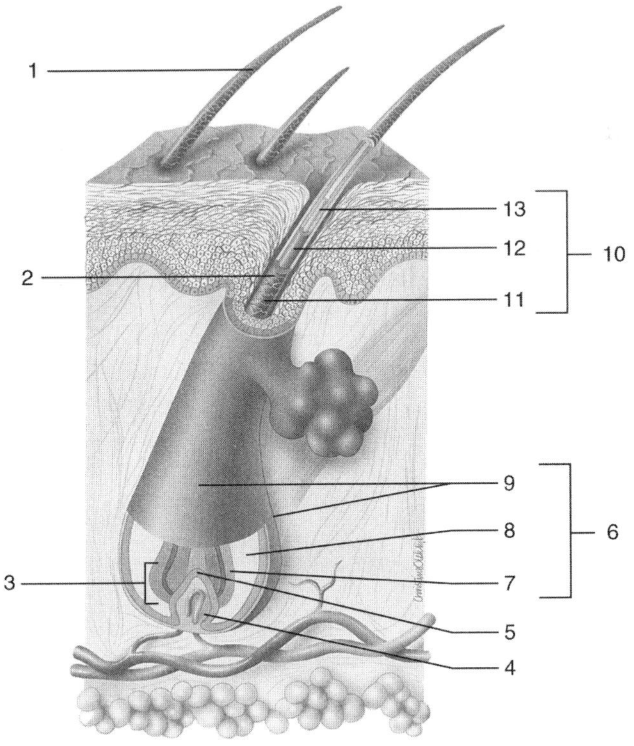

Figure 5.2

Chapter 5

Glands

A. Match these terms with the correct statement or definition:

Apocrine sweat gland
Ceruminous gland
Merocrine sweat gland
Sebaceous gland
Sebum

_____ 1. White substance rich in lipids; oils the hair and skin surface, prevents drying, and protects against bacteria.

_____ 2. Gland that opens into a hair follicle; produces sebum.

_____ 3. Two types of sweat, or sudoriferous, glands.

_____

_____ 4. Gland that opens to the surface of the skin and secretes an isotonic fluid that is mostly water; involved with temperature regulation.

_____ 5. Gland that usually opens into a hair follicle; secretes an organic substance that is metabolized by bacteria to produce body odor.

_____ 6. Gland that produces earwax.

B. Match these terms with the glands labeled in figure 5.3:

Apocrine sweat gland
Merocrine sweat gland
Sebaceous gland

1. _____
2. _____
3. _____

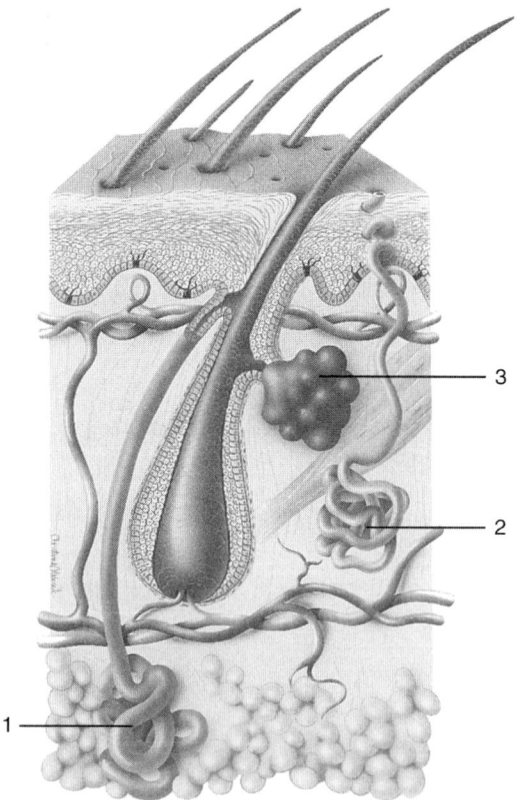

Figure 5.3

Chapter 5

Nails

A. Match these terms with the correct statement or definition:

Eponychium
Hyponychium
Lunula
Nail bed
Nail body
Nail fold
Nail groove
Nail matrix
Nail root

_____ 1. Proximal portion of the nail that is covered by skin.

_____ 2. Portion of the skin that covers the lateral and proximal edges of the nail.

_____ 3. Holds the edges of the nail in place.

_____ 4. Cuticle; the stratum corneum of the nail fold that grows onto the nail body.

_____ 5. Nail root and nail body attach to this structure.

_____ 6. Proximal portion of the nail bed; produces most of the nail.

_____ 7. Whitish, crescent-shaped area at the base of the nail; part of the nail matrix.

B. Match these terms with the parts of the nail labeled in figure 5.4:

Eponychium (cuticle)
Hyponychium
Lunula
Nail bed
Nail body
Nail matrix
Nail root

1. _____
2. _____
3. _____
4. _____
5. _____
6. _____
7. _____
8. _____
9. _____
10. _____

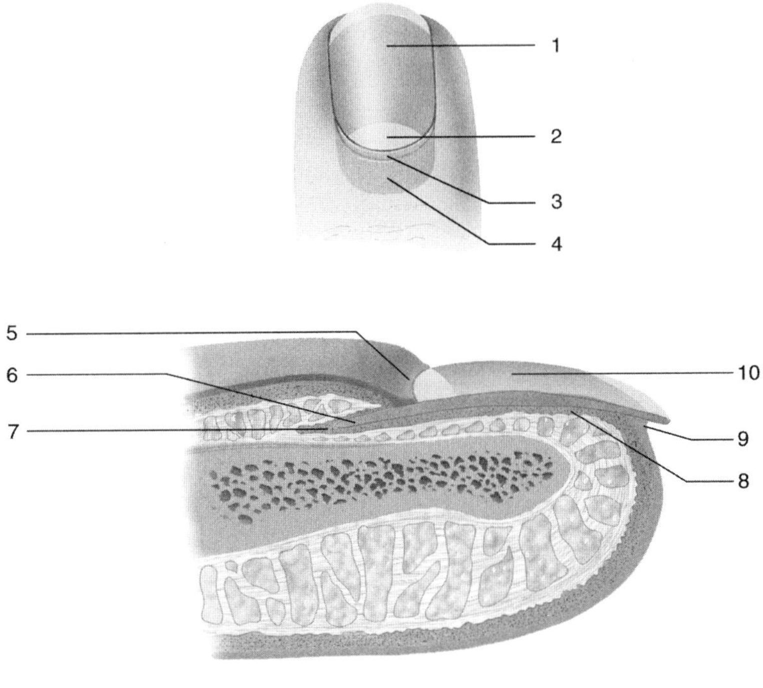

Figure 5.4

Chapter 5

Functions of the Integumentary System

Match these terms with the correct statement or definition:

Excretion
Protection
Sensation
Temperature regulation
Vitamin D production

_____ 1. Accomplished by the skin as a physical barrier, as a permeability barrier, as a barrier against ultraviolet light, and as a barrier against abrasion.

_____ 2. Detection of touch, temperature, and pain.

_____ 3. Carried out by producing sweat and increasing or decreasing blood vessel diameter.

_____ 4. Begins when a precursor molecule in the skin is exposed to ultraviolet light and is converted to cholecalciferol.

_____ 5. Occurs to a very slight degree with sweat production when some urea, uric acid, and ammonia are lost.

The Effects of Aging on the Integumentary System

Using the terms provided, complete the following statements:

Decrease(s)   Increase(s)

As the body ages, blood flow to the skin _(1)_, and the thickness of skin _(2)_. Elastic fibers in the dermis _(3)_, and the skin tends to sag. A(n) _(4)_ in the activity of sebaceous and sweat glands results in dry skin and a(n) _(5)_ in thermoregulatory ability. The _(6)_ in ability to sweat can contribute to death from heat prostration in elderly individuals. The number of functioning melanocytes _(7)_, but in some localized areas, especially the hands and face, melanocytes _(8)_ to produce age spots. White or gray hairs occur because of a(n) _(9)_ in melanin production.

1. _____
2. _____
3. _____
4. _____
5. _____
6. _____
7. _____
8. _____
9. _____

Integumentary Disorders

A. Match these terms with the correct statement or definition:

Acne
Decubitus ulcers
Ringworm and athlete's foot
Warts

_____ 1. Disorder of the hair follicles and sebaceous glands that involves testosterone and bacteria.

_____ 2. Viral infection of the epidermis.

_____ 3. Fungal infections that affect the keratinized portion of the skin.

_____ 4. Disorder caused by ischemia and necrosis of the hypodermis.

B. Match these terms with the correct statement or definition:

Basal cell carcinoma Squamous cell carcinoma
Malignant melanoma

_____ 1. Cancer that begins in the stratum basale and extends into the dermi; the most frequent type of skin cancer.

_____ 2. Cancer that typically produces a nodular, keratinized tumor confined to the epidermis.

_____ 3. Less common form of skin cancer that usually arises from a preexisting mole; most often fatal.

C. Match these terms with the correct statement or definition:

First degree burn Second degree burn
Full-thickness burn Third degree burn
Partial-thickness burn

_____ 1. Involves only the epidermis; burned area is red and painful and swelling may occur.

_____ 2. Damages the epidermis and dermis, but not the hypodermis; repair is from the wound edges and epithelial tissue in hair follicles and sweat glands.

_____ 3. Complete destruction of the epidermis and dermis; may involve deeper tissues; repair is only from the wound edges.

_____ 4. These two are also called partial-thickness burns.

_____

D. Match these terms with the correct statement or definition:

Major burn 1%
Minor burn 9%
Moderate burn 18%

_____ 1. Surface area of the head and neck of an adult; surface area of one entire upper limb of an adult.

_____ 2. Surface area of one entire lower limb of an adult; surface area of the anterior or posterior trunk of an adult.

_____ 3. Surface area of the genitalia of an adult.

_____ 4. Third degree burn over 10% or more of the BSA, a second degree burn over 25% or more of the BSA, or a second or third degree burn of the hands, feet, face, genitals, or anal region.

_____ 5. Third degree burn of 2%–10% of the BSA or a second degree burn of 15%–25% of the BSA.

_____ 6. Third degree burn of less than 2% or a second degree burn of less than 15% of the BSA.

Chapter 5

Quick Recall

1. List the five strata of the epidermis from the deepest to the most superficial.

2. Name the two layers of the dermis.

3. List three types of hair located at different stages of development in humans.

4. Name the two stages in the hair production cycle.

5. List the three major types of glands associated with the skin.

6. List four protective functions of the integumentary system.

7. List two ways the integumentary system regulates body temperature.

8. List four effects of aging on the integumentary system.

ANSWERS TO CHAPTER 5

CONTENT LEARNING ACTIVITY

Skin
1. Dermis; 2. Epidermis; 3. Epidermis

Epidermis
A. 1. Keratinocytes; 2. Langerhans cells; 3. Merkel cells; 4. Desquamate; 5. Keratinization; 6. Strata
B. 1. Stratum basale; 2. Stratum spinosum; 3. Stratum granulosum; 4. Keratohyalin; 5. Lamellar bodies; 6. Stratum lucidum; 7. Stratum corneum
C. 1. Epidermis; 2. Dermis; 3. Stratum basale; 4. Stratum spinosum; 5. Stratum granulosum; 6. Stratum lucidum; 7. Stratum corneum

Thick Skin and Thin Skin
1. Thick skin; 2. Thin skin; 3. Thin skin; 4. Callus; 5. Corn

Skin Color
1. Melanin; 2. Melanocytes; 3. Melanosomes; 4. Albinism; 5. Ultraviolet light; 6. Carotene; 7. Cyanosis

Dermis
1. Reticular layer; 2. Striae; 3. Papillae

Hypodermis
1. Hypodermis; 2. Loose connective tissue; 3. Hypodermis

Hair
A. 1. Lanugo; 2. Terminal hairs; 3. Vellus hairs; 4. Terminal hairs
B. 1. Shaft; 2. Hair bulb; 3. Medulla; 4. Cortex; 5. Cuticle
C. 1. Dermal root sheath; 2. Epithelial root sheath; 3. Matrix; 4. Melanocytes; 5. Arrector pili
D. 1. Hair shaft; 2. Hair root; 3. Hair bulb; 4. Dermal papilla; 5. Matrix; 6. Hair follicle; 7. Internal epithelial root sheath; 8. External epithelial root sheath; 9. Dermal root sheath; 10. Hair; 11. Cuticle; 12. Cortex; 13. Medulla

Glands
A. 1. Sebum; 2. Sebaceous gland; 3. Merocrine sweat gland and apocrine sweat gland; 4. Merocrine sweat gland; 5. Apocrine sweat gland; 5. Ceruminous gland
B. 1. Apocrine sweat gland; 2. Merocrine sweat gland; 3. Sebaceous gland

Nails
A. 1. Nail root; 2. Nail fold; 3. Nail groove; 4. Eponychium; 5. Nail bed; 6. Nail matrix; 7. Lunula
B. 1. Nail body; 2. Lunula; 3. Eponychium (cuticle); 4. Nail root; 5. Eponychium (cuticle); 6. Nail root; 7. Nail matrix; 8. Nail bed; 9. Hyponychium; 10. Nail body

Functions of the Integumentary System
1. Protection; 2. Sensation; 3. Temperature regulation; 4. Vitamin D production; 5. Excretion

The Effects of Aging on the Integumentary System
1. Decreases; 2. Decreases; 3. Decrease; 4. Decrease; 5. Decrease; 6. Decrease; 7. Decreases; 8. Increase; 9. Decrease

Integumentary Disorders
A. 1. Acne; 2. Warts; 3. Ringworm and athlete's foot; 4. Decubitus ulcers
B. 1. Basal cell carcinoma; 2. Squamous cell carcinoma; 3. Malignant melanoma
C. 1. First degree burn; 2. Second degree burn 3. Third degree burn; 4. First and second degree burns
D. 1. 9%; 2. 18%; 3. 1%; 4. Major burn; 5. Moderate burn; 6. Minor burn

QUICK RECALL

1. Stratum basale, stratum spinosum, stratum granulosum, stratum lucidum, and stratum corneum
2. Reticular and papillary layers
3. Lanugo, vellus hairs, and terminal hairs
4. Growth stage and resting stage
5. Sebaceous glands, apocrine sweat glands, and merocrine sweat glands
6. Skin: prevents water loss, physical barrier to microorganisms, protection against mechanical damage, protection against ultraviolet light; hair (eyebrows and eyelashes): protection against damage to eyes; nails: protect digits
7. Dilation/constriction of blood vessels and sweat production
8. Decreased blood flow to skin, thinner skin, decreased elastic fibers, decreased activity of sweat and sebaceous glands, generally decreased melanocyte activity but age spots in some areas

Skeletal System: Bones and Bone Tissue

CONTENT LEARNING ACTIVITY

Functions of the Skeletal System

Match these terms with the correct statement or definition:

Bone
Cartilage
Ligament
Tendon

_____ 1. Very rigid tissue that bears weight and supports the body, protects internal organs, and provides attachment for muscles to produce body movement.

_____ 2. Tissue that provides flexible support and forms a smooth surface at some joints.

_____ 3. Strong band of fibrous connective tissue that attaches muscle to bone.

_____ 4. Strong band of fibrous connective tissue that attaches bone to bone; allows movement but prevents excessive movement.

_____ 5. Tissue that stores minerals and fat.

_____ 6. Contains marrow that gives rise to blood cells and platelets.

Cartilage

A. Match these terms with the correct statement or definition:

Articular cartilage
Chondroblast
Chondrocyte
Perichondrium

_____ 1. Cell that produces new matrix on the outside of cartilage.

_____ 2. Mature cartilage cell, in a lacuna that is surrounded by matrix.

_____ 3. Double-layered connective tissue sheath around cartilage.

_____ 4. Covers the ends of bones where they come together to form joints.

B. Match these terms with the correct statement or definition:

Appositional growth Interstitial growth

1. Cartilage growth that occurs when chondroblasts in the perichondrium lay down new matrix and add chondrocytes to the outside of the tissue.

2. Cartilage growth that occurs when chondrocytes within the tissue divide and add more matrix between the cells.

Bone Matrix

Match these terms with the correct statement or definition:

Collagen and proteoglycans
Hydroxyapatite

1. Major organic component of bone; lends flexible strength to bones.

2. Primary mineral in bone; gives bone matrix compression (weight-bearing) strength.

Bone Cells

Match these terms with the correct statement or definition:

Canaliculi Osteoclast
Lacunae Osteocytes
Ossification (osteogenesis) Ruffled border
Osteoblast Stem cells
Osteochondral progenitor cells

1. Cell that produces mineralized bone matrix but is not surrounded by matrix.

2. Formation of bone by osteoblasts.

3. Osteoblast that has become surrounded by bone matrix.

4. Spaces occupied by osteocyte cell bodies.

5. Spaces occupied by osteocyte cell processes.

6. Large cell with several nuclei; responsible for the resorption (breakdown) of mineralized bone matrix.

7. Projections from the plasma membrane of an osteoclast where it contacts bone matrix.

8. Mesenchymal cells that give rise to more specialized cell types.

9. Cells that have the ability to become osteoblasts or chondroblasts.

Chapter 6

Woven and Lamellar Bone

Match these terms with the correct statement or definition:

Lamellae
Lamellar (mature) bone
Remodeling
Woven bone

_____ 1. Type of bone in which collagen fibers are randomly oriented in many directions; formed during fetal development or during fracture repair.

_____ 2. Type of bone that is organized into thin sheets or layers; within a layer, the collagen fibers are parallel to each other.

_____ 3. Thin layers or sheets of bone matrix.

_____ 4. Process of changing woven bone to form lamellar bone.

Cancellous and Compact Bone

A. Using the terms provided, complete these statements:

Central (haversian) canals
Circumferential lamellae
Concentric lamellae
Interstitial lamellae
Osteon (haversian system)
Perforating (Volkmann's) canals
Trabeculae

Cancellous bone consists of interconnecting rods or plates of bone called __(1)__. Compact bone is more dense, with blood vessels that run parallel to the long axis of the bone through __(2)__, surrounded by __(3)__. A(n) __(4)__ consists of a single central (haversian) canal, its contents, and associated concentric lamellae and osteocytes. In between osteons are __(5)__, which are remnants of older osteons. The outer surfaces of compact bone are covered by thin plates of bone called __(6)__. The blood vessels from the periosteum or medullary cavity are interconnected by a network of vessels contained within __(7)__, which run perpendicular to the long axis of the bone.

1. _____
2. _____
3. _____
4. _____
5. _____
6. _____
7. _____

B. Match these terms with the correct parts labeled in figure 6.1:

Central (haversian) canal
Circumferential lamellae
Concentric lamellae
Osteocyte in lacuna
Osteon (haversian system)
Interstitial lamellae
Perforating (Volkmann's) canal
Periosteum

1. _____
2. _____
3. _____
4. _____
5. _____
6. _____
7. _____
8. _____

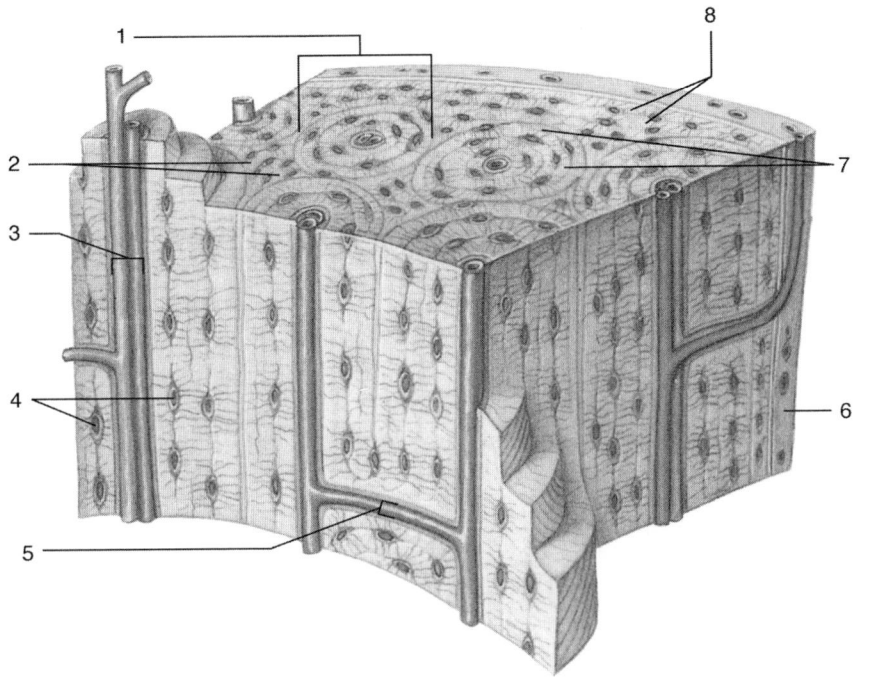

Figure 6.1

Bone Anatomy

A. Match these bone shapes with the correct bones:

Flat bones
Irregular bones
Long bones
Short bones

_____

_____

_____

_____

1. Limb bones.

2. Carpals and tarsals.

3. Some skull bones, ribs, sternum, and scapula.

4. Vertebrae and facial bones.

B. Match these terms with the correct statement or definition as it applies to the structure of a long bone:

Articular cartilage
Diaphysis
Epiphyseal line
Epiphyseal plate
Epiphysis
Medullary cavity

_____

_____

_____

1. Shaft of a long bone; composed primarily of compact bone.

2. Hyaline cartilage that covers the end of a long bone.

3. End of a long bone; composed of cancellous (spongy) bone, with an outer layer of compact bone.

Chapter 6   79

_____ 4. Hyaline cartilage between the diaphysis and epiphysis; growth in length of the bone occurs here.

_____ 5. Ossified epiphyseal plate.

_____ 6. Large internal space within the diaphysis.

C. Match these terms with the correct statement or definition:

Endosteum
Periosteum
Perforating (Sharpey's) fibers
Red marrow
Sinus
Yellow marrow

_____ 1. Site of blood cell formation.

_____ 2. Adipose tissue in medullary cavity of long bones.

_____ 3. Double-layered connective tissue membrane that covers the outer surface of a bone.

_____ 4. Collagen fibers that penetrate the periosteum and outer part of the bone.

_____ 5. Single layer of cells that lines the internal surfaces of all cavities within the bone, including the medullary cavity.

_____ 6. Air space located in some flat and irregular skull bones.

D. Match these terms with the correct parts labeled in figure 6.2:

Articular cartilage
Cancellous bone
Compact bone
Diaphysis
Endosteum
Epiphyseal line
Epiphysis
Medullary cavity
Periosteum

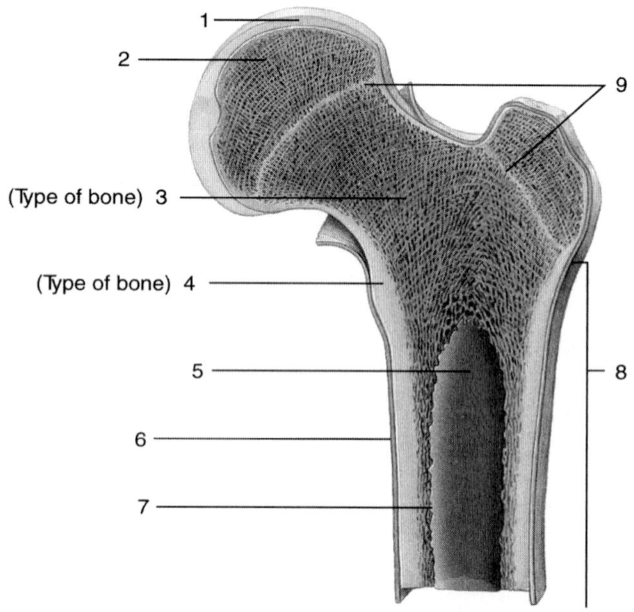

1. _____
2. _____
3. _____
4. _____
5. _____
6. _____
7. _____
8. _____
9. _____

Figure 6.2

Intramembranous Ossification

Using the terms provided, complete these statements:

Centers of ossification
Fontanels
Membrane
Mesenchyme
Osteochondral progenitor cells
Remodeling
Skull
Woven bone

Intramembranous ossification produces many bones of the (1). At the site of intramembranous ossification, embryonic (2) condenses around the brain to form a(n) (3). (4) cells become osteoblasts and produce tiny trabeculae of (5). Woven bone is converted to lamellar bone by (6), which contributes to the final shape of the bone. These locations of bone formation are called (7). At birth, some membrane is not ossified and these regions are called (8).

1. _____
2. _____
3. _____
4. _____
5. _____
6. _____
7. _____
8. _____

Endochondral Ossification

Using the terms provided, complete these statements:

Bone collar
Calcified cartilage
Chondrocytes
Epiphyses
Hyaline cartilage
Hypertrophy
Osteoblasts
Osteochondral progenitor
Osteoclasts
Primary ossification center

The first phase of endochondral ossification is the formation of a(n) (1) model by chondroblasts, which become (2) when they are surrounded by cartilage matrix. When blood vessels invade the perichondrium surrounding the cartilage, (3) cells within the perichondrium become osteoblasts. The osteoblasts produce compact bone on the surface of the cartilage model, forming a(n) (4). Chondrocytes in the center of the cartilage model (5), or enlarge, and the matrix between the enlarged cells becomes mineralized to form (6). When chondrocytes in this calcified area die, blood vessels grow into the enlarged lacunae and bring in (7), which produce bone trabeculae. This area of bone formation is called a(n) (8). At the same time, (9) remove bone from the diaphysis to produce the medullary cavity. Secondary ossification centers appear in the (10) about 1 month before birth.

1. _____
2. _____
3. _____
4. _____
5. _____
6. _____
7. _____
8. _____
9. _____
10. _____

Chapter 6

Bone Growth

A. Using the terms provided, complete these statements:

Appositional growth   Length
Epiphyses   Width

Bone growth can occur only by (1) , the formation of new bone on the surface of older bone or cartilage. Appositional growth beneath the periosteum is responsible for the increase in (2) of long bones and the increase in size o thickness of other bones. Growth at the epiphyseal plate is responsible for the increase in (3) of long bones, or the elongation of projections on some bones. Growth also occurs in articular cartilage and is responsible for the growth of the (4) .

1. _____
2. _____
3. _____
4. _____

B. Using the terms provided, complete these statements:

Appositional growth   Zone of
Remodeled   proliferation
Zone of calcification   Zone of resting
Zone of hypertrophy   cartilage

In the epiphyseal plate, randomly arranged chondrocytes nearest the epiphysis are in the (1) and divide slowly. Chondrocytes in the (2) divide rapidly and produce new cartilage by interstitial growth. Chondrocytes produced in the zone of proliferation mature and enlarge in the (3) . Hypertrophied chondrocytes are surrounded by calcium carbonate and eventually die in the (4) , a thin layer close to the diaphysis. Osteoblasts line up on the surface of the calcified cartilage and deposit bone matrix through (5) to produce new bone matrix, which is later (6) .

1. _____
2. _____
3. _____
4. _____
5. _____
6. _____

Factors Affecting Bone Growth

A. Match these terms with the correct statement or definition:

Osteomalacia   Vitamin C
Rickets   Vitamin D
Scurvy

_____ 1. Necessary for the normal absorption of calcium from the intestines.

_____ 2. Can occur when children have insufficient vitamin D.

_____ 3. Softening of the bones because of calcium depletion in adults.

_____ 4. Necessary for normal collagen synthesis by osteoblasts.

_____ 5. Results from vitamin C deficiency in adults or children.

B. Match these terms with the correct statement or definition:

Estrogen Testosterone
Growth hormone Thyroid hormone

_____ 1. Increase general tissue growth (two).

_____

_____ 2. Sex hormone that causes the earliest closure of the epiphyseal plate.

Bone Remodeling

Using the terms provided, complete these statements:

Basic multicellular unit (BMU)
Interstitial lamellae
Medullary cavity
Osteoblasts
Osteoclasts
Osteons
Trabecula

Remodeling occurs constantly in bones. For example, as long bones increase in length and diameter, the size of the _(1)_ also increases. A(n) _(2)_ is a temporary assembly of osteoclasts and osteoblasts that travels through or across the surface of bone, removing old bone matrix and replacing it with new bone matrix. This process produces new _(3)_ in compact bone. However, portions of older osteons, called _(4)_, are left between the newly developed osteons. In cancellous bone, the BMU removes bone matrix from the surface of a(n) _(5)_, forming a cavity, which is then filled in by the BMU with new bone matrix. Mechanical stress increases the activity of _(6)_; in addition, pressure in bone causes an electric change that increases the activity of _(7)_ and speeds the healing of broken bones.

1. _____
2. _____
3. _____
4. _____
5. _____
6. _____
7. _____

Bone Repair

Using the terms provided, complete these statements:

Cancellous
Chondroblasts
Compact
External callus
Fibroblasts
Hematoma
Internal callus
Osteoblasts

When bone is fractured, blood vessels in the periosteum and in the bone bleed, and a(n) _(1)_ is formed. Bone tissue adjacent to the fracture site dies. Blood vessels grow into the clot produced by the hematoma. When the clot dissolves, _(2)_ produce a fibrous network. _(3)_ invade the fibrous network and form a(n) _(4)_, which is located *between* the ends of the broken bone. Osteochondral progenitor cells from the periosteum become _(5)_, which form bone, and _(6)_, which produce cartilage. Together these cells form a bone;ndcartilage collar, called a(n) _(7)_, *around* opposing ends of the bone fragments. Cartilage in the internal and external calluses is replaced by woven, _(8)_ bone. Finally, woven bone is remodeled to form _(9)_ bone, in which osteons extend across the fracture line to "peg" the bone fragments together.

1. _____
2. _____
3. _____
4. _____
5. _____
6. _____
7. _____
8. _____
9. _____

Calcium Homeostasis

Using the terms provided, complete these statements:

Decreases
Increases
Osteoblasts
Osteoclast precursor cells
Osteoclasts
Osteoprotegerin (OPG)
PTH

When blood Ca^{2+} levels are too low, osteoclast activity _(1)_ and more Ca^{2+} are released into the blood. If blood Ca^{2+} levels are too high, osteoclast activity _(2)_, which results in a net movement of Ca^{2+} from blood to bone. _(3)_ from the parathyroid glands is the major regulator of blood Ca^{2+} levels. If the blood Ca^{2+} level _(4)_, the secretion of PTH increases, resulting in increased numbers of _(5)_. Increased PTH promotes an increase in osteoclast numbers by increasing *receptor for activation of nuclear factor kappa B ligand (RANKL)* production on the surface of _(6)_ and stromal (stem) cells. RANKL binds to *receptor for activation of nuclear factor kappaB ligand (RANK)* on the surface of _(7)_ and stimulates them to become osteoclasts. _(8)_, which is also secreted from osteoblasts and stromal cells, inhibits osteoclast production. When PTH secretion increases, OPG secretion _(9)_, resulting in less inhibition of osteoclast precursor cells, as well as more osteoclasts. PTH also _(10)_ Ca^{2+} uptake in the small intestine, _(11)_ the formation of vitamin D in the kidneys, and _(12)_ the reabsorption of Ca^{2+} from the urine, which _(13)_ the Ca^{2+} lost in the urine. Calcitonin, secreted from the thyroid gland, _(14)_ osteoclast activity.

1. _____
2. _____
3. _____
4. _____
5. _____
6. _____
7. _____
8. _____
9. _____
10. _____
11. _____
12. _____
13. _____
14. _____

Effects of Aging on the Skeletal System

Using the terms provided, complete these statements:

Cancellous
Collagen
Compact
Endosteum
Fracture
Remodeling
Stress
Trabeculae

The bone matrix in an older bone is more brittle than a younger bone because decreased _(1)_ production results in a matrix that has relatively more mineral and less collagen fibers. After 35, both men and women have an age-related loss of bone. _(2)_ bone is lost at first as the _(3)_ become thinner and weaker. Trabecular bone loss is greatest in the trabeculae that are under the least _(4)_. A slow loss of _(5)_ bone begins about 40 and increases after age 45. Bones become thinner, but their outer dimensions change little, because most loss of compact bone occurs under the _(6)_ on the inner surface of bone. In addition, the bone becomes weaker because of incomplete bone _(7)_. Significant loss of bone increases the likelihood of bone _(8)_.

1. _____
2. _____
3. _____
4. _____
5. _____
6. _____
7. _____
8. _____

QUICK RECALL

1. List three important components of the extracellular matrix of cartilage.

2. List the major organic and inorganic compounds in the extracellular matrix of bone.

3. List three types of bone cells, depending on their function.

4. List three types of lamellae visible in bone matrix.

5. Name, and distinguish between, two types of bone marrow.

6. List two types of bone according to their internal structure.

7. List three important components of an osteon (haversian system).

8. Name two types of bone ossification.

9. Name three types of bone growth.

10. List three factors that affect bone growth and calcium homeostasis.

11. Name the abnormal bone conditions that occur with a dietary deficiency of vitamin D.

12. List five hormones that influence bone growth.

ANSWERS TO CHAPTER 6

CONTENT LEARNING ACTIVITY

Functions of the Skeletal System
1. Bone; 2. Cartilage; 3. Tendon; 4. Ligament;
5. Bone; 6. Bone

Cartilage
- A. 1. Chondroblast; 2. Chondrocyte;
3. Perichondrium; 4. Articular cartilage
- B. 1. Appositional growth; 2. Interstitial growth

Bone Matrix
1. Collagen and proteoglycans;
2. Hydroxyapatite

Bone Cells
1. Osteoblast; 2. Ossification (osteogenesis);
3. Osteocyte; 4. Lacunae; 5. Canaliculi;
6. Osteoclast; 7. Ruffled border; 8. Stem cells;
9. Osteochondral progenitor cells

Woven and Lamellar Bone
1. Woven bone; 2. Lamellar (mature) bone;
3. Lamellae; 4. Remodeling

Cancellous and Compact Bone
- A. 1. Trabeculae; 2. Central (haversian) canals;
3. Concentric lamellae; 4. Osteon (haversian system); 5. Interstitial lamellae;
6. Circumferential lamellae; 7. Perforating (Volkmann's) canals
- B. 1. Osteon (haversian system); 2. Interstitial lamellae; 3. Central (haversian) canal;
4. Osteocyte in lacuna; 5. Perforating (Volkmann's) canal; 6. Periosteum;
7. Concentric lamellae; 8. Circumferential lamellae

Bone Anatomy
- A. 1. Long bones; 2. Short bones; 3. Flat bones;
4. Irregular bones
- B. 1. Diaphysis; 2. Articular cartilage;
3. Epiphysis; 4. Epiphyseal plate;
5. Epiphyseal line; 6. Medullary cavity
- C. 1. Red marrow; 2. Yellow marrow;
3. Periosteum; 4. Perforating (Sharpey's) fibers; 5. Endosteum; 6. Sinus
- D. 1. Articular cartilage; 2. Epiphysis;
3. Cancellous bone; 4. Compact bone;
5. Medullary cavity; 6. Periosteum;
7. Endosteum; 8. Diaphysis; 9. Epiphyseal lines

Intramembranous Ossification
1. Skull; 2. Mesenchyme; 3. Membrane;
4. Osteochondral progenitor cells; 5. Woven bone;
6. Remodeling; 7. Centers of ossification;
8. Fontanels

Endochondral Ossification
    1. Hyaline cartilage; 2. Chondrocytes;
    3. Osteochondral progenitor; 4. Bone collar;
    5. Hypertrophy; 6. Calcified cartilage;
    7. Osteoblasts; 8. Primary ossification center;
    9. Osteoclasts; 10. Epiphyses

Bone Growth
    A. 1. Appositional growth; 2. Width;
       3. Length; 4. Epiphyses
    B. 1. Zone of resting cartilage; 2. Zone of proliferation; 3. Zone of hypertrophy; 4. Zone of calcification; 5. Appositional growth; 6. Remodeled

Factors Affecting Bone Growth
    A. 1. Vitamin D; 2. Rickets; 3. Osteomalacia;
       4. Vitamin C; 5. Scurvy
    B. 1. Growth hormone and thyroid hormone;
       2. Estrogen

Bone Remodeling
    1. Medullary cavity; 2. Basic multicellular unit (BMU); 3. Osteons; 4. Interstitial lamellae; 5. Trabecula; 6. Osteoblasts; 7. Osteoblasts

Bone Repair
    1. Hematoma; 2. Fibroblasts; 3. Chondroblasts;
    4. Internal callus; 5. Osteoblasts; 6. Chondroblasts;
    7. External callus; 8. Cancellous; 9. Compact

Calcium Homeostasis
    1. Increases; 2. Decreases; 3. PTH; 4. Decreases;
    5. Osteoclasts; 6. Osteoblasts; 7. Osteoclast precursor cells; 8. Osteoprotegerin (OPG);
    9. Decreases; 10. Increases; 11. Increases;
    12. Increases; 13. Decreases; 14. Decreases

Effects of Aging on the Skeletal System
    1. Collagen; 2. Cancellous; 3. Trabeculae;
    4. Stress; 5. Compct; 6. Endosteum;
    7. Remodeling; 8. Fracture

QUICK RECALL

1. Collagen, proteoglycans, and water
2. Organic: collagen and proteoglycans; inorganic: hydroxyapatite
3. Osteoblasts, osteocytes, and osteoclasts
4. Concentric, circumferential, and interstitial
5. Yellow marrow: mostly fat; red marrow: site of blood cell formation
6. Compact bone and cancellous (spongy) bone
7. Central (haversian) canal, concentric lamellae, and osteocytes in lacunae
8. Intramembranous and endochondral
9. Growth at the epiphyseal plate, growth at the articular cartilage, and increase in diameter
10. Genetics, vitamin deficiencies, and hormones
11. Rickets in children and osteomalacia in adults
12. Growth hormone, thyroid hormone, sex hormones, parathyroid hormone, and calciton

7 Skeletal System: Gross Anatomy

CONTENT LEARNING ACTIVITY

Introduction

Match these skeletal subdivisions with the correct bones:

Appendicular skeleton
Axial skeleton

_____ 1. Vertebral column.

_____ 2. Skull.

_____ 3. Hyoid bone.

_____ 4. Limbs.

_____ 5. Thoracic cage.

_____ 6. Limb girdles.

General Considerations

Match these bone features with the correct definition:

Canal (meatus)　　Foramen
Condyle　　　　　Fossa
Facet　　　　　　Process

_____ 1. Smooth, rounded articular surface.

_____ 2. Small, flattened articular surface.

_____ 3. Projection.

_____ 4. Hole in a bone for passage of blood vessels or nerves.

_____ 5. Tunnel.

_____ 6. Depression.

Skull

A. Match these structures with the correct description or definition:

Auditory ossicles
Cranium
Crista galli
Hard palate
Hyoid bone
Nasal conchae
Nasal septum
Neurocranium (braincase)
Orbit
Paranasal sinuses
Sella turcica
Viscerocranium

_____ 1. Skull.

_____ 2. Six bones that function in hearing; located within cavities of the temporal bones.

_____ 3. Subdivision of the skull that protects the brain.

_____ 4. Subdivision of the skull that forms most of the facial bones.

_____ 5. Neck and tongue muscles attach to it; "floats" in the neck.

_____ 6. Structure in the skull that encloses and protects the eye.

_____ 7. Bony shelves of the nasal cavity; warm and moisten the air.

_____ 8. Air-filled cavities that open into the nasal cavity.

_____ 9. Connective tissue membrane (one of the meninges) that holds the brain in place attaches to this prominent ridge.

_____ 10. Saddle-shaped structure occupied by the pituitary gland.

_____ 11. Divides the nasal cavity into two parts.

_____ 12. Separates the nasal cavity from the oral cavity.

B. Match these bone parts with structures to which they contribute:

Horizontal plate of palatine bone
Palatine process of maxilla
Perpendicular plate of ethmoid bone
Temporal process of zygomatic bone
Vomer
Zygomatic process of temporal bone

_____
_____ 1. Two structures that form the hard palate.

_____
_____ 2. Two structures that form the nasal septum.

_____
_____ 3. Two structures that form the zygomatic arch.

C. Match these terms with the correct parts of the skull labeled in figure 7.1.

- Hard palate
- Horizontal plate of palatine bone
- Nasal septum
- Palatine process of maxillary bone
- Perpendicular plate of ethmoid bone
- Septal cartilage
- Temporal process of zygomatic bone
- Vomer
- Zygomatic arch
- Zygomatic process of temporal bone

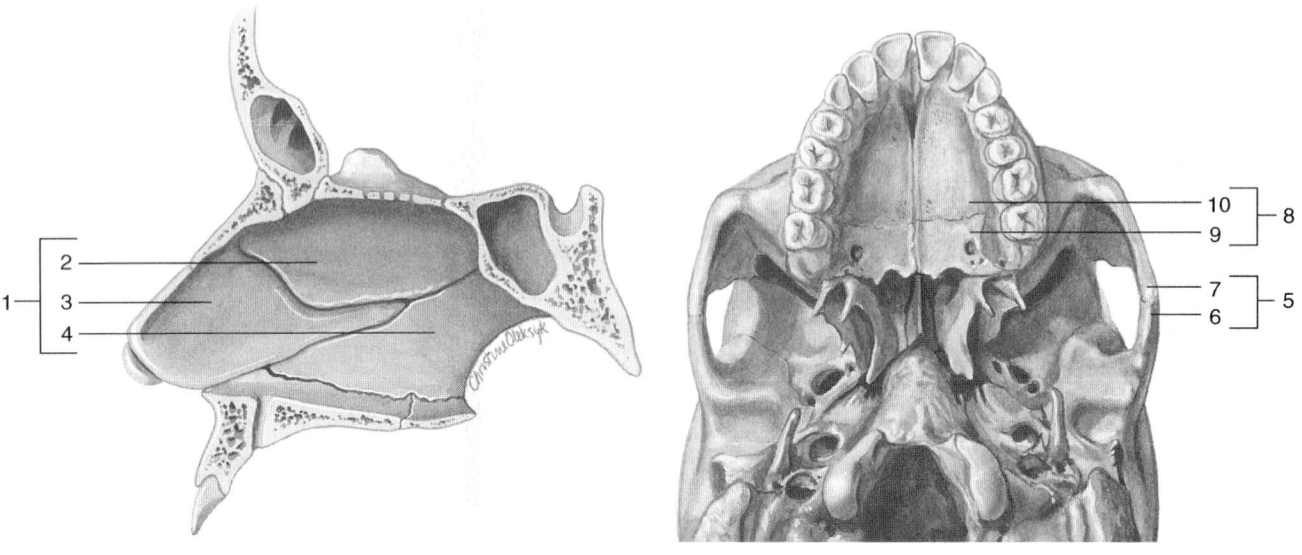

Figure 7.1

1. _____
2. _____
3. _____
4. _____
5. _____
6. _____
7. _____
8. _____
9. _____
10. _____

D. Match these bones with the correct structures that make up part of that bone:

- Ethmoid
- Hyoid
- Mandible
- Occipital
- Sphenoid
- Temporal

_____ 1. Coronoid process

_____ 2. Cribriform plate

_____ 3. Crista galli

_____ 4. External acoustic meatus

_____ 5. Foramen magnum

_____ 6. Greater cornu

_____ 7. Lateral pterygoid plate

_____ 8. Lesser wing

_____ 9. Mandibular fossa

_____ 10. Mastoid process

_____ 11. Middle nasal concha

_____ 12. Perpendicular plate

_____ 13. Sella turcica

_____ 14. Styloid process

Muscles of the Skull

Match these muscles with their correct attachments:

Eye movement muscles   Muscles of mastication
Muscles of facial expression   Neck muscles

_____ 1. Attach to the temporal bones, zygomatic arches, lateral pterygoid plates of the sphenoid, and mandible.

_____ 2. Attach to the nuchal lines, external occipital protuberance, and mastoid process.

_____ 3. Attach to the bones of the orbits.

Openings of the Skull

Match the skull opening with its function or with the structures it contains:

Carotid canals   Jugular foramen
External acoustic meatus   Nasolacrimal canal
Foramen magnum   Olfactory foramina
Internal acoustic meatus   Optic canal

_____ 1. Cribriform plate holes; contain nerves for the sense of smell.

_____ 2. Contains a duct carrying tears from the eye to the nasal cavity.

_____ 3. Contains the nerve for the sense of vision.

_____ 4. Transmits sound waves toward the eardrum.

_____ 5. Contains the nerve for the sense of hearing.

_____ 6. Most blood leaves the skull through this opening.

_____ 7. Openings through which most blood reaches the brain.

_____ 8. Opening through which the spinal cord connects with the brain.

Joints of the Skull

Match these skull joints with their correct description:

Coronal suture   Occipital condyle
Lambdoid suture   Sagittal suture
Mandibular fossa Squamous suture

_____ 1. Located between the parietal bones.

_____ 2. Located between the parietal bones and the frontal bone.

_____ 3. Located between the parietal bones and the occipital bone.

_____ 4. Located between the parietal bone and the temporal bone.

_____ 5. Articulation point between the skull and mandible.

_____ 6. Articulation point between the skull and the vertebral column.

Skull Diagrams

A. Match these terms with the correct parts of the skull labeled in figure 7.2:

Coronal suture   Occipital bone
Frontal bone   Parietal bone
Lambdoid suture   Sagittal suture

1. _____
2. _____
3. _____
4. _____
5. _____
6. _____

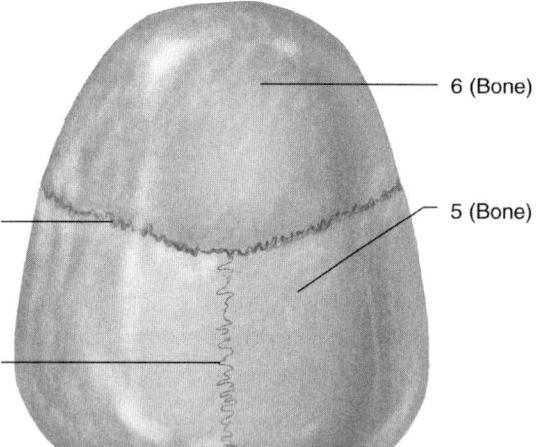

Figure 7.2

92   Chapter 7

B. Match these terms with the correct parts of the orbit labeled in figure 7.3:

Ethmoid bone
Frontal bone
Inferior orbital fissure
Lacrimal bone
Maxilla
Nasolacrimal canal
Optic canal
Palatine bone
Sphenoid bone
Superior orbital fissure
Zygomatic bone

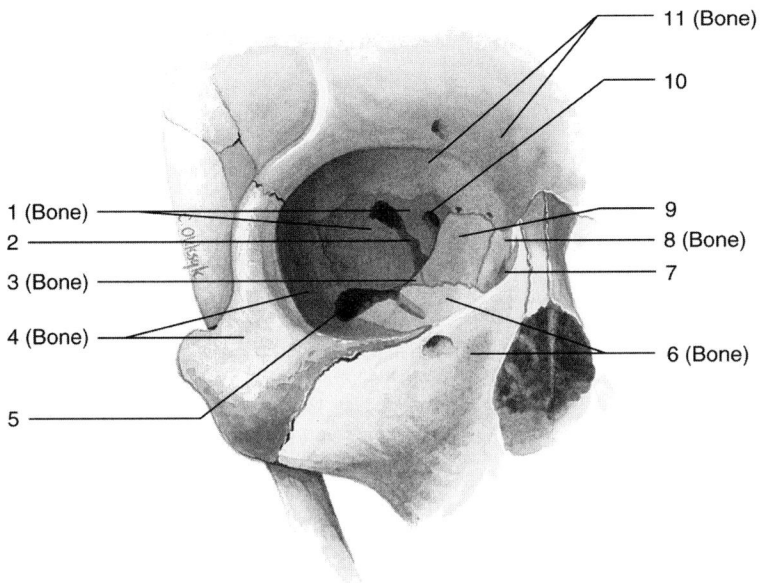

Figure 7.3

1. _____
2. _____
3. _____
4. _____
5. _____
6. _____
7. _____
8. _____
9. _____
10. _____
11. _____

Chapter 7

C. Match these terms with the correct parts of the skull labeled in figure 7.4:

Coronal suture
External acoustic meatus
Frontal bone
Greater wing
Lambdoid suture
Mandible
Mastoid process
Maxilla
Nasal bone
Occipital bone
Parietal bone
Sphenoid bone
Squamous portion
Squamous suture
Styloid process
Temporal bone
Temporal lines
Zygomatic bone

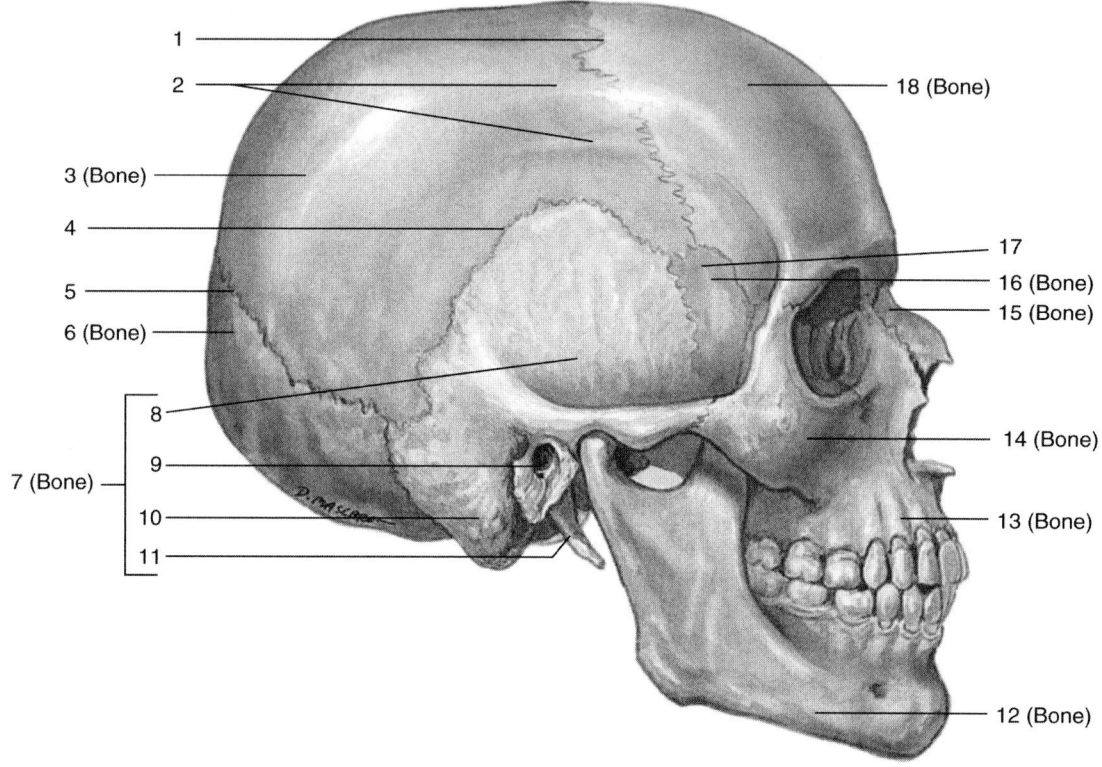

Figure 7.4

1. _____
2. _____
3. _____
4. _____
5. _____
6. _____
7. _____
8. _____
9. _____
10. _____
11. _____
12. _____
13. _____
14. _____
15. _____
16. _____
17. _____
18. _____

D. Match these terms with the correct parts of the skull labeled in figure 7.5:

Ethmoid bone
Frontal bone
Inferior nasal conchae
Infraorbital foramen
Mandible
Mandibular symphysis
Maxilla
Middle nasal concha
Nasal bone
Nasal cavity
Nasal septum
Orbit
Perpendicular plate
Supraorbital foramen
Supraorbital margin
Vomer
Zygomatic bone

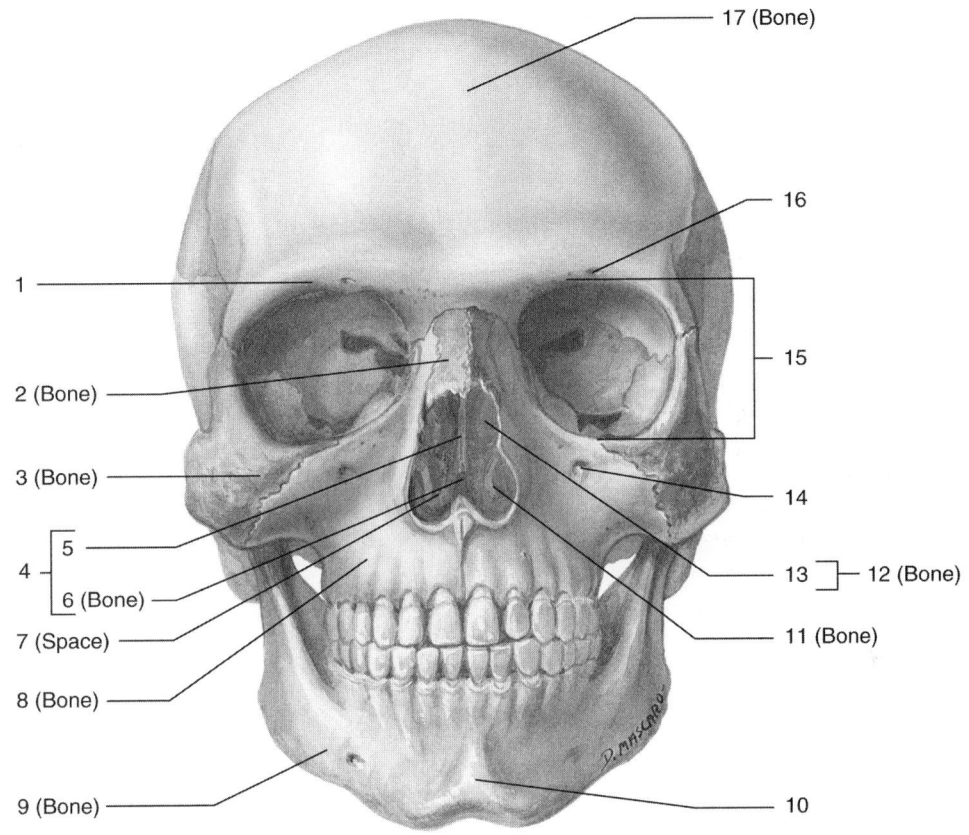

Figure 7.5

1. _____
2. _____
3. _____
4. _____
5. _____
6. _____
7. _____
8. _____
9. _____
10. _____
11. _____
12. _____
13. _____
14. _____
15. _____
16. _____
17. _____

Chapter 7

E. Match these terms with the correct parts of the skull labeled in figure 7.6:

Carotid canal
Cribriform plate
Crista galli
Ethmoid bone
Foramen magnum
Frontal bone
Frontal sinuses
Greater wing
Internal acoustic meatus
Jugular foramen
Lesser wing
Occipital bone
Optic canal
Petrous portion
Sella turcica
Sphenoid bone
Squamous portion
Temporal bone

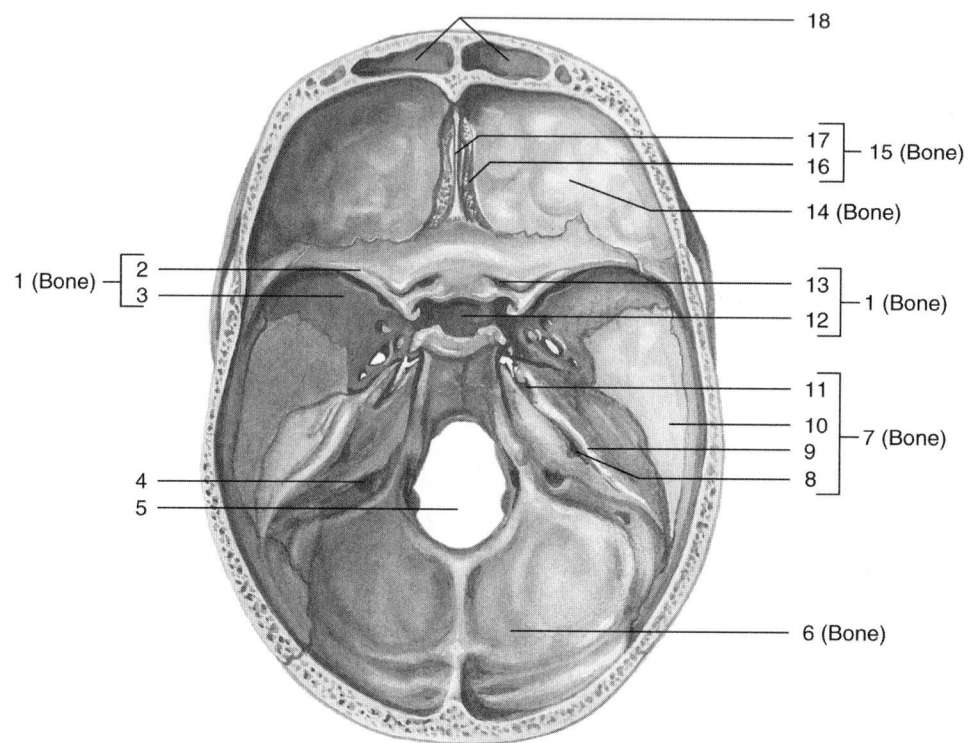

Figure 7.6

1. _____   7. _____   13. _____
2. _____   8. _____   14. _____
3. _____   9. _____   15. _____
4. _____   10. _____   16. _____
5. _____   11. _____   17. _____
6. _____   12. _____   18. _____

F. Match these terms with the correct parts of the skull labeled in figure 7.7:

Carotid canal
External occipital protuberance
Foramen magnum
Incisive fossa
Jugular foramen
Lateral pterygoid plate
Mandibular fossa
Mastoid process
Maxilla
Medial pterygoid plate
Nuchal lines
Occipital bone
Occipital condyle
Palatine bone
Sphenoid bone
Styloid process
Temporal bone
Vomer

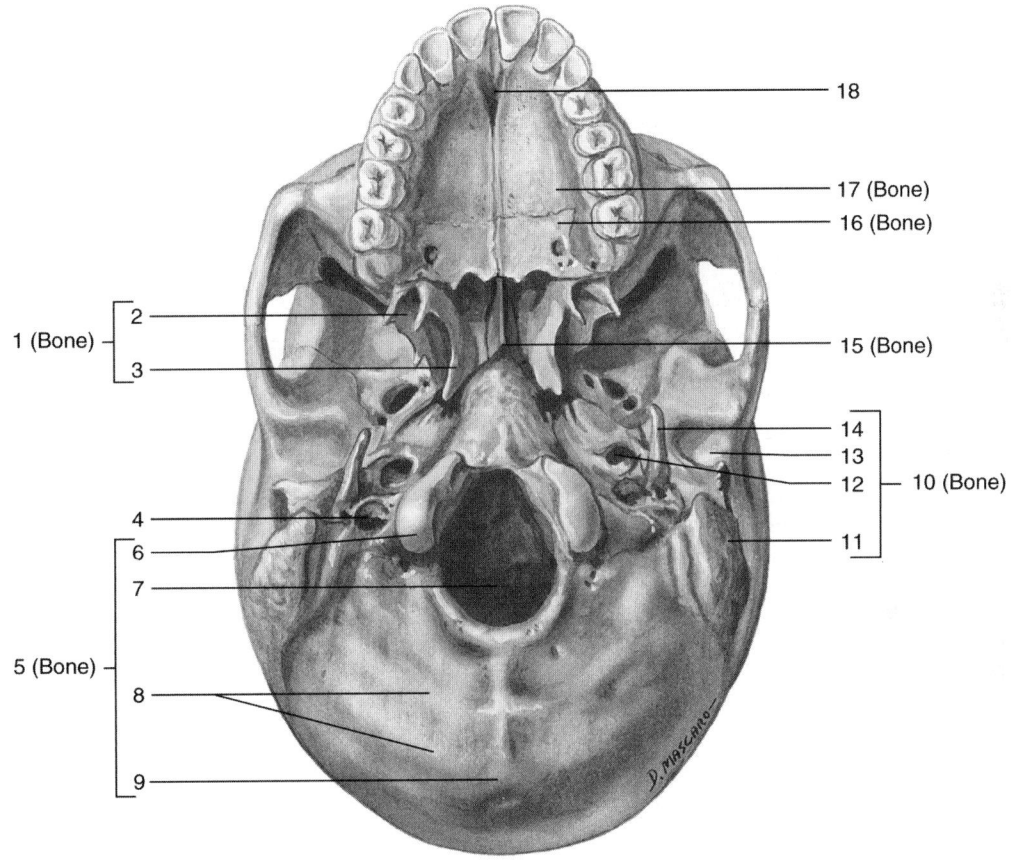

Figure 7.7

1. _____
2. _____
3. _____
4. _____
5. _____
6. _____
7. _____
8. _____
9. _____
10. _____
11. _____
12. _____
13. _____
14. _____
15. _____
16. _____
17. _____
18. _____

Chapter 7

G. Match these terms with the correct parts labeled in figure 7.8:

Ethmoidal labyrinth (sinuses)
Frontal sinus
Maxillary sinus
Sphenoidal sinus

1. _____
2. _____
3. _____
4. _____

Figure 7.8

H. Match these terms with the correct parts of the mandible labeled in figure 7.9:

Angle
Alveolar process
Body
Coronoid process
Mandibular condyle
Mandibular foramen
Mental foramen
Ramus

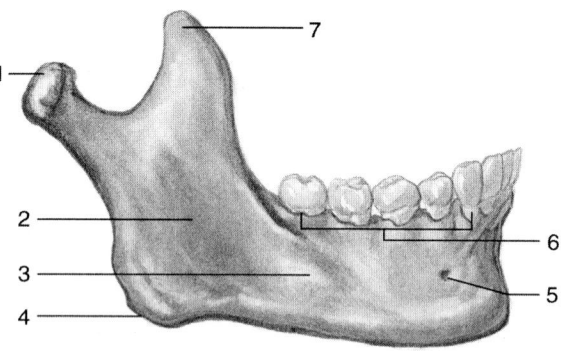
Right half of the mandible
Lateral view

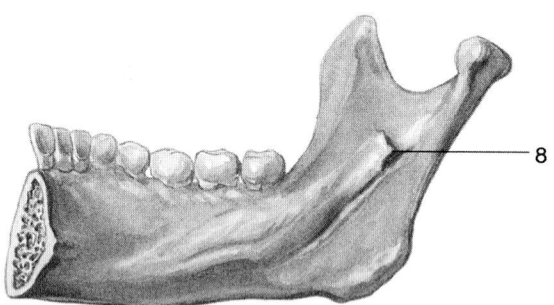
Right half of the mandible
Medial view

Figure 7.9

1. _____
2. _____
3. _____
4. _____
5. _____
6. _____
7. _____
8. _____

Vertebral Column

A. Match these terms with the correct statement or definition:

Cervical Sacral and coccygeal
Kyphosis Scoliosis
Lordosis Thoracic
Lumbar

_____ 1. Two sections of the vertebral column that become convex anteriorly after birth.

_____ 2. Exaggerated lumbar curvature.

_____ 3. Exaggerated thoracic curvature.

_____ 4. Abnormal bending of the spine to one side.

B. Match these terms with the correct statement or definition:

Anulus fibrosus Nucleus pulposus
Intervertebral disk

_____ 1. Structure located between the bodies of adjacent vertebrae.

_____ 2. External portion of an intervertebral disk.

_____ 3. Inner portion of an intervertebral disk.

_____ 4. Herniated disk results from damage to this structure.

General Plan of the Vertebrae

A. Match these parts of a vertebra with their correct function or description:

Articular process Pedicle
Body Spinous process
Intervertebral foramina Transverse process
Lamina Vertebral foramen

_____ 1. Main weight-bearing portion of the vertebra.

_____ 2. Two parts that form the vertebral arch, which protects the spinal cord.

_____ 3. Contains the spinal cord; all of them together form the vertebral canal.

_____ 4. Where the spinal nerves exit the vertebrae.

_____ 5. Connects one vertebra to another, increasing the rigidity of the vertebral column while allowing movement.

_____ 6. Midline point of muscle attachment.

_____ 7. Muscles and the tubercle of the rib attach to this structure.

Chapter 7

B. Match the structures that articulate with each other.

Articular facet on rib head Intervertebral disk
Articular facet on rib tubercle Occipital condyle
Body of vertebra Pelvic bone (ilium)
Dens Superior articular facet

_____ 1. Body of vertebra.

_____ 2. Articular facet on the body of the vertebra.

_____ 3. Articular facet on the transverse process.

_____ 4. Inferior articular facet.

_____ 5. Superior articular facet of atlas.

_____ 6. Facet on the anterior arch of the atlas.

_____ 7. Ala.

C. Match these terms with the correct parts of the vertebra labeled in figure 7.10:

Body Superior articular process
Lamina Transverse process
Pedicle Vertebral arch
Spinous process Vertebral foramen

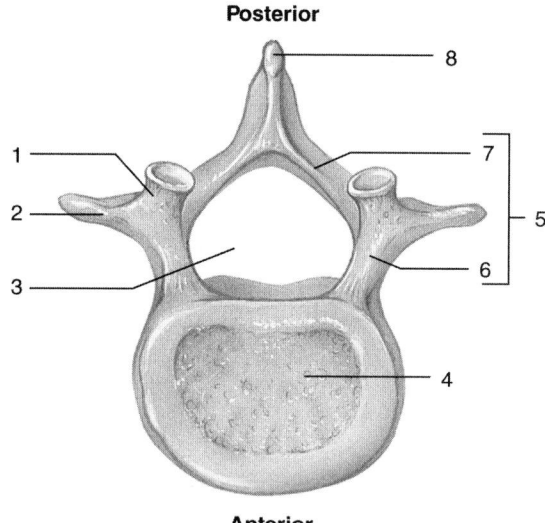

Figure 7.10

1. _____

2. _____

3. _____

4. _____

5. _____

6. _____

7. _____

8. _____

D. Match these terms with the correct parts of the vertebra labeled in figure 7.11:

Articular facet for rib head
Articular facet for rib tubercle
Body
Inferior articular process
Inferior intervertebral notch
Lamina
Pedicle
Spinous process
Superior articular facet
Superior articular process
Superior intervertebral notch
Transverse process

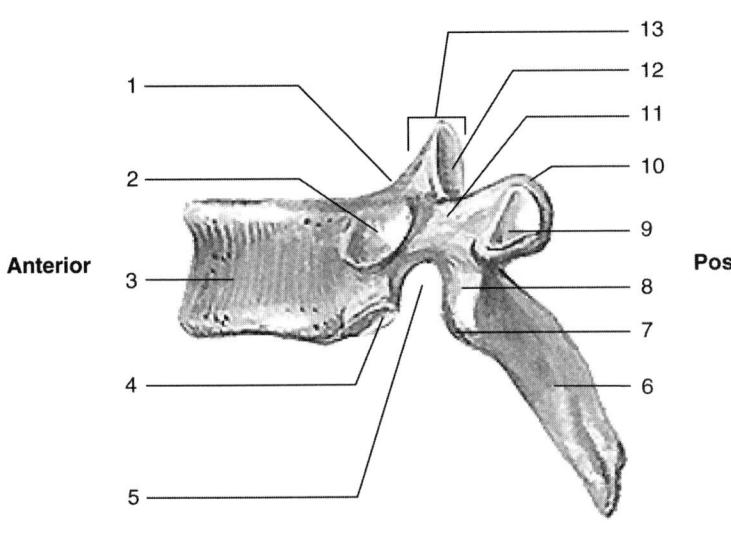

Figure 7.11

1. _____
2. _____
3. _____
4. _____
5. _____
6. _____
7. _____
8. _____
9. _____
10. _____
11. _____
12. _____
13. _____

Regional Differences in Vertebrae

A. Match the type of vertebra with the characteristic unique to it:

Cervical vertebra
Coccygeal vertebra
Lumbar vertebra
Sacrum
Thoracic vertebra

_____ 1. Transverse foramina through which the vertebral arteries going to the brain pass; spinous process is usually bifid (split).

_____ 2. Articular facet on the transverse process and on the body for the attachment of the rib.

_____ 3. Thick bodies and rectangular transverse and spinous processes; superior articular processes face medially, limiting movement.

_____ 4. Transverse processes are fused to form the alae, which attach to the pelvic bones.

_____ 5. No vertebral foramen.

Chapter 7

B. Match these terms with the the correct statement and the structure labeled in figure 7.12:

Articular facet for rib head
Articular facet for rib tubercle
Atlas
Axis
Bifid
Blunt
Body
Cervical vertebra
Dens
Lumbar vertebra
Medial
Slender and pointed
Spinous process
Thoracic vertebra
Transverse foramen

1. Type of vertebra. _____

2. Characteristic common to this type of vertebrae. _____

3. Name of this vertebra. _____

4. Two parts that all other vertebrae have but is missing from the atlas. _____

5. Type of vertebra. _____

6. Part of vertebra. _____

7. Name of this vertebra. _____

8. Type of vertebra. _____

9. Type of spinous process in most cervical vertebrae. _____

10. Type of vertebra. _____

11. Characteristic common to all vertebrae of this type. _____

12. Characteristic common to all vertebrae of this type. _____

13. Type of spinous process in most thoracic vertebrae. _____

14. Type of vertebra. _____

15. Superior articular process faces in this direction. _____

16. Type of spinous process in most lumbar vertebrae. _____

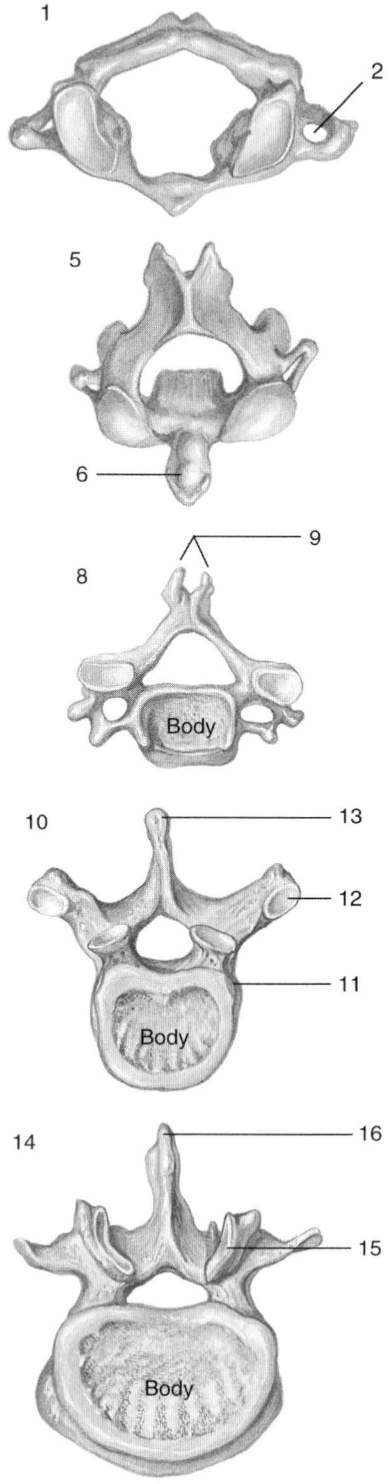

Figure 7.12

C. Match these terms with the correct parts of the sacrum labeled in figure 7.13:

Ala
Articular surface
   (joins coxal bone)
Coccyx
Median sacral crest
Sacral canal
Sacral foramina
Sacral hiatus
Superior articular facet

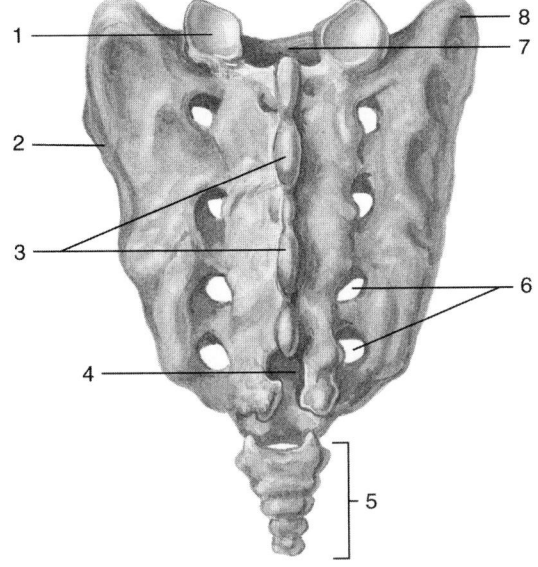

Figure 7.13

1. _____
2. _____
3. _____
4. _____
5. _____
6. _____
7. _____
8. _____

Thoracic Cage

A. Match these terms with the correct statement or definition:

Body
False (vertebrochondral) ribs
Floating (vertebral) ribs
Head
Jugular notch
Manubrium
True (vertebrosternal) ribs
Tubercle
Sternal angle
Xiphoid process

_____ 1. First seven pairs of ribs, which articulate with both the vertebrae and the sternum.

_____ 2. Ribs that are attached to a common cartilage that, in turn, is attached to the sternum.

_____ 3. Eleventh and twelfth ribs, which have no attachment to the sternum.

_____ 4. Rounded process on a rib that articulates with the transverse process of one vertebra.

_____ 5. Middle part of the sternum.

_____ 6. Most inferior portion of the sternum.

_____ 7. Marks the attachment site of the second ribs.

Chapter 7

B. Match these terms with the correct parts of the thoracic cage labeled in figure 7.14:

Body
Costal cartilage
False ribs
Floating ribs
Jugular notch

Manubrium
Sternal angle
Sternum
True ribs
Xiphoid process

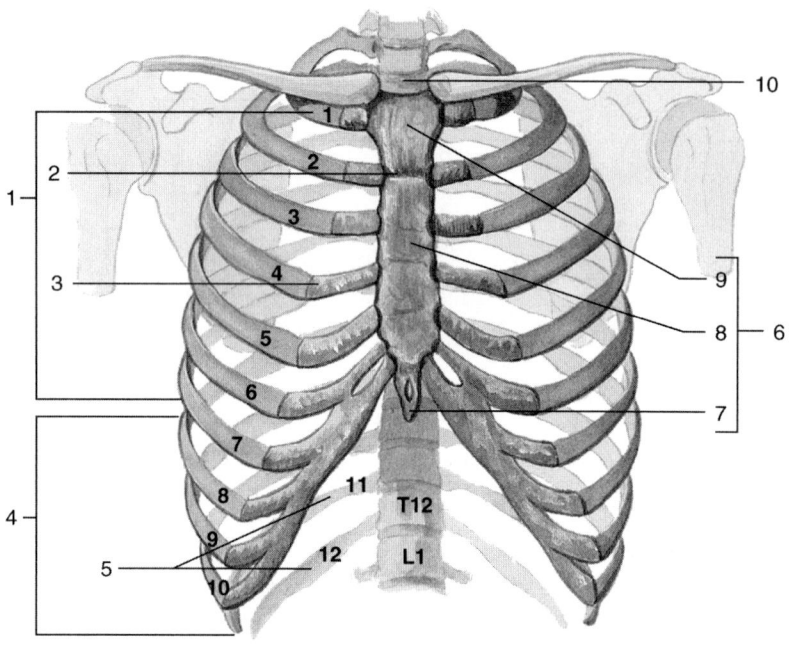

Figure 7.14

1. _____
2. _____
3. _____
4. _____
5. _____
6. _____
7. _____
8. _____
9. _____
10. _____

C. Match these terms with the correct parts of the rib labeled in figure 7.15:

Angle
Articular facet for body of vertebrae
Articular facet for transverse process of vertebra

Body
Head
Neck
Tubercle

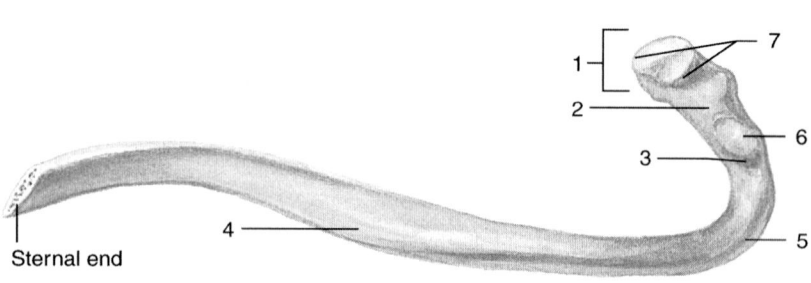

Figure 7.15

1. _____
2. _____
3. _____
4. _____
5. _____
6. _____
7. _____

Pectoral Girdle

A. Match these terms with the correct description:

Acromion process
Clavicle
Coracoid process
Glenoid cavity
Inferior angle
Infraspinous fossa
Scapular spine
Subscapular fossa
Superior angle
Superior border
Supraspinous fossa

_____ 1. Apex of the scapula.

_____ 2. Base of the scapula.

_____ 3. Projection of scapula over the shoulder joint; articulates with the clavicle.

_____ 4. Large ridge extending from the acromion process across the scapula.

_____ 5. Projection on the anterior side of the scapula that provides a point of attachment for shoulder and arm muscles.

_____ 6. Shallow depression in the scapula, where it articulates with the humerus.

_____ 7. Deep, anterior surface of the scapula.

_____ 8. Depression inferior to the spine.

_____ 9. Holds the upper limb away from the body.

B. Match these terms with the correct parts of the clavicle labeled in figure 7.16:

Acromial (lateral) end
Body
Sternal (medial) end

1. _____

2. _____

3. _____

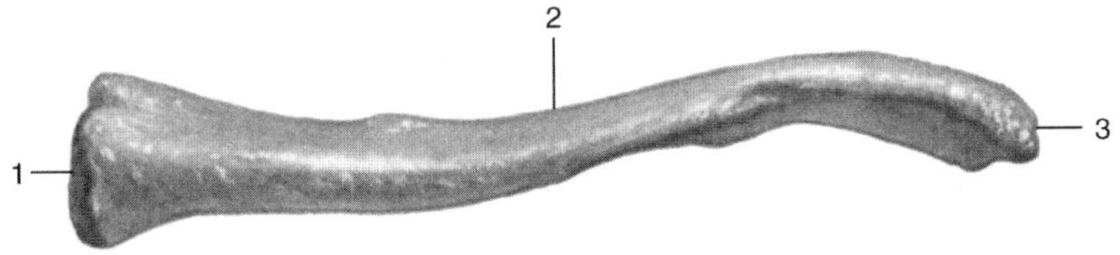

Figure 7.16

Chapter 7

C. Match these terms with the correct parts of the scapula labeled in figure 7.17:

Acromion process
Coracoid process
Glenoid cavity
Inferior angle
Infraspinous fossa
Lateral (axillary) border
Medial (vertebral) border
Spine
Subscapular fossa
Superior angle
Superior border
Supraspinous fossa

1. _____
2. _____
3. _____
4. _____
5. _____
6. _____
7. _____
8. _____
9. _____
10. _____
11. _____
12. _____

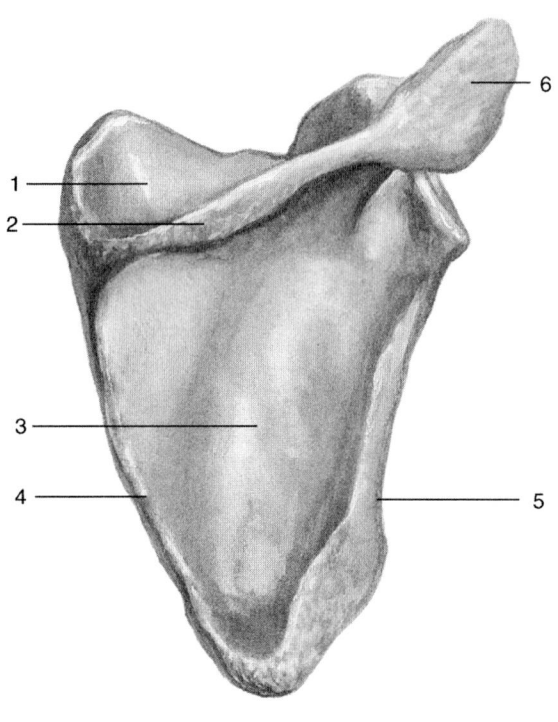

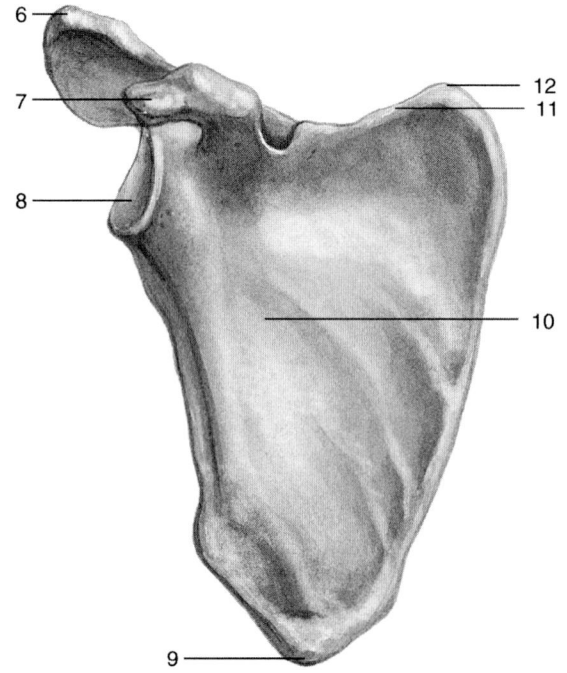

Figure 7.17

Upper Limb

A. Match the bone parts that articulate with each other:

Head of humerus
Head of radius
Head of ulna
Trochlea

_____ 1. Glenoid cavity.

_____ 2. Capitulum.

_____ 3. Trochlear (semilunar) notch.

_____ 4. Radial notch.

_____ 5. Carpals.

B. Match these terms with the correct parts of the humerus labeled in figure 7.18:

Anatomical neck
Capitulum
Coronoid fossa
Deltoid tuberosity
Greater tubercle
Head
Intertubercular groove
Lateral epicondyle
Lesser tubercle
Medial epicondyle
Olecranon fossa
Radial fossa
Surgical neck
Trochlea

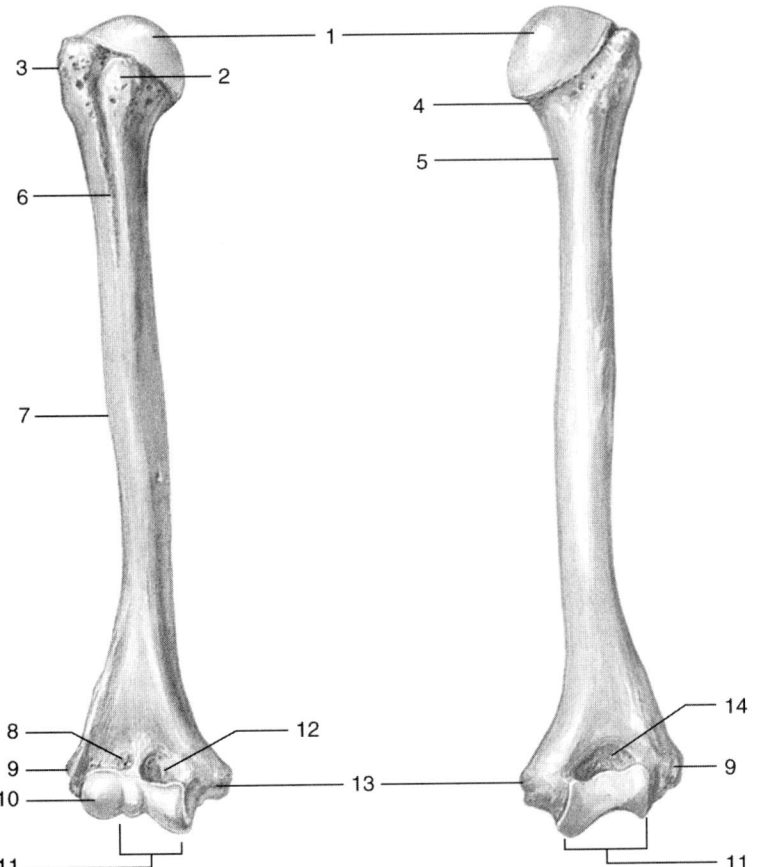

1. _____
2. _____
3. _____
4. _____
5. _____
6. _____
7. _____
8. _____
9. _____
10. _____
11. _____
12. _____
13. _____
14. _____

Figure 7.18

Chapter 7

C. Match the bony part with its function:

- Deltoid tuberosity
- Epicondyles
- Greater tubercle
- Intertubercular groove
- Lesser tubercle
- Olecranon process
- Radial tuberosity
- Styloid processes

_____ 1. Three places that shoulder muscles attach to the humerus.

_____

_____

_____ 2. Groove between the greater and lesser tubercle where one tendon of the biceps brachii muscle passes.

_____ 3. Location where forearm muscles attach to the humerus.

_____ 4. Two places that arm muscles attach to the forearm.

_____

_____ 5. Location where ligaments of the wrist attach to the forearm bones.

D. Match these terms with the correct parts of the forearm labeled in figure 7.19:

- Coronoid process
- Head
- Neck
- Olecranon process
- Radial notch
- Radial tuberosity
- Radius
- Styloid process
- Trochlear notch
- Ulna

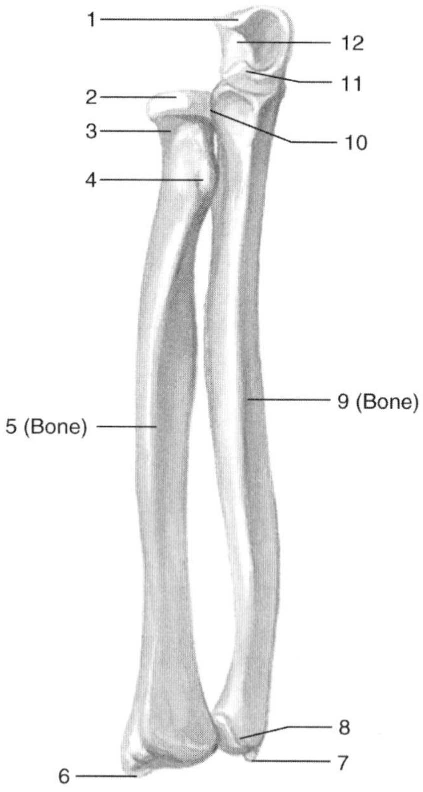

1. _____
2. _____
3. _____
4. _____
5. _____
6. _____
7. _____
8. _____
9. _____
10. _____
11. _____
12. _____

Figure 7.19

E. Match these groups of bones with the correct descriptions:

Carpals
Metacarpals
Phalanges

_____ 1. Five bones that form the hand.

_____ 2. Eight bones in two rows, forming the wrist.

_____ 3. Three of these bones in each finger; the thumb has two.

F. Match these terms with the correct parts of the wrist and hand labeled in figure 7.20:

Capitate
Carpals
Distal phalanx
Hamate
Lunate
Metacarpals
Middle phalanx
Pisiform
Proximal phalanx
Scaphoid
Trapezium
Trapezoid
Triquetrum

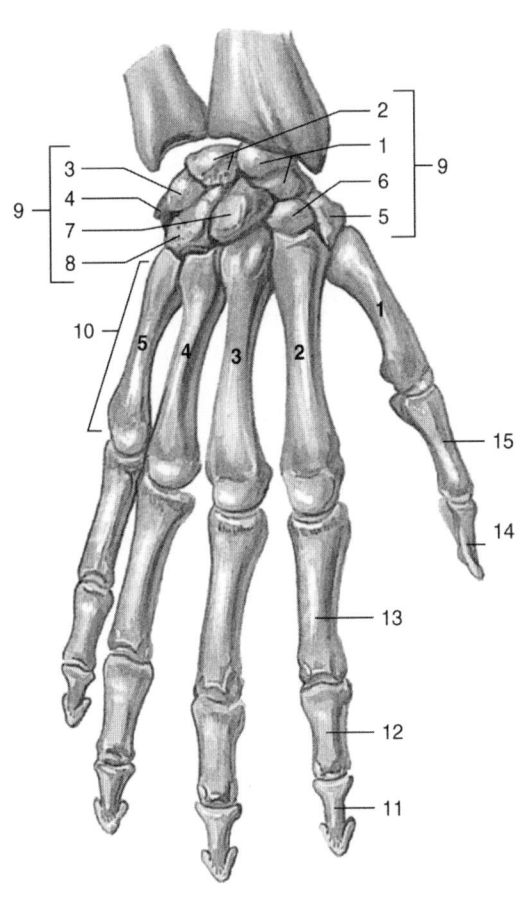

Figure 7.20

1. _____
2. _____
3. _____
4. _____
5. _____
6. _____
7. _____
8. _____
9. _____
10. _____
11. _____
12. _____
13. _____
14. _____
15. _____

Pelvic Girdle

A. Using the terms provided, complete these statements:

    Articular surface   Iliac spines
    Coxal bone   Ischial tuberosity
    False pelvis   Obturator foramen
    Iliac crest   Symphysis pubis
    Iliac fossa   True pelvis

Each (1) is formed by the fusion of the ilium, ischium, and pubis. A large hole in the inferior half of the coxal bone is the (2). The superior portion of the ilium is called the (3), whereas the large depression on the medial side of the ilium is the (4). The anterior and posterior ends of the iliac crest are referred to as (5), which serve as points of muscle attachment. A heavy (6) on the ischium provides a place for muscle attachment and a place to sit. The two coxal bones are joined at the (7), and the (8) of each coxal bone joins the sacrum (ala) to form the sacroiliac joints. The (9) is superior to the pelvic brim and is partially surrounded by bone, whereas the (10) is inferior to the pelvic brim and is completely surrounded by bone.

1. _____
2. _____
3. _____
4. _____
5. _____
6. _____
7. _____
8. _____
9. _____
10. _____

B. Match these terms with the correct parts of the pelvic girdle labeled in figure 7.21:

    Acetabulum   Pubis
    Anterior superior iliac spine   Sacral promontory
    Coxal bone   Sacroiliac joint
    Ilium   Sacrum
    Ischium   Subpubic angle
    Obturator foramen   Symphysis pubis

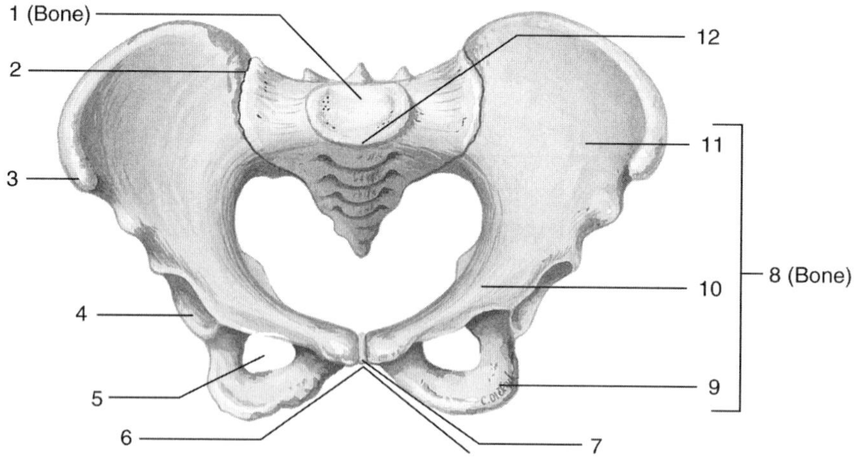

Figure 7.21

1. _____
2. _____
3. _____
4. _____
5. _____
6. _____
7. _____
8. _____
9. _____
10. _____
11. _____
12. _____

C. Match these terms with the correct parts of the coxa labeled in figure 7.22:

Acetabulum
Anterior inferior iliac spine
Anterior superior iliac spine
Articular surface
Greater sciatic notch
Iliac crest
Iliac fossa
Ilium
Ischial spine
Ischial tuberosity
Ischium
Lesser sciatic notch
Obturator foramen
Posterior inferior iliac spine
Posterior superior iliac spine
Pubis
Symphysis pubis

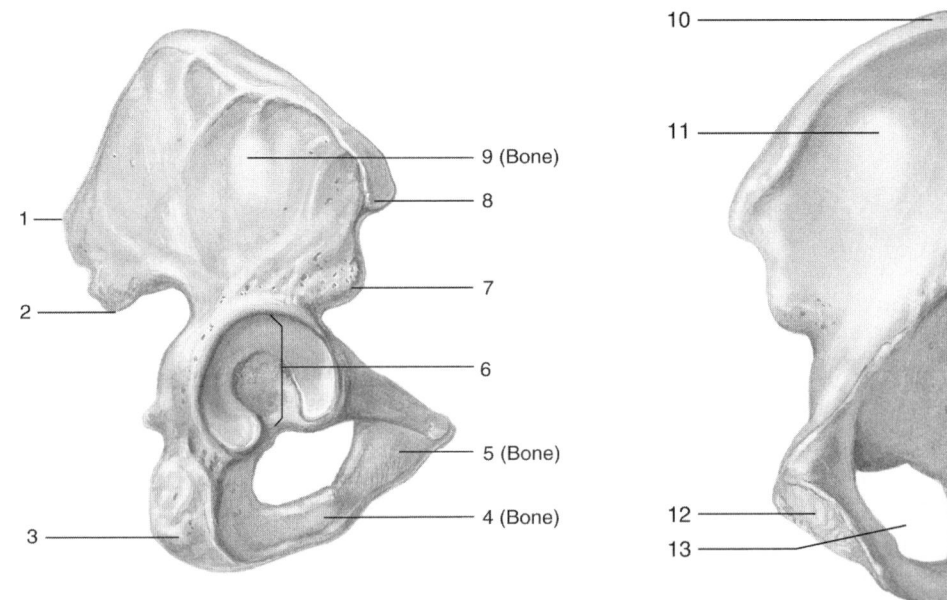

Figure 7.22

1. _____
2. _____
3. _____
4. _____
5. _____
6. _____
7. _____
8. _____
9. _____
10. _____
11. _____
12. _____
13. _____
14. _____
15. _____
16. _____
17. _____

D. Match these terms with the correct descriptions:

Female pelvis
Male pelvis

_____ 1. Pelvic inlet tends to be heart-shaped.

_____ 2. Subpubic angle is more than 90 degrees.

_____ 3. Ischial spines closer together.

_____ 4. Pelvic outlet is broader.

Chapter 7   111

Lower Limb

A. Match the bone parts that articulate with each other:

    Acetabulum   Patellar groove
    Condyles   Talus
    Head of fibula

_____ 1. Head of the femur articulates with the coxal bone.

_____ 2. Femur and tibia.

_____ 3. Tibia and proximal fibula.

_____ 4. Tibia and fibula with the ankle bone.

B. Match these terms with the correct parts of the femur labeled in figure 7.23:

    Fovea capitis   Linea aspera
    Greater trochanter   Medial condyle
    Head   Medial epicondyle
    Intercondylar fossa   Neck
    Lateral condyle   Patella
    Lateral epicondyle   Patellar groove
    Lesser trochanter

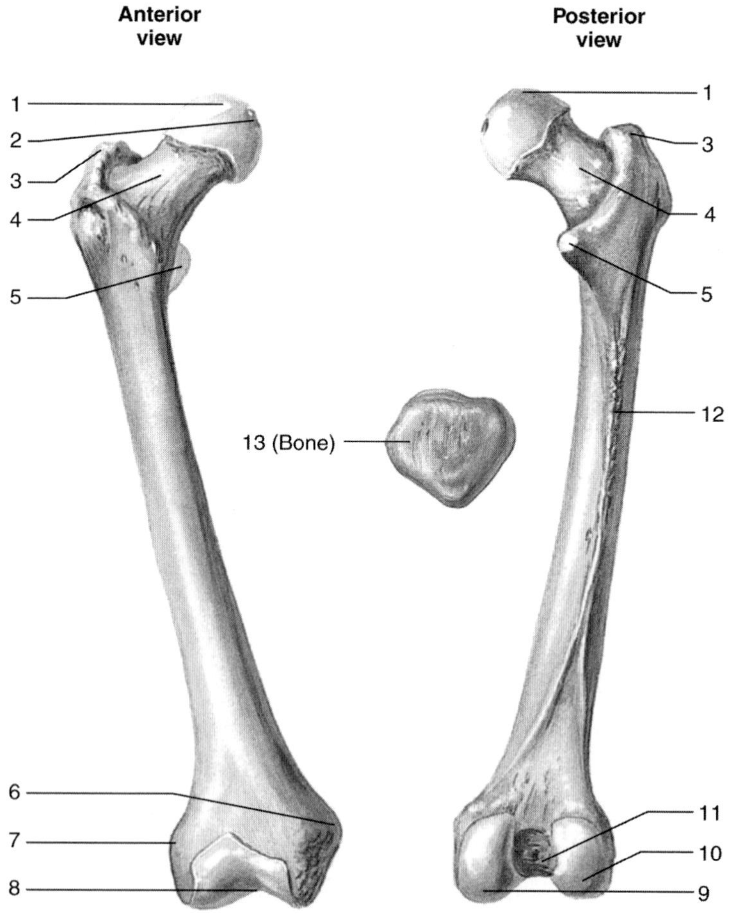

Anterior view   **Posterior view**

1. _____
2. _____
3. _____
4. _____
5. _____
6. _____
7. _____
8. _____
9. _____
10. _____
11. _____
12. _____
13. _____

Figure 7.23

112   Chapter 7

C. Match the bony part with its function:

Calcaneus Patella
Condyles Tibial tuberosity
Epicondyles Trochanters (lesser and greater)

_____ 1. Site of hip muscle attachment lateral to the neck of the femur.

_____ 2. Muscle attachment sites next to the condyles of the femur.

_____ 3. Bone within an anterior tendon of the thigh muscles.

_____ 4. Point of attachment on the tibia for anterior thigh muscles.

_____ 5. Attachment site for the calf muscles; the heel bone.

D. Match these terms with the correct parts of the leg labeled in figure 7.24:

Anterior crest Lateral malleolus
Fibula Medial condyle
Head Medial malleolus
Intercondylar eminence Tibia
Lateral condyle Tibial tuberosity

1. _____
2. _____
3. _____
4. _____
5. _____
6. _____
7. _____
8. _____
9. _____
10. _____

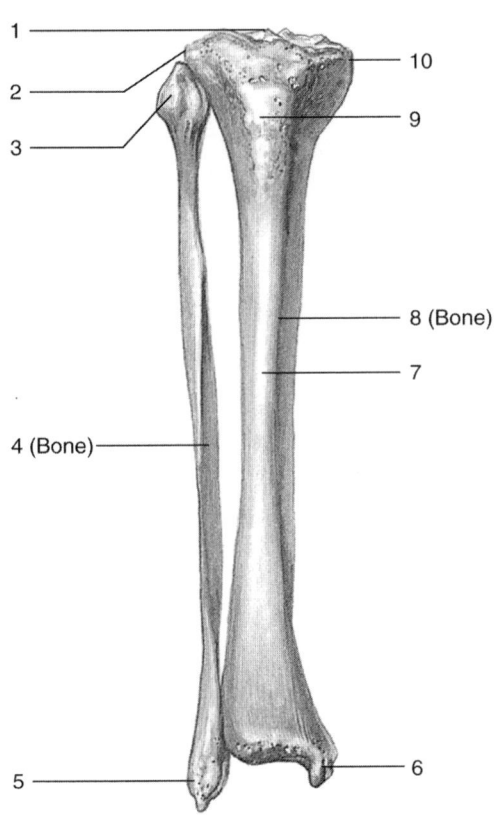

Figure 7.24

Chapter 7

E. Match these terms with the correct description:

    Lateral malleolus   Phalanges
    Medial malleolus   Talus
    Metatarsals   Tarsals

_____ 1. Located on the distal end of the tibia; forms the ankle.

_____ 2. Seven of these bones form the proximal portion of the foot.

_____ 3. Ankle bone.

_____ 4. Five of these bones form the middle of the foot.

_____ 5. Three of these bones are in each toe; but the big toe has two.

F. Match these terms with the correct parts of the foot labeled in figure 7.25:

    Calcaneus   Metatarsals
    Cuboid   Navicular
    Distal phalanx   Middle phalanx
    Intermediate cuneiform   Proximal phalanx
    Lateral cuneiform   Talus
    Medial cuneiform   Tarsals

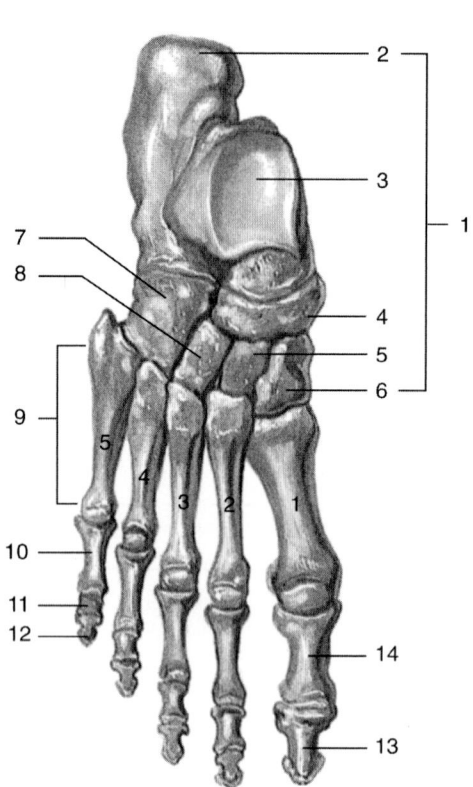

Figure 7.25

1. _____
2. _____
3. _____
4. _____
5. _____
6. _____
7. _____
8. _____
9. _____
10. _____
11. _____
12. _____
13. _____
14. _____

Quick Recall

1. List the two major subdivisions of the skull.

2. List the four major sutures in the neurocranium (braincase).

3. List the four major curvatures of the vertebral column of the adult, and give the direction in which they curve.

4. Name the five types of vertebrae, and give the number of each found in the vertebral column.

5. Name the three types of ribs according to their attachment, and give the number of each type.

6. List the three parts of the sternum.

7. Name the bones of the pectoral girdle, pelvic girdle, and pelvis.

8. State the number of carpals, metacarpals, and phalanges in the upper limb, and state the number of tarsals, metatarsals, and phalanges in the lower limb.

Name the bone or part of the bone responsible for the described bony landmark:

_____ 9. Bump posterior and inferior to the ear.

_____ 10. Bump that can be felt anterior to the ear when the jaw is moved side to side.

_____ 11. Bridge of the nose (two bones).

_____ 12. Tip of the shoulder.

_____ 13. Elbow projection on the medial side of the upper limb.

_____ 14. Elbow projection along the midline of the posterior surface of the upper limb.

_____ 15. Bump at the distal end of the forearm on the medial, posterior surface.

_____ 16. Knuckles.

_____ 17. Ridge of bone felt when the hands are placed on the hips.

_____ 18. Bony prominence that one sits upon.

_____ 19. Kneecap.

_____ 20. Shinbone.

_____ 21. Bump on the anterior leg just distal to the kneecap.

_____ 22. Large protuberance on the lateral surface of the ankle.

ANSWERS TO CHAPTER 7

Content Learning Activity

Introduction
1. Axial skeleton; 2. Axial skeleton; 3. Axial skeleton; 4. Appendicular skeleton; 5. Axial skeleton; 6. Appendicular skeleton

General Considerations
1. Condyle; 2. Facet; 3. Process; 4. Foramen; 5. Canal (meatus); 6. Fossa

The Skull
A. 1. Cranium; 2. Auditory ossicles; 3. Neurocranium (braincase); 4. Viscerocranium; 5. Hyoid bone; 6. Orbit; 7. Nasal conchae; 8. Paranasal sinuses; 9. Crista galli; 10. Sella turcica; 11. Nasal septum; 12. Hard palate
B. 1. Horizontal plate of palatine bone and palatine process of maxilla; 2. Perpendicular plate of ethmoid bone and vomer; 3. Temporal process of zygomatic bone and zygomatic process of temporal bone
C. 1. Nasal septum; 2. Perpendicular plate of ethmoid bone; 3. Septal cartilage; 4. Vomer; 5. Zygomatic arch; 6. Zygomatic process of temporal bone; 7. Temporal process of zygomatic bone; 8. Hard palate; 9. Horizontal plate of palatine bone; 10. Palatine process of maxillary bone
D. 1. Mandible; 2. Ethmoid; 3. Ethmoid; 4. Temporal; 5. Occipital; 6. Hyoid; 7. Sphenoid; 8. Sphenoid; 9. Temporal; 10. Temporal; 11. Ethmoid 12. Ethmoid; 13. Sphenoid; 14. Temporal

Muscles of the Skull
1. Muscles of mastication; 2. Neck muscles; 3. Eye movement muscles

Openings of the Skull
1. Olfactory foramina; 2. Nasolacrimal canal; 3. Optic canal; 4. External acoustic meatus; 5. Internal acoustic meatus; 6. Jugular foramen; 7. Carotid canals and foramen magnum; 8. Foramen magnum

Joints of the Skull
1. Sagittal suture; 2. Coronal suture; 3. Lambdoid suture; 4. Squamous suture; 5. Mandibular fossa; 6. Occipital condyle.

Skull Diagrams
A. 1. Coronal suture; 2. Sagittal suture; 3. Lambdoid suture; 4. Occipital bone; 5. Parietal bone; 6. Frontal bone
B. 1. Sphenoid bone; 2. Superior orbital fissure; 3. Palatine bone; 4. Zygomatic bone; 5. Inferior orbital fissure; 6. Maxilla; 7. Nasolacrimal canal; 8. Lacrimal bone; 9. Ethmoid bone; 10. Optic canal; 11. Frontal bone
C. 1. Coronal suture; 2. Temporal lines; 3. Parietal bone; 4. Squamous suture; 5. Lambdoid suture; 6. Occipital bone; 7. Temporal bone; 8. Squamous portion; 9. External acoustic meatus; 10. Mastoid process; 11. Styloid process; 12. Mandible; 13. Maxilla; 14. Zygomatic bone; 15. Nasal bone; 16. Sphenoid bone; 17. Greater wing; 18. Frontal bone
D. 1. Supraorbital margin; 2. Nasal bone; 3. Zygomatic bone; 4. Nasal septum; 5. Perpendicular plate; 6. Vomer; 7. Nasal cavity; 8. Maxilla; 9. Mandible; 10. Mandibular symphysis; 11. Inferior nasal concha; 12. Ethmoid bone; 13. Middle nasal concha; 14. Infraorbital foramen; 15. Orbit; 16. Supraorbital foramen; 17. Frontal bone
E. 1. Sphenoid bone; 2. Lesser wing; 3. Greater wing; 4. Jugular foramen; 5. Foramen magnum; 6. Occipital bone; 7. Temporal bone; 8. Internal acoustic meatus; 9. Petrous portion; 10. Squamous portion; 11. Carotid canal; 12. Sella turcica; 13. Optic canal; 14. Frontal bone; 15. Ethmoid bone; 16. Cribriform plate; 17. Crista galli; 18. Frontal sinus
F. 1. Sphenoid bone; 2. Lateral pterygoid plate; 3. Medial pterygoid plate; 4. Jugular foramen; 5. Occipital bone; 6. Occipital condyle; 7. Foramen magnum; 8. Nuchal lines; 9. External occipital protuberance; 10. Temporal bone; 11. Mastoid process; 12. Carotid canal; 13. Mandibular fossa; 14. Styloid process; 15. Vomer; 16. Palatine bone; 17. Maxilla; 18. Incisive fossa
G. 1. Frontal sinus; 2. Ethmoidal labyrinth (sinuses); 3. Sphenoidal sinus; 4. Maxillary sinus
H. 1. Mandibular condyle; 2. Ramus; 3. Body; 4. Angle; 5. Mental foramen; 6. Alveolar process; 7. Coronoid process; 8. Mandibular foramen

Vertebral Column
A. 1. Cervical and lumbar; 2. Lordosis; 3. Kyphosis; 4. Scoliosis
B. 1. Intervertebral disk; 2. Anulus fibrosus; 3. Nucleus pulposus; 4. Anulus fibrosus

General Plan of the Vertebrae
A. 1. Body; 2. Lamina and pedicle; 3. Vertebral foramen; 4. Intervertebral foramina; 5. Articular process; 6. Spinous process; 7. Transverse process
B. 1. Intervertebral disk; 2. Articular facet on rib head; 3. Articular facet on rib tubercle; 4. Superior articular facet; 5. Occipital condyle; 6. Dens; 7. Pelvic bone (ilium)
C. 1. Superior articular process; 2. Transverse process; 3. Vertebral foramen; 4. Body; 5. Vertebral arch; 6. Pedicle; 7. Lamina; 8. Spinous process
D. 1. Superior intervertebral notch; 2. Articular facet for rib head; 3. Body; 4. Articular facet for rib head; 5. Inferior intervertebral notch; 6. Spinous process; 7. Inferior articular process; 8. Lamina; 9. Articular facet for rib tubercle; 10. Transverse process; 11. Pedicle; 12. Superior articular facet; 13. Superior articular process

Regional Differences in Vertebrae
A. 1. Cervical vertebra; 2. Thoracic vertebra; 3. Lumbar vertebra; 4. Sacrum; 5. Coccygeal vertebra
B. 1. Cervical vertebra; 2. Transverse foramen; 3. Atlas; 4. Body and spinous process; 5. Cervical vertebra; 6. Dens; 7. Axis; 8. Cervical vertebra; 9. Bifid; 10. Thoracic vertebra; 11. Articular facet for rib head; 12. Articular facet for rib tubercle; 13. Slender and pointed; 14. Lumbar; 15. Medial; 16. Blunt
C. 1. Superior articular facet; 2. Articular surface (joins coxal bone); 3. Median sacral crest; 4. Sacral hiatus; 5. Coccyx; 6. Sacral foramina; 7. Sacral canal; 8. Ala

Thoracic Cage
A. 1. True (vertebrosternal) ribs; 2. False (vertebrochondral) ribs; 3. Floating vertebral) ribs; 4. Tubercle; 5. Body; 6. Xiphoid process; 7. Sternal angle
B. 1. True ribs; 2. Sternal angle; 3. Costal cartilage; 4. False ribs; 5. Floating ribs; 6. Sternum; 7. Xiphoid process; 8. Body; 9. Manubrium; 10. Jugular notch
C. 1. Head; 2. Neck; 3. Tubercle; 4. Body; 5. Angle; 6. Articular facet for transverse process of vertebrae; 7. Articular facet for body of vertebra

Pectoral Girdle
A. 1. Inferior angle; 2. Superior border; 3. Acromion process; 4. Spine; 5. Coracoid process; 6. Glenoid cavity; 7. Subscapular fossa; 8. Infraspinous fossa; 9. Clavicle
B. 1. Medial Sternal (medial) end; 2. Body; 3. Acromial (lateral) end
C. 1. Supraspinous fossa; 2. Spine; 3. Infraspinous fossa; 4. Medial (vertebral) border; 5. Lateral (axillary) border; 6. Acromion process; 7. Coracoid process; 8. Glenoid cavity; 9. Inferior angle; 10. Subscapular fossa; 11. Superior border; 12. Superior angle

Upper Limb
A. 1. Head of humerus; 2. Head of radius; 3. Trochlea; 4. Head of radius; 5. Head of ulna
B. 1. Head; 2. Lesser tubercle; 3. Greater tubercle; 4. Anatomical neck; 5. Surgical neck; 6. Intertubercular groove; 7. Deltoid tuberosity; 8. Radial fossa; 9. Lateral epicondyle; 10. Capitulum; 11. Trochlea; 12. Coronoid fossa; 13. Medial epicondyle; 14. Olecranon fossa
C. 1. Greater tubercle; lesser tubercle; deltoid tuberosity; 2. Intertubercular groove; 3. Epicondyles; 4. Olecranon process and radial tuberosity; 5. Styloid processes
D. 1. Olecranon process; 2. Head; 3. Neck; 4. Radial tuberosity; 5. Radius; 6. Styloid process; 7. Styloid process; 8. Head; 9. Ulna; 10. Radial notch; 11. Coronoid process; 12. Trochlear notch
E. 1. Metacarpals; 2. Carpals; 3. Phalanges

F. 1. Scaphoid bone; 2. Lunate bone;
    3. Triquetrum bone; 4. Pisiform bone;
    5. Trapezium bone; 6. Trapezoid bone;
    7. Capitate bone; 8. Hamate bone; 9. Carpals;
    10. Metacarpals; 11. Distal phalanx;
    12. Middle phalanx; 13. Proximal phalanx;
    14. Distal phalanx; 15. Proximal phalanx

Pelvic Girdle
A. 1. Coxal bone; 2. Obturator foramen; 3. Iliac crest; 4. Iliac fossa; 5. Iliac spines; 6. Ischial tuberosity; 7. Symphysis pubis; 8. Articular surface; 9. False pelvis; 10. True pelvis
B. 1. Sacrum; 2. Sacroiliac joint; 3. Anterior superior iliac spine; 4. Acetabulum; 5. Obturator foramen; 6. Subpubic angle; 7. Symphysis pubis; 8. Coxal bone; 9. Ischium; 10. Pubis; 11. Ilium; 12. Sacral promontory
C. 1. Posterior superior iliac spine; 2. Posterior inferior iliac spine; 3. Ischial tuberosity; 4. Ischium; 5. Pubis; 6. Acetabulum; 7. Anterior inferior iliac spine; 8. Anterior superior iliac spine; 9. Ilium; 10. Iliac crest; 11. Iliac fossa; 12. Symphysis pubis; 13. Obturator foramen; 14. Lesser sciatic notch; 15. Ischial spine; 16. Greater sciatic notch; 17. Articular surface
D. 1. Male pelvis; 2. Female pelvis; 3. Male pelvis; 4. Female pelvis

Lower Limb
A. 1. Acetabulum; 2. Condyles; 3. Head of fibula; 4. Talus
B. 1. Head; 2. Fovea capitis; 3. Greater trochanter; 4. Neck; 5. Lesser trochanter; 6. Medial epicondyle; 7. Lateral epicondyle; 8. Patellar groove; 9. Medial condyle; 10. Lateral condyle; 11. Intercondylar fossa; 12. Linea aspera; 13. Patella
C. 1. Trochanters (lesser and greater); 2. Epicondyles; 3. Patella; 4. Tibial tuberosity; 5. Calcaneus
D. 1. Intercondylar eminence; 2. Lateral condyle; 3. Head; 4. Fibula; 5. Lateral malleolus; 6. Medial malleolus; 7. Anterior crest; 8. Tibia; 9. Tibial tuberosity; 10. Medial condyle
E. 1. Medial malleolus; 2. Tarsals; 3. Talus; 4. Metatarsals; 5. Phalanges
F. 1. Tarsals; 2. Calcaneus; 3. Talus; 4. Navicular; 5. Intermediate cuneiform; 6. Medial cuneiform; 7. Cuboid; 8. Lateral cuneiform; 9. Metatarsals; 10. Proximal phalanx; 11. Middle phalanx; 12. Distal phalanx; 13. Distal phalanx; 14. Proximal phalanx

QUICK RECALL

1. Neurocranium (braincase) and viscerocranium (facial bones)
2. Coronal, squamous, lambdoid, and sagittal sutures
3. Cervical: convex anteriorly; thoracic: concave anteriorly; lumbar: convex anteriorly; sacrum and coccyx: concave anteriorly
4. Cervical: 7; thoracic: 12; lumbar: 5; sacrum: 1; coccyx: 1
5. True (vertebrosternal) ribs: 7 pairs; false ribs: 5 pairs; 3 pairs of the false ribs are vertebrochondral ribs and 2 pairs are floating (vertebral) ribs
6. Manubrium, body, and xiphoid process
7. Pectoral girdle: the two scapulae and the two clavicles; pelvic girdle: right and left coxal bones and sacrum; pelvis; pelvic girdle and coccyx
8. Upper limb: carpals 8; metacarpals 5; phalanges 14; lower limb: tarsals 7; metatarsals 5; phalanges 14
9. Mastoid process
10. Mandibular condyle
11. Nasal bones and maxillae
12. Acromion process of scapula
13. Medial epicondyle of humerus
14. Olecranon process of ulna
15. Head of ulna
16. Distal end of metacarpals
17. Iliac crest
18. Ischial tuberosity
19. Patella
20. Anterior crest of tibia
21. Tibial tuberosity
22. Lateral malleolus of fibula

Articulations and Movement

CONTENT LEARNING ACTIVITY

Fibrous and Cartilaginous Joints

A. Match the type of joint with its definition:

    Fontanel   Symphysis
    Gomphosis   Synchondrosis
    Periodontal ligament Syndesmosis
    Sutural ligament   Synostosis
    Suture

_____ 1. Seams between skull bones; the bones are joined by short, dense regular collagenous connective tissue.

_____ 2. Two layers of periosteum plus dense fibrous connective tissue between adjacent skull bones.

_____ 3. Membranous area within some sutures in the newborn.

_____ 4. Result when bones of a joint grow together to form a single bone.

_____ 5. Joint with bones joined by ligaments; farther apart than sutures.

_____ 6. Specialized joints with pegs that fit into sockets; bundles of regular collagenous connective tissue join teeth to sockets.

_____ 7. Connective tissue bundle between teeth and sockets.

_____ 8. Joint with two bones joined by hyaline cartilage; most become synostoses.

_____ 9. Joint with bones joined by fibrocartilage.

B. Match the type of joint with the correct example:

    Gomphosis   Synchondrosis
    Suture   Syndesmosis
    Symphysis   Synostosis

_____ 1. Between the temporal and parietal bones.

_____ 2. Epiphyseal line.

Chapter 8

_____ 3. Between the ulna and the radius.

_____ 4. Between the mandible and the teeth.

_____ 5. Epiphyseal plate.

_____ 6. Intervertebral discs.

Synovial Joints

A. Match these terms with the correct statement or definition:

Articular cartilage   Meniscus
Articular disk   Synovial fluid
Bursa   Synovial membrane
Bursitis   Tendon sheath
Fibrous capsule

_____ 1. Thin layer of hyaline cartilage that covers the articular surface of bones in synovial joints.

_____ 2. Fibrocartilage structures that provide extra strength and support for joints such as the knee and temporomandibular joint.

_____ 3. Incomplete, crescent-shaped fibrocartilage in joints such as the knee and wrist.

_____ 4. Portion of the joint capsule that is continuous with the fibrous layer of the periosteum.

_____ 5. Portion of the joint capsule; an inner membrane that secretes fluid.

_____ 6. Fluid that consists of serum filtrate and secretions from the synovial cells; forms lubricating film on joint surfaces.

_____ 7. Pocket or sac of the synovial membrane extending away from the rest of the joint cavity.

_____ 8. Pocket or sac containing synovial fluid extending along a tendon for some distance.

_____ 9. Inflammation of a bursa; results in pain and restriction of movement.

B. Match these terms with the correct statement or definition:

Biaxial   Uniaxial
Multiaxial

_____ 1. Synovial joint movement occurring around one axis.

_____ 2. Synovial joint movement occurring around two axes situated at right angles to each other.

_____ 3. Synovial joint movement occurring around several axes.

C. Match these terms with the correct parts labeled in figure 8.1:

Articular cartilage
Bursa
Fibrous capsule
Joint capsule

Joint cavity
Synovial membrane
Tendon sheath

1. _____
2. _____
3. _____
4. _____
5. _____
6. _____
7. _____

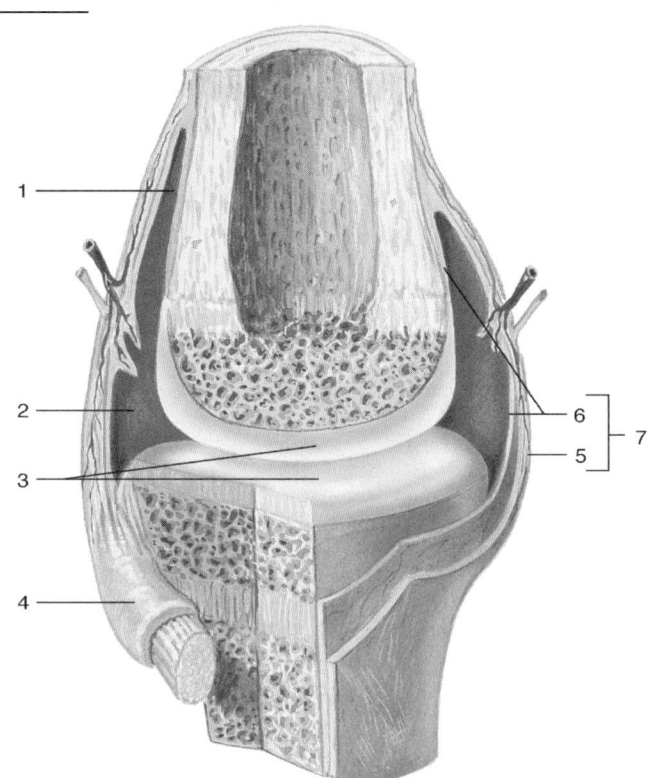

Figure 8.1

D. Match the type of synovial joint with the correct definition and type of movement:

Ball-and-socket joint
Ellipsoid joint
Hinge joint

Pivot joint
Plane (gliding) joint
Saddle joint

_____
_____
_____
_____
_____
_____

1. Two opposed flat surfaces approximately equal in size; uniaxial movement is gliding or slightly rotating.
2. Two saddle-shaped surfaces at right angles; biaxial movement.
3. Convex cylinder applied to a concavity; uniaxial movement.
4. Cylindrical process rotates within a ring; uniaxial rotation.
5. Round head fitting into a round depression; multiaxial.
6. Modified ball-and-socket joint; range of motion is biaxial.

Chapter 8

121

E. Match the type of synovial joint with the correct example:

Ball-and-socket joint   Pivot joint
Ellipsoid joint   Plane (gliding) joint
Hinge joint   Saddle joint

_____ 1. Articular process of adjacent vertebrae.

_____ 2. Carpal and metacarpal of thumb.

_____ 3. Between phalanges.

_____ 4. Between the atlas and axis.

_____ 5. Hip joint.

_____ 6. Atlas and occipital bone.

F. Match these terms with the correct type of synovial joint in figure 8.2:

Ball-and-socket joint   Pivot joint
Ellipsoid joint   Plane (gliding) joint
Hinge joint   Saddle joint

1. _____   3. _____   5. _____

2. _____   4. _____   6. _____

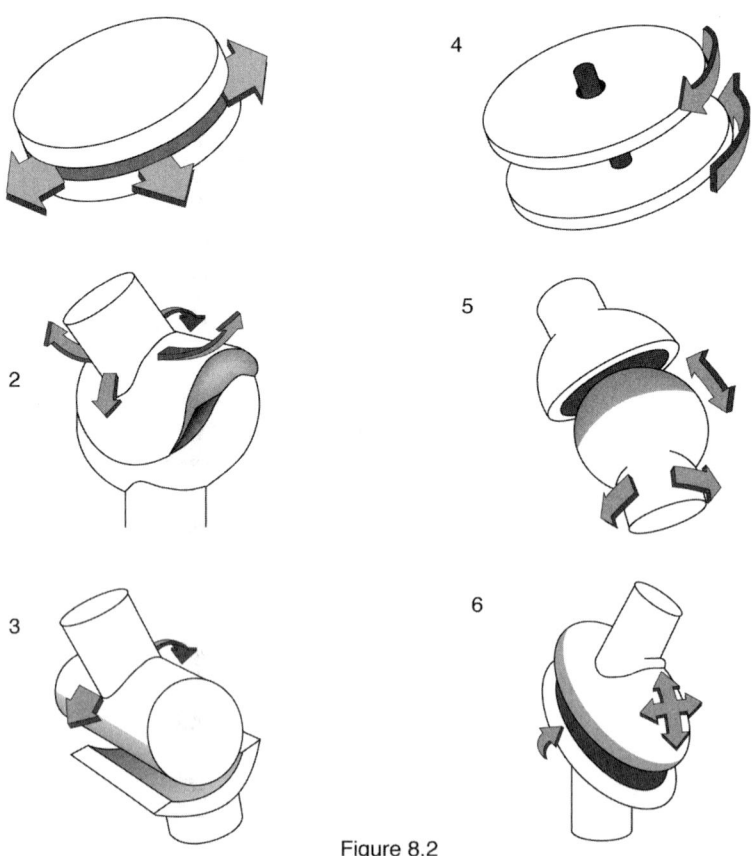

Figure 8.2

Gliding and Angular Movements

Match these terms with the correct statement or definition:

Abduction
Adduction
Dorsiflexion
Extension
Flexion
Gliding
Lateral flexion
Plantar flexion

_____ 1. Simplest type of movement; occurs in plane joints between two flat or nearly flat surfaces.

_____ 2. Moving a part of the body in an anterior or ventral direction (except the knee).

_____ 3. Movement of the leg in an anterior direction.

_____ 4. Movement of the foot toward the plantar surface.

_____ 5. Movement away from the median plane.

_____ 6. Movement toward the median plane.

_____ 7. Bending at the waist to one side.

Circular Movements

Match these terms with the correct statement or definition:

Circumduction
Pronation
Rotation
Supination

_____ 1. Turning of a structure around its long axis.

_____ 2. Rotation of the palm so that it faces anteriorly in relation to the anatomical position.

_____ 3. Combination of flexion, extension, abduction, and adduction.

Special Movements

Match these terms with the correct statement or definition:

Depression
Elevation
Eversion
Inversion
Lateral excursion
Medial excursion
Opposition
Protraction
Reposition
Retraction

_____ 1. Movement of a structure superiorly—e.g., the mandible.

_____ 2. Movement of a structure anteriorly—e.g., the scapula.

_____ 3. Movement of the mandible to the side.

_____ 4. Movement of the thumb and little finger together.

_____ 5. Turning the ankle so that the plantar surface faces laterally.

Movement Diagrams

Match these terms with the correct parts labeled in figure 8.3:

Abduct
Adduct
Circumduct
Dorsiflex
Extend
Flex
Plantar flex
Pronate
Supinate

1. _____
2. _____
3. _____
4. _____
5. _____
6. _____
7. _____
8. _____
9. _____
10. _____
11. _____

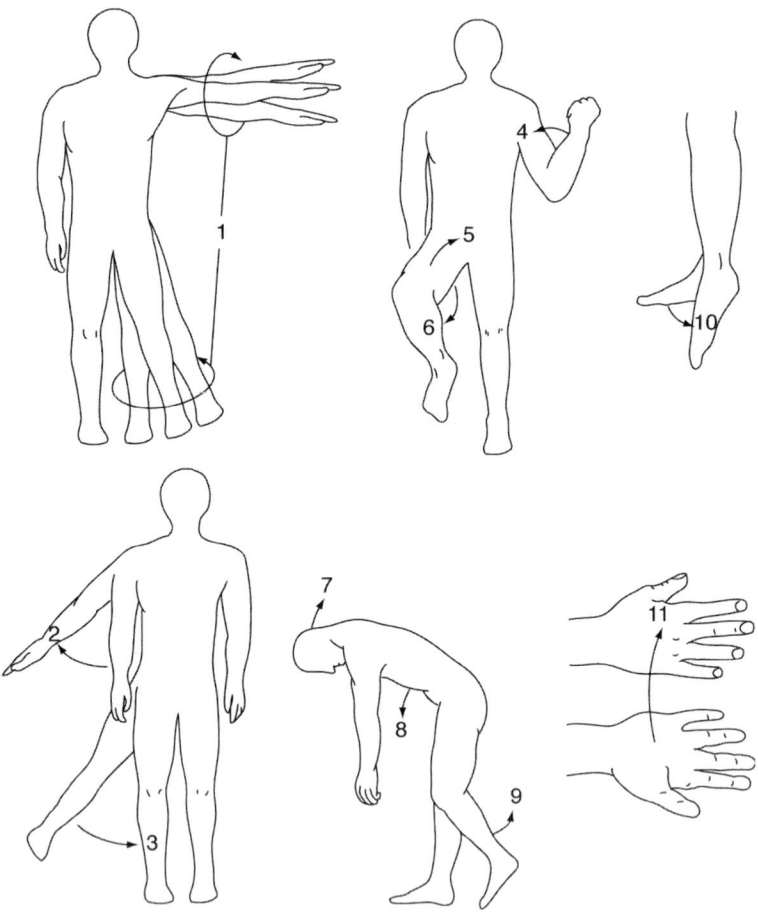

Figure 8.3

Range of Motion

Using the terms provided, complete these statements:

Active range of motion   Ligaments and tendons
Cartilage   Passive range of motion
Fluid   Use or disuse

Range of motion is an expression of the amount of mobility that can be demonstrated in a given joint. _(1)_ is the amount of movement that can be accomplished by contraction of the muscles that normally act across a joint, whereas _(2)_ is the amount of movement that can be accomplished at a joint when the structures that meet at that joint are moved by outside force. The range of motion for a given joint is influenced by a number of factors, including the shape of the articular surfaces of the bones forming the joint, the amount and shape of the _(3)_ covering those articular surfaces, the strength and location of _(4)_ surrounding the joint, the strength and location of muscles associated with the joint, the amount of _(5)_ in and around the joint, the amount of pain in and around the joint, and the amount of _(6)_ the joint has received over time.

1. _____
2. _____
3. _____
4. _____
5. _____
6. _____

Temporomandibular Joint

Using the terms provided, complete these statements:

Articular disk   Mandibular fossa
Condyle   Plane and ellipsoid
Ligaments

In the temporomandibular joint, the _(1)_ of the mandible articulates with the _(2)_ of the temporal bone. A fibrocartilage _(3)_ is interposed between the mandible and temporal bone, and the joint is surrounded by a fibrous capsule. The joint is also strengthened by lateral and accessory _(4)_. The temporomandibular joint is a combination _(5)_ joint, which allows depression, protraction, and excursion of the mandible.

1. _____
2. _____
3. _____
4. _____
5. _____

Shoulder Joint

Match these terms with the correct statement or definition:

Biceps brachii Rotator cuff
Glenoid fossa Subacromial bursa
Glenoid labrum Subscapular bursa

_____ 1. Fibrocartilage ring that builds up the rim of the glenoid fossa.

_____ 2. Sac with synovial fluid that opens into the shoulder joint cavity.

_____ 3. Sac with synovial fluid near the joint cavity, but separated from the cavity by the joint capsule.

_____ 4. Four muscles, collectively, that hold the humeral head tightly within the glenoid fossa.

_____ 5. Tendon of this muscle passes through the articular capsule and helps hold the humerus against the glenoid fossa.

Elbow Joint

Using the terms provided, complete these statements:

Humeroradial Proximal radioulnar
Humeroulnar Radial annular ligament
Joint capsule Radial collateral ligament
Olecranon bursa Ulnar collateral ligament

The elbow joint consists of a _(1)_ joint, between the humerus and ulna, and a _(2)_ joint, between the humerus and radius. The _(3)_ joint between the proximal radius and ulna is also closely related. The elbow joint is surrounded by a _(4)_. The humeroulnar joint is reinforced by the _(5)_, the humeroradial joint is reinforced by the _(6)_, and the proximal radioulnar joint is reinforced by the _(7)_. A subcutaneous _(8)_ covers the proximal and posterior surfaces of the olecranon process.

1. _____
2. _____
3. _____
4. _____
5. _____
6. _____
7. _____
8. _____

Hip Joint

Match these terms with the correct statement or definition:

Acetabular labrum Ligament of the head of the
Iliofemoral ligament   femur
Joint capsule Transverse acetabular ligament

_____ 1. Lip of fibrocartilage; strengthens and deepens the acetabulum.

_____ 2. Ligament that passes inferior to and strengthens the acetabulum.

_____ 3. Strong connective tissue; extends from the rim of the acetabulum to the neck of the femur, completely surrounding the femur head.

_____ 4. Strong, thick ligament that most people use to support the body's weight.

_____ 5. Located inside the hip joint between the femoral head and the acetabulum.

Knee Joint

Match these terms with the correct statement or definition:

Collateral and popliteal ligaments
Cruciate ligaments
Menisci
Suprapatellar bursa

_____ 1. Thick articular disks at the margins of the tibia; deepen the articular surface.

_____ 2. Ligaments that extend between the fossa of the femur and the intercondylar eminence of the tibia; limit anterior and posterior displacement of the joint.

_____ 3. Ligaments that surround and strengthen the knee.

_____ 4. Superior extension of the joint capsule; allows movement of the anterior thigh muscles over the distal end of the femur.

Ankle Joint and Arches of the Foot

Using the terms provided, complete these statements:

Dorsiflexion
Calcaneus
Hinge joint
Talocrural
Talus

The ankle, or _(1)_ joint, is a highly modified _(2)_ formed by the articulation of the tibia and fibula with the _(3)_. Plantar flexion and _(4)_, as well as limited inversion and eversion, can occur at this joint.

1. _____
2. _____
3. _____
4. _____

Effects of Aging on the Joints

Using the terms provided, complete these statements:

Decrease(s)
Increase(s)

In general, as a person ages, tissues become less flexible and less elastic as protein cross-linking _(1)_. With age, the rate of tissue repair and new blood vessel development _(2)_ and, as a result, the rate of repair of worn cartilage covering articular surfaces _(3)_. With increased age, production of synovial fluid _(4)_, which also results in increased wear of the articular cartilage. In addition, the ligaments and tendons surrounding a joint shorten and become less flexible with age, resulting in a(n) _(5)_ in the range of motion of a joint. A lack of activity in older people also contributes to a(n) _(6)_ in the range of motion of their joints.

1. _____
2. _____
3. _____
4. _____
5. _____
6. _____

Quick Recall

1. List the three major classes of joints.

2. Name the three types of fibrous joint and give an example of each.

3. Name the two types of cartilaginous joints; give an example of each.

4. Name the six types of synovial joints; give an example of each.

5. Name six types of angular movement.

6. List four types of circular movement.

7. Name 10 types of special movement.

ANSWERS TO CHAPTER 8

CONTENT LEARNING ACTIVITY

Fibrous and Cartilaginous Joints
  A. 1. Suture; 2. Sutural ligament; 3. Fontanel; 4. Synostosis; 5. Syndesmosis; 6. Gomphosis; 7. Periodontal ligament; 8. Synchondrosis; 9. Symphysis
  B. 1. Suture; 2. Synostosis; 3. Syndesmosis; 4. Gomphosis; 6. Synchondrosis; 7. Symphysis

Synovial Joints
  A. 1. Articular cartilage; 2. Articular disk; 3. Meniscus; 4. Fibrous capsule; 5. Synovial membrane; 6. Synovial fluid; 7. Bursa; 8. Tendon sheath; 9. Bursitis
  B. 1. Uniaxial; 2. Biaxial; 3. Multiaxial
  C. 1. Bursa; 2. Joint cavity; 3. Articular cartilage; 4. Tendon sheath; 5. Fibrous capsule; 6. Synovial membrane; 7. Joint capsule
  D. 1. Plane (gliding) joint; 2. Saddle joint; 3. Hinge joint; 4. Pivot joint; 5. Ball-and-socket joint; 6. Ellipsoid joint
  E. 1. Plane joint; 2. Saddle joint; 3. Hinge joint; 4. Pivot joint; 5. Ball-and-socket joint; 6. Ellipsoid joint
  F. 1. Plane joint; 2. Saddle joint; 3. Hinge joint; 4. Pivot joint; 5. Ball-and-socket joint; 6. Ellipsoid joint

Gliding and Angular Movements
  1. Gliding; 2. Flexion; 3. Extension; 4. Plantar flexion; 5. Abduction; 6. Adduction; 7. Lateral flexion

Circular Movements
  1. Rotation; 2. Supination; 3. Circumduction

Special Movements
  1. Elevation; 2. Protraction; 3. Lateral excursion; 4. Opposition; 5. Eversion

Movement Diagrams
  1. Circumduct; 2. Abduct; 3. Adduct; 4. Flex; 5. Flex; 6. Extend; 7. Extend; 8. Flex; 9. Flex; 10. Plantar flex; 11. Pronate

Range of Motion
  1. Active range of motion; 2. Passive range of motion; 3. Cartilage; 4. Ligaments and tendons; 5. Fluid; 6. Use or disuse

Temporomandibular Joint
  1. Condyle; 2. Mandibular fossa; 3. Articular disk; 4. Ligaments; 5. Plane (gliding) and ellipsoid

Shoulder Joint
  1. Glenoid labrum; 2. Subscapular bursa; 3. Subacromial bursa; 4. Rotator cuff; 5. Biceps brachii

Elbow Joint
  1. Humeroulnar; 2. Humeroradial; 3. Proximal radioulnar; 4. Joint capsule; 5. Ulnar collateral ligament; 6. Radial collateral ligament; 7. Radial annular ligament; 8. Olecranon bursa

Hip Joint
  1. Acetabular labrum; 2. Transverse acetabular ligament; 3. Joint capsule; 4. Iliofemoral ligament; 5. Ligamentum teres

Knee Joint
  1. Menisci; 2. Cruciate ligaments; 3. Collateral and popliteal ligaments; 4. Suprapatellar bursa

Ankle Joint and Arches of the Foot
  1. Talocrural; 2. Hinge joint; 3. Talus; 4. Dorsiflexion

Effects of Aging on the Joints
  1. Increases; 2. Decreases; 3. Decreases; 4. Decreases; 5. Decrease; 6. Decrease

QUICK RECALL

1. Fibrous, cartilaginous, and synovial joints
2. Sutures: see table 8.1 for examples; syndesmosis: see table 8.1 for examples; gomphosis: between the teeth and alveolar process in mandible or maxilla
3. Synchondrosis: see table 8.1 for examples; symphysis: see table 8.1 for examples
4. Plane joint: see table 8.2 for examples; saddle joint: between carpal and metacarpal of thumb; hinge joint: see table 8.2 for examples; pivot joint: between atlas and axis, or between radius and ulna; ball-and-socket joint: between coxal bone and femur, or between the scapula and humerus; ellipsoid joint: see table 8.2 for examples
5. Flexion, extension, plantar flexion, dorsiflexion, abduction, and adduction
6. Rotation, circumduction, pronation, and supination
7. Elevation, depression, protraction, retraction, lateral excursion, medial excursion, opposition, reposition, inversion, and eversion

Muscular System: Histology and Physiology

CONTENT LEARNING ACTIVITY

General Functional Characteristics of Muscle

Match these terms with the correct statement or definition:

Cardiac muscle Extensibility
Contractility Skeletal muscle
Elasticity Smooth muscle
Excitability

_____ 1. Ability to shorten forcefully.

_____ 2. Ability to be stretched to normal resting length or beyond.

_____ 3. Ability to recoil to original resting length after being stretched.

_____ 4. Comprises about 40% of the body; under voluntary control; responsible for locomotion, facial expressions, posture, respiratory movements, and other body movements.

_____ 5. Located in the walls of hollow organs, the internal eye muscles, and the walls of blood vessels; under involuntary control and capable of spontaneous contractions; regulates urine and blood flow and movement and mixing of food.

_____ 6. Located only in the heart; under involuntary control, capable of spontaneous contractions; pumps blood throughout the circulatory system.

Skeletal Muscle Structure

A. Match these terms with the correct statement or definition:

Branched Many
Cylindrical Myoblasts
Increase in number One
Increase in size Striated

_____ 1. Shape of skeletal muscle fibers.

_____ 2. Number of nuclei within a skeletal muscle fiber.

_____ 3. Cells from which muscle fibers develop.

_____ 4. Responsible for most enlargement (hypertrophy) of skeletal muscles after birth.

_____ 5. Appearance of skeletal muscle fibers as seen in longitudinal sections.

B. Match these terms with the correct statement or definition:

Endomysium   Muscular fascia
Epimysium   Perimysium
External lamina   Sarcolemma

_____ 1. Muscle fiber cell membrane.

_____ 2. Layer of reticular fibers just outside the sarcolemma.

_____ 3. Layer of reticular tissue just outside the external lamina.

_____ 4. Surrounds a bundle of muscle fibers (fasciculus).

_____ 5. Surrounds groups of fasciculi (a whole muscle); consists of dense collagenous connective tissue.

_____ 6. Connective tissue superficial to the epimysium; separates and compartmentalizes individual muscles or groups of muscles.

C. Match these terms with the correct parts labeled in figure 9.1:

Endomysium
Epimysium (fascia)
Fasciculus (bundle)
Muscle fiber (cell)
Muscular fascia
Perimysium
Sarcolemma

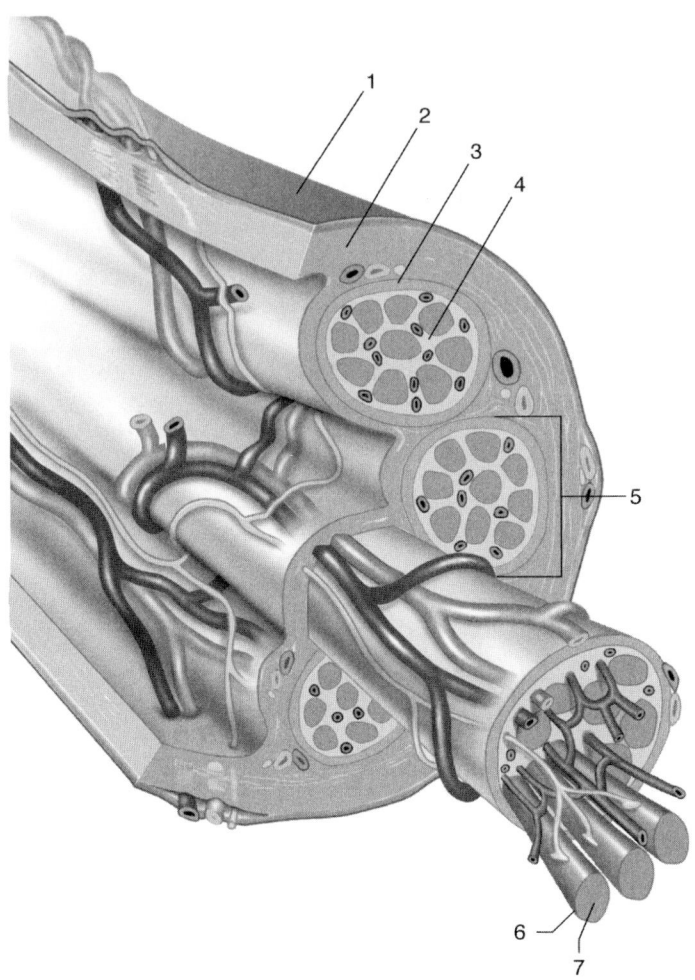

1. _____
2. _____
3. _____
4. _____
5. _____
6. _____
7. _____

Figure 9.1

Chapter 9

Muscle Fibers, Myofilaments, and Sarcomeres

A. Match these terms with the correct statement or definition:

    Motor neurons   Sarcolemma
    Myofibril   Sarcomeres
    Myofilaments   Sarcoplasm

_____ 1. Nerve cells that stimulate muscles to contract.

_____ 2. Cytoplasmic material of a muscle cell (without the myofibrils).

_____ 3. Threadlike structure that extends from one end of a muscle fiber to the other; composed of myofilaments.

_____ 4. Two types of proteins, actin and myosin.

_____ 5. Highly ordered units of actin and myosin myofilaments, which are joined end to end to form myofibrils.

B. Match these terms with the correct statement or definition:

    Cross-bridge   Myosin head
    F-actin   Tropomyosin
    G-actin   Troponin

_____ 1. Molecule with two strands coiled to form a double helix; extend the length of the actin myofilament.

_____ 2. Small, globular units that combine to form F-actin; each unit has an active site to which myosin can bind.

_____ 3. Elongated protein that covers active sites on the actin molecules.

_____ 4. Composed of three subunits, each of which binds with actin, tropomyosin, or Ca^{2+}.

_____ 5. Structure formed when a myosin head binds with an active site on actin.

_____ 6. Attaches to the rodlike portion of myosin by a hinge region that can bend during contraction.

_____ 7. Has ATPase, which splits ATP to ADP + P and releases energy.

C. Match these terms with the correct statement or definition:

A band
Actin myofilaments
H zone
I band
M line
Myosin myofilaments
Titin
Z disk

_____ 1. Filamentous network of protein forming a disklike structure to which actin myofilaments attach; the boundary of a sarcomere.

_____ 2. Area that extends between two myosin myofilaments and includes a Z disk; the light band.

_____ 3. Area extending the length of a myosin myofilament within a sarcomere; the dark band.

_____ 4. Area that extends between two myofilaments in the center of a sarcomere; no overlap of actin and myosin myofilaments occurs here.

_____ 5. Dark band in the middle of the H zone composed of delicate filaments that attach to myosin myofilaments and hold them in place.

_____ 6. Large protein that helps hold myosin myofilaments in place; allows the sarcomere to stretch and recoil.

D. Match these terms with the correct parts labeled in figure 9.2:

A band
Actin
Active site
F-actin molecule
H zone
I band
Myosin
Myosin head
Sarcomere
Tropomyosin
Troponin
Z disk

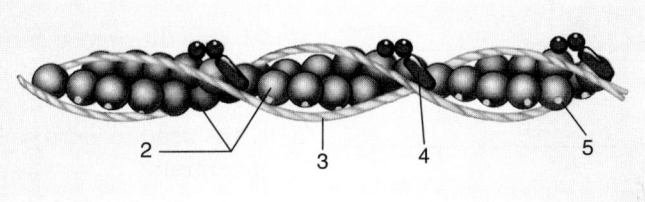

1. Type of myofilament

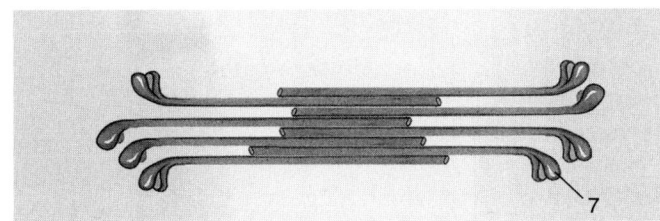

6. Type of myofilament

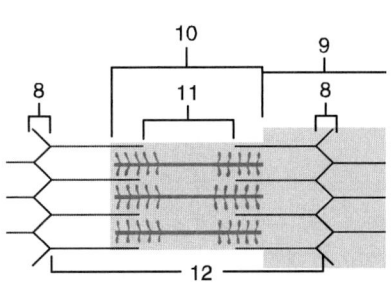

Figure 9.2

1. _____ 5. _____ 9. _____
2. _____ 6. _____ 10. _____
3. _____ 7. _____ 11. _____
4. _____ 8. _____ 12. _____

Chapter 9

Sliding Filament Model

Using the terms provided, complete these statements:

A bands	H zones
Actin	I bands
Do	M line
Do not	Myosin
Gravity	Sarcomeres

During contraction, the actin and myosin molecules __(1)__ change length. During contraction, the __(2)__ myofilaments slide past the __(3)__ myosin myofilaments toward each other. As a consequence, the __(4)__ and __(5)__ become more narrow, but the __(6)__ remain constant in length. As the actin myofilaments slide over the myosin myofilaments, the Z lines are brought closer together, and the __(7)__ are shortened. During relaxation, sarcomeres lengthen passively from the force of other muscles or __(8)__.

1. _____
2. _____
3. _____
4. _____
5. _____
6. _____
7. _____
8. _____

Physiology of Skeletal Muscle Fibers

A. Match these terms with the correct statement or definition:

Millivolts	Positive
Negatives	Resting membrane potential

_____ 1. Charge difference across the plasma membrane of an unstimulated cell.

_____ 2. Units used to express the charge difference across a plasma membrane.

_____ 3. Compared with the outside of the plasma membrane, the type of charge on the inside.

B. Match these terms with the correct statement or definition:

Into cells	Out of cells
Ligand	Receptor
Ligand-gated	Voltage-gated

_____ 1. Molecule that binds to a receptor.

_____ 2. Protein or glycoprotein to which a ligand can bind.

_____ 3. Type of ion channel that opens when a neurotransmitter binds to it.

_____ 4. Type of ion channel that opens in response to a change in the charge difference across the plasma membrane.

_____ 5. Direction of movement of Na^+ when gated Na^+ channels open.

_____ 6. Direction of movement of K^+ when gated K^+ channels open.

C. Match these terms with the correct statement or definition:

Action potentials
All-or-none principal
Decrease
Depolarization
Increase
Propagate
Repolarization
Threshold

_____ 1. The nervous system controls the contraction of skeletal muscles through these.

_____ 2. Occurs when the inside of the plasma membrane becomes less negative.

_____ 3. Results in an action potential when the depolarization caused by stimulating a cell reaches this level.

_____ 4. Results when gated Na^+ channels open and Na^+ move into the cell.

_____ 5. Return of the membrane potential to its resting value.

_____ 6. Results when gated K^+ channels open and K^+ move out of the cell.

_____ 7. An action potential always occurs with the same magnitude, or does not occur.

_____ 8. Spread of an action potential across the plasma membrane.

_____ 9. Effect of increasing stimulus strength on action potential frequency.

Neuromuscular Junction

A. Match these terms with the correct statement or definition:

Acetylcholinesterase
Neurotransmitter
Postsynaptic membrane
   (motor end plate)
Presynaptic terminal
Synaptic cleft
Synaptic vesicles
Voltage-gated Ca^{2+} channels

_____ 1. Enlarged axon terminal that rests in an invagination of the sarcolemma at the neuromuscular junction.

_____ 2. Space between the presynaptic terminal and the muscle fiber.

_____ 3. Muscle cell membrane in the neuromuscular junction.

_____ 4. Spherical sacs in presynaptic terminal; contain acetylcholine.

_____ 5. Substance (such as acetylcholine) released from a presynaptic terminal, diffuses across the synaptic cleft, and stimulates (or inhibits) an action potential in the postsynaptic membrane.

_____ 6. Open when action potential reaches presynaptic terminal; Ca^{2+} enter cell and acetylcholine is released from vesicles.

_____ 7. Enzyme that breaks acetylcholine into acetic acid and choline; prevents accumulation of acetylcholine in the synaptic cleft.

B. Match these terms with the parts labeled in figure 9.3:

Postsynaptic membrane Synaptic cleft
Presynaptic terminal Synaptic vesicles

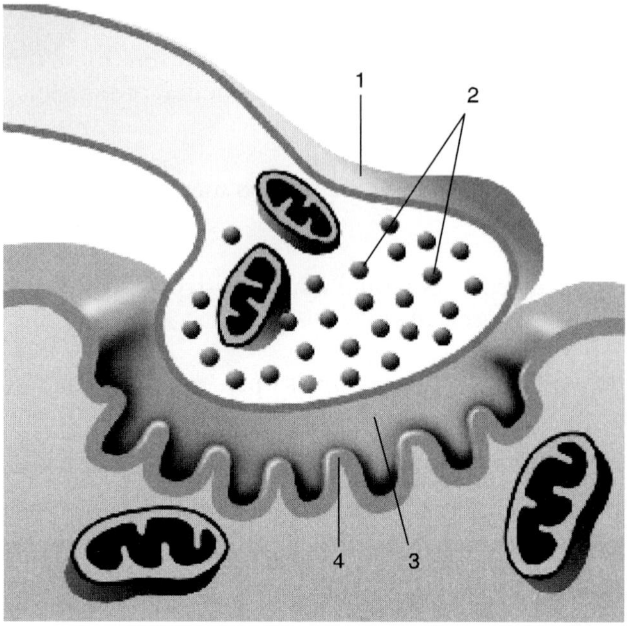

1. _____
2. _____
3. _____
4. _____

Figure 9.3

Excitation–Contraction Coupling

A. Match these terms with the correct statement or definition:

Sarcoplasmic reticulum Terminal cisterna
T tubules Triad

_____ 1. Tubelike invaginations of the sarcolemma that project into the sarcoplasm and wrap around where actin and myosin overlap.

_____ 2. Smooth endoplasmic reticulum, the membrane of which actively transports Ca^{2+} from the sarcoplasm into its lumen.

_____ 3. Enlargement of the sarcoplasmic reticulum next to a T tubule.

_____ 4. Grouping of a T tubule and its two adjacent terminal cisternae.

B. Using the terms provided, complete these statements:

Cross-bridge T tubules
Myosin Tropomyosin
Sarcolemma Troponin
Sarcoplasmic reticulum

Action potentials are propagated along the _(1)_ and into the interior of the muscle fiber by _(2)_. The action potentials cause Ca^{2+} channels in the terminal cisternae to open, and Ca^{2+} are released from the _(3)_. Calcium ions diffuse into the sarcoplasm surrounding the myofilaments and bind to _(4)_ of the actin myofilaments. This causes _(5)_ to move and expose the active sites of actin to _(6)_. As a result, a _(7)_ is formed.

1. _____
2. _____
3. _____
4. _____
5. _____
6. _____
7. _____

C. Using the terms provided, complete these statements:

Actin
ATP
ATPase
Head
Power
Recovery

After a cross-bridge has formed, the myosin head moves at its hinge region and the actin myofilament is pulled past the myosin. Release of the myosin head from actin requires (1) to bind to the head of the myosin molecule. ATP is broken down by (2) in the head of the myosin molecule and energy is stored in the (3) of the myosin molecule. When the myosin molecule binds to (4) to form another cross-bridge, much of the energy is used for cross-bridge formation and movement. Before the cross-bridge can be released for another cycle, once again an ATP molecule must bind to the head of myosin molecule. Movement of the myosin molecule while the cross-bridge is attached to actin is a(n) (5) stroke, whereas return of the myosin head to its original position after cross-bridge release is a(n) (6) stroke.

1. _____
2. _____
3. _____
4. _____
5. _____
6. _____

D. Using the terms provided, complete these statements:

Active site
Ca^{2+}
Sarcoplasmic reticulum

Relaxation occurs as a result of the active transport of (1) back into the (2), which allows the troponin–tropomyosin complex to block the (3) on the actin molecules and cross-bridges cannot reform.

1. _____
2. _____
3. _____

Muscle Twitch

Match these terms with the correct description:

Contraction phase
Lag phase
Relaxation phase

_____ 1. An action potential causes the presynaptic terminal to release acetylcholine. Acetylcholine crosses the synaptic cleft and binds to postsynaptic receptors, causing an action potential.

_____ 2. An action potential propagates down the T tubules, causing the release of Ca^{2+} from the sarcoplasmic reticulum.

_____ 3. Calcium ions bind with troponin, the troponin–tropomyosin complex changes position, and active sites on actin molecules are exposed to the heads of the myosin molecules.

_____ 4. Cross-bridges between actin and myosin molecules form, move, release, and reform, causing sarcomeres to shorten.

_____ 5. Calcium ions are actively transported into the sarcoplasmic reticulum, troponin–tropomyosin complexes inhibit cross-bridge formation, and muscle fibers lengthen passively.

Stimulus Strength and Muscle Contraction

A. Match these terms with the correct statement or definition:

Maximal stimulus
Multiple motor unit summation
Submaximal stimulus
Subthreshold stimulus
Supramaximal stimulus
Threshold stimulus

_____ 1. Stimulus just strong enough to produce an action potential in a single motor unit.

_____ 2. Stimulus strength between threshold and maximal values.

_____ 3. Stimulus that is stronger than necessary to activate all the motor units in a muscle.

_____ 4. Increasing stimulus strength, between threshold and maximum values, produces a graded increase in force of contraction of a muscle.

B. Match these terms with the correct parts of the graph in figure 9.4:

Maximal stimulus
Submaximal stimulus
Subthreshold stimulus
Supramaximal stimulus
Threshold stimulus

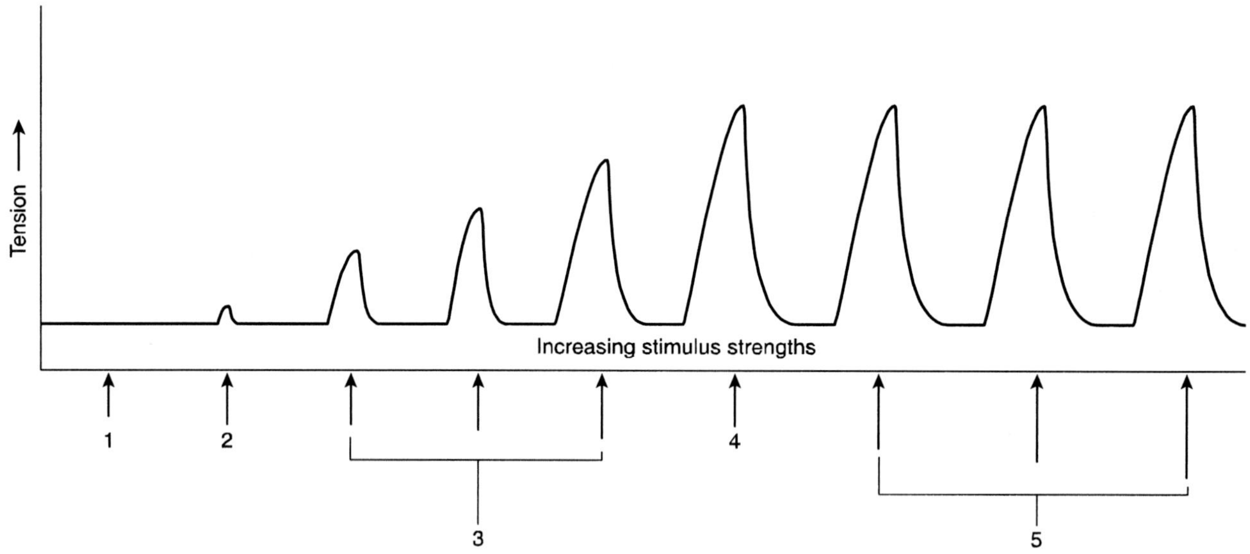

Figure 9.4

1. _____
2. _____
3. _____
4. _____
5. _____

Stimulus Frequency and Muscle Contraction

A. Match these terms with the correct statement or definition:

Complete tetanus
Incomplete tetanus
Multiple wave summation
Treppe

_____ 1. Increase in the force of a muscle contraction caused by an increased frequency of stimulation.

_____ 2. Muscle fibers partially relax between contractions.

_____ 3. Stimuli occur so frequently that there is no muscle relaxation.

_____ 4. Graded response occurring in muscle that has rested for a prolonged period of time. If the muscle is stimulated with a maximal stimulus at a frequency that allows complete relaxation between stimuli, the second contraction is of a slightly greater magnitude than the first, and the third is greater than the second. After a few stimuli, all contractions are of equal magnitude.

B. Match these terms with the correct parts labeled in figure 9.5:

Complete tetanus
Incomplete tetanus
Multiple wave summation
Treppe

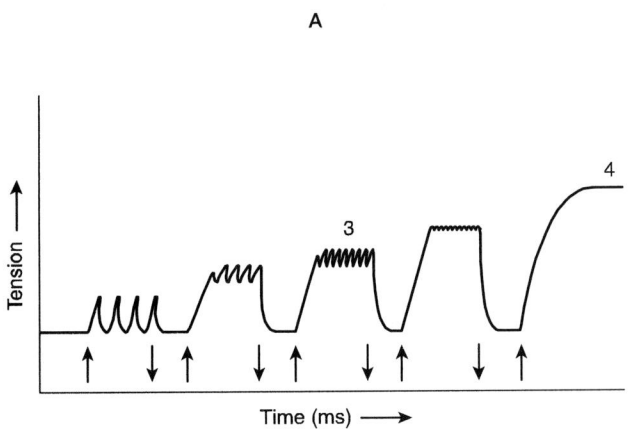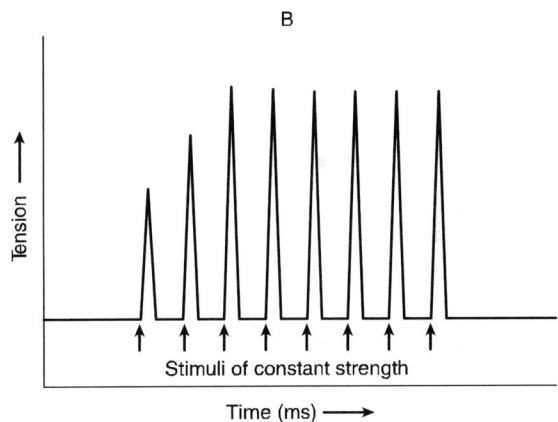

Figure 9.5

_____ 1. Demonstrated by graph A.

_____ 2. Demonstrated by graph B

_____ 3. Name part 3 of graph A.

_____ 4. Name part 4 of graph A.

Chapter 9

C. Using the terms provided, complete these statements:

Active sites
Ca^{2+}
Decreases
Elasticity
Increases
Muscle twitch
Tension
Tetanus

Multiple wave summation is increased tension that is apparent when a muscle is exhibiting incomplete or complete _(1)_. Two factors play a role in this increased tension. First, as the action potential frequency increases, the concentration of _(2)_ around the myofibrils becomes greater than during a single _(3)_, causing a greater degree of contraction. The additional Ca^{2+} cause the exposure of additional _(4)_ on the actin filaments. Second, the sarcoplasm and the connective tissue components of muscle have some _(5)_. In a muscle stimulated at high frequency, the elastic elements are stretched during the early part of the prolonged contraction. The stretching allows all the _(6)_ produced by the muscle to be applied to the load to be lifted, and the observed tension produced by the muscle _(7)_.

1. _____
2. _____
3. _____
4. _____
5. _____
6. _____
7. _____

Types of Muscle Contractions

Match these terms with the correct statement or definition:

Concentric
Eccentric
Isometric
Isotonic
Muscle tone

_____ 1. Contractions in which muscle length is the same but tension changes.

_____ 2. Contractions in which muscle tension is the same but muscle length changes.

_____ 3. Muscle tension is constant and the muscle decreases in length.

_____ 4. Muscle tension is constant and the muscle increases in length.

_____ 5. Maintenance of constant tension for long periods of time.

Length vs. Tension

Match these terms with the correct statement or definition:

Active tension
Passive tension
Total tension

_____ 1. Produced when a muscle contracts.

_____ 2. Produced when a muscle is stretched but is not stimulated.

_____ 3. Sum of active and passive tension.

Fatigue

Match these terms with the correct statement or definition:

Muscular fatigue   Synaptic fatigue
Psychologic fatigue

_____ 1. Involves the central nervous system; muscles are capable of functioning but the person "perceives" work is not possible.

_____ 2. Result of ATP depletion; without adequate ATP levels in muscle fibers, cross-bridges cannot function normally.

_____ 3. Occurs in the neuromuscular junction when the rate of acetylcholine release is greater than the rate of acetylcholine synthesis; rare but can occur after extreme exertion.

Physiologic Contracture and Rigor Mortis

Match these terms with the correct statement or definition:

Physiologic contracture
Rigor mortis

_____ 1. Extreme muscular fatigue caused by a lack of ATP in which a muscle can neither contract nor relax.

_____ 2. Rigid muscles that occur after death; caused by Ca^{2+} leakage from sarcoplasmic reticulum and no ATP to allow relaxation.

Energy Sources and Oxygen Debt

A. Match these terms with the correct statement or definition:

Aerobic respiration
Anaerobic respiration

_____ 1. Does not require oxygen and results in the breakdown of glucose to yield ATP and lactic acid.

_____ 2. Requires oxygen and breaks down glucose, fatty acids, or amino acids to produce ATP, carbon dioxide, and water.

_____ 3. More efficient (produces the most ATP for each molecule of glucose used) of the two types of respiration.

_____ 4. More suited to short periods of intense exercise.

B. Using the terms provided, complete these statements:

- ADP
- Aerobic
- Anaerobic
- ATP
- Creatine phosphate
- Lactic acid
- Liver
- Oxygen deficit (debt)
- Recovery oxygen consumption

The immediate source of energy for muscle contractions is _(1)_. During resting conditions, only a small amount of ATP is present in muscle cells. Energy is stored when ATP transfers a high-energy phosphate to creatine to form _(2)_. During exercise, creatine phosphate releases a phosphate that combines with _(3)_ to produce ATP. Resting muscles or muscles undergoing long-term exercise depend primarily upon _(4)_ respiration for ATP synthesis. On the other hand, during short periods of intense exercise _(5)_ respiration combined with the breakdown of creatine phosphate provides enough ATP for 1–3 minutes. These processes are limited by the depletion of creatine phosphate and glucose and the buildup of _(6)_ within muscle fibers. The lack of increased oxygen consumption relative to increased activity of oxygen consumption creates a(n) _(7)_. The elevated level of oxygen consumption relative to activity level after exercise has ended is called _(8)_. It repays the oxygen deficit and supports metabolic processes that restore homeostasis.

1. _____
2. _____
3. _____
4. _____
5. _____
6. _____
7. _____
8. _____

Slow and Fast Fibers

Match these terms with the correct statement or definition:

- Fast-twitch muscle fibers
- Slow-twitch muscle fibers

1. Also called type I muscle fibers.
2. Also called type II muscle fibers.
3. Have a better developed blood supply.
4. Have very little myoglobin and fewer and smaller mitochondria.
5. More fatigue-resistant.
6. Have large deposits of glycogen and are well adapted to perform anaerobic respiration.
7. There is a greater concentration of this type of fiber in large postural muscles.

Effects of Exercise

A. Using the terms provided, complete these statements:

Aerobic
Anaerobic
Fatigue-resistant
Type IIa
Type IIb

Intense exercise resulting in _(1)_ respiration has the greatest effect on fast-twitch muscle fibers, causing them to increase in strength and mass. The blood supply to both fast-twitch and slow-twitch muscle fibers is increased by endurance exercise requiring _(2)_ respiration, making both types of fibers more _(3)_. Fast-twitch muscle fibers that fatigue readily are called _(4)_, whereas aerobically trained fast-twitch fibers that are more resistant to fatigue are called _(5)_.

1. _____
2. _____
3. _____
4. _____
5. _____

B. Using the terms provided, complete these statements:

Atrophy
Cardiovascular
Fat
Hypertrophy
Motor units
Number
Size

In response to exercise, muscles increase in size, strength, and endurance. The increase in size is called _(1)_. Conversely, muscles that are not used _(2)_. The increase or decrease in size of individual muscles is caused by a change in the _(3)_ of muscle fibers. The increased strength of a trained muscles also occurs because of the recruitment of more _(4)_, the reduction of excess _(5)_, greater ATP production, and increased efficiency of the _(6)_ and respiratory systems.

1. _____
2. _____
3. _____
4. _____
5. _____
6. _____

Heat Production

Match these terms with the correct statement or definition:

Decrease
Increase
Oxygen deficit
Shivering

_____ 1. Change in the amount of heat produced in cells as a result of exercise.

_____ 2. Helps keep body temperature elevated after exercise stops.

_____ 3. Heat production involving rapid skeletal muscle contractions that produce shaking rather than coordinated movements.

Chapter 9

Smooth Muscle

Using the terms provided, complete these statements:

Actin myofilaments
Ca^{2+}
Calmodulin
Caveolae
Dense bodies
Intermediate filaments
Myosin
Myosin phosphatase
Single
Spindle-shaped
Striated

Skeletal muscle fibers are cylindrical cells, whereas smooth muscle fibers are (1), with a(n) (2) nucleus per cell. Skeletal muscle fibers appear to be (3), but smooth muscle fibers do not. Instead of Z disks, smooth muscle has (4), to which (5) and noncontractile (6) are attached. Although smooth muscle does not have a T tubule system, it does have (7), which are invaginations along the plasma membrane. Smooth muscle lacks an extensive sarcoplasmic reticulum, so (8) must enter the cell from the extracellular fluid to initiate contraction. Smooth muscle has no troponin. Instead of binding to troponin, Ca^{2+} binds to (9) and activates myosin kinase, an enzyme that transfers a phosphate group from ATP to (10) and initiates cross-bridge formation. Another enzyme, (11), removes the phosphate from myosin molecules, releasing the cross-bridges.

1. _____
2. _____
3. _____
4. _____
5. _____
6. _____
7. _____
8. _____
9. _____
10. _____
11. _____

Types of Smooth Muscle

Match the smooth muscle type with the correct statement:

Multiunit smooth muscle
Visceral (unitary) smooth muscle

_____

_____

_____

_____

1. Often autorhythmic; has many gap junctions; acts as a single unit.

2. Usually found in sheets (digestive, respiratory, and urinary tracts).

3. Occurs in sheets (blood vessels), small bundles (arrector pili), or single cells (spleen capsule).

4. Has few gap junctions and normally contracts only when stimulated by nerves or hormones.

Electrical and Functional Properties of Smooth Muscle

Using the terms provided, complete these statements:

Do
Do not
Gradual
Hormones
Involuntary
Pacemaker cells
Prostaglandin
Smooth muscle tone
Sudden

Spontaneous generation of action potentials in smooth muscle occurs because of the (1) leakage of Na^+ and Ca^{2+} into the cell. Smooth muscle cells (2) respond in an all-or-none fashion to action potentials. Certain smooth muscle cells in the uterus, ureter, and digestive tract tend to develop action potentials more rapidly than other cells and are called (3). Due to a(n) (4) stretch, some smooth muscle contracts. Despite a (5) increase or decrease in length, smooth muscle maintains a constant tension and amplitude of contraction. This constant tension is called (6). Smooth muscle is innervated by the autonomic nervous system and is therefore under (7) control. Some (8) are also important in regulating smooth muscle—e.g., epinephrine and oxytocin.

1. _____
2. _____
3. _____
4. _____
5. _____
6. _____
7. _____
8. _____

Cardiac Muscle

Match these terms with the correct statement or definition:

Autorhythmic
Intercalated disks
Involuntary
Many
Na^+ and Ca^{2+}
One
Refractory period

_____ 1. Cell-to-cell attachments between cardiac muscle cells.

_____ 2. Spontaneous, repetitive contraction of cardiac muscle cells.

_____ 3. Usual number of nuclei in a cardiac muscle cell.

_____ 4. Responsible for depolarization of cardiac muscle.

_____ 5. Much longer in cardiac muscle than in skeletal muscle.

Effects of Aging on Skeletal Muscle

Match these terms with the correct statement or definition:

Decrease
Increase

_____ 1. Effect of age on the number of fast-twitch muscle fibers.

_____ 2. Effect of age on the surface area of the synapse.

_____ 3. Effect of age on the number of motor units.

_____ 4. Effect of age on the density of capillaries in skeletal muscles.

Chapter 9

Quick Recall

1. List seven major functions of muscle.

2. List the four functional characteristics of muscle.

3. List the five connective tissue structures associated with skeletal muscle.

4. List the parts of a sarcomere found in the I band, A band, and H zone.

5. List the three substances with which troponin can combine.

6. List three important properties of the myosin head.

7. Name two parts of an action potential and describe ion movement during each part.

8. Describe the all-or-none principle of action potentials.

9. List the events that result in the transfer of an action potential from a neuron to a skeletal muscle.

10. List three roles of ATP in muscle contraction and relaxation.

11. List the three phases of a muscle twitch.

12. State the all-or-none law of skeletal muscle contraction.

13. Explain how the force of contraction increases during multiple motor unit summation and multiple wave summation.

14. List four types of muscle contraction.

15. List three types of muscle fatigue.

16. Name two ways that ATP is produced in a muscle fiber during short periods of intense exercise.

17. List two types of skeletal muscle fibers and two types of smooth muscle.

ANSWERS TO CHAPTER 9

CONTENT LEARNING ACTIVITY

General Functional Characteristics of Muscle
1. Contractility; 2. Extensibility; 3. Elasticity;
4. Skeletal muscle; 5. Smooth muscle; 6. Cardiac muscle

Skeletal Muscle Structure
- A. 1. Cylindrical; 2. Many; 3. Myoblasts;
  4. Increase in size; 5. Striated
- B. 1. Sarcolemma; 2. External lamina;
  3. Endomysium; 4. Perimysium;
  5. Epimysium; 6. Muscular fascia
- C. 1. Muscular fascia; 2. Epimysium;
  3. Perimysium; 4. Endomysium;
  5. Fasciculus (bundle); 6. Sarcolemma;
  7. Muscle fiber (cell)

Muscle Fibers, Myofilaments, and Sarcomeres
- A. 1. Motor neurons; 2. Sarcoplasm;
  3. Myofibril; 4. Myofilaments; 5. Sarcomere
- B. 1. F-actin; 2. G-actin; 3. Tropomyosin;
  4. Troponin; 5. Cross-bridge; 6. Myosin head;
  7. Myosin head
- C. 1. Z disk; 2. I band; 3. A band; 4. H zone;
  5. M line; 6. Titin
- D. 1. Actin; 2. F-actin molecule; 3. Tropomyosin;
  4. Troponin; 5. Active site; 6. Myosin;
  7. Myosin head; 8. Z disk; 9. I band;
  10. A band; 11. H zone; 12. Sarcomere

Sliding Filament Model
1. Do not; 2. Actin myofilaments; 3. Myosin myofilaments; 4. H zones; 5. I bands; 6. A bands;
7. Sarcomere; 8. Gravity

Physiology of Skeletal Muscle Fibers
- A. 1. Resting membrane potential; 2. Millivolts;
  3. Negative
- B. 1. Ligand; 2. Receptor; 3. Ligand-gated;
  4. Voltage-gated; 5. Into cells; 6. Out of cells
- C. 1. Action potentials; 2. Depolarization;
  3. Threshold; 4. Depolarization;
  5. Repolarization; 6. Repolarization;
  7. All-or-none principle; 8. Propagate;
  9. Increase

Neuromuscular Junction
- A. 1. Presynaptic terminal; 2. Synaptic cleft;
  3. Postsynaptic membrane (motor end plate);
  4. Synaptic vesicles; 5. Neurotransmitter;
  6. Voltage-gated Ca^{2+} channels;
  7. Acetylcholinesterase
- B. 1. Presynaptic terminal; 2. Synaptic vesicles;
  3. Synaptic cleft; 4. Postsynaptic membrane

Excitation-Contraction Coupling
- A. 1. T tubules; 2. Sarcoplasmic reticulum;
  3. Terminal cisterna; 4. Triad
- B. 1. Sarcolemma; 2. T tubules; 3. Sarcoplasmic reticulum; 4. Troponin; 5. Tropomyosin;
  6. Myosin; 7. Cross-bridge
- C. 1. ATP; 2. ATPase; 3. Head; 4. Actin; 5. Power;
  6. Recovery
- D. 1. Ca^{2+}; 2. Sarcoplasmic reticulum;
  3. Active site

Muscle Twitch
1. Lag phase; 2. Lag phase; 3. Lag phase;
4. Contraction phase; 5. Relaxation phase

Stimulus Strength and Muscle Contraction
- A. 1. Threshold stimulus; 2. Submaximal stimulus; 3. Supramaximal stimulus;
  4. Multiple motor unit summation
- B. 1. Subthreshold stimulus; 2. Threshold stimulus; 3. Submaximal stimulus;
  4. Maximal stimulus; 5. Supramaximal stimulus

Stimulus Frequency and Muscle Contraction
- A. 1. Multiple wave summation; 2. Incomplete tetanus; 3. Complete tetanus; 4. Treppe
- B. 1. Multiple wave summation; 2. Treppe;
  3. Incomplete tetanus; 4. Complete tetanus
- C. 1. Tetanus; 2. Ca^{2+}; 3. Muscle twitch;
  4. Active sites; 5. Elasticity; 6. Tension;
  7. Increases

Types of Muscle Contractions
1. Isometric; 2. Isotonic; 3. Concentric;
4. Eccentric; 5. Muscle tone

Length vs. Tension
1. Active tension; 2. Passive tension; 3. Total tension

Fatigue
1. Psychologic fatigue; 2. Muscular fatigue;
3. Synaptic fatigue

Physiologic Contracture and Rigor Mortis
1. Physiologic contracture; 2. Rigor mortis

Energy Sources and Oxygen Debt
- A. 1. Anaerobic respiration; 2. Aerobic respiration; 3. Aerobic respiration;
  4. Anaerobic respiration
- B. 1. ATP; 2. Creatine phosphate; 3. ADP;
  4. Aerobic; 5. Anaerobic; 6. Lactic acid;
  7. Oxygen deficit; 8. Recovery oxygen consumption

Slow and Fast Fibers
1. Slow-twitch muscle fibers; 2. Fast-twitch muscle fibers; 3. Slow-twitch muscle fibers;
4. Fast-twitch muscle fibers; 5. Slow-twitch muscle fibers; 6. Fast-twitch muscle fibers;
7. Slow-twitch muscle fibers

Effects of Exercise
- A. 1. Anaerobic; 2. Aerobic; 3. Fatigue resistant;
  4. Type IIb; 5. Type IIa
- B. 1. Hypertrophy; 2. Atrophy; 3. Size;
  4. Motor units; 5. Fat; 6. Cardiovascular

Heat Production
1. Increase; 2. Oxygen deficit; 3. Shivering

Smooth Muscle
1. Spindle-shaped; 2. Single; 3. Striated;
4. Dense bodies; 5. Actin myofilaments;
6. Intermediate filaments; 7. Caveolae; 8. Ca^{2+};
9. Calmodulin; 10. Myosin; 11. Myosin phosphatase

Types of Smooth Muscle
1. Visceral (unitary) smooth muscle; 2. Visceral (unitary) smooth muscle; 3. Multiunit smooth muscle; 4. Multiunit smooth muscle

Electrical and Functional Properties of Smooth Muscle
1. Gradual; 2. Do not; 3. Pacemaker cells;
4. Sudden; 5. Gradual; 6. Smooth muscle tone;
7. Involuntary; 8. Hormones

Cardiac Muscle
1. Intercalated disks; 2. Autorhythmic; 3. One;
4. Na^+ and Ca^{2+}; 5. Refractory period

Effects of Aging on Skeletal Muscle
1. Decrease; 2. Decrease; 3. Decrease; 4. Decrease

QUICK RECALL

1. Body movement, maintenance of posture, respiration, production of body heat, communication, constriction of organs and vessels, and heart beat (movement of blood)
2. Contractility, excitability, extensibility, and elasticity
3. External lamina, endomysium, perimysium, epimysium, and muscular fascia
4. I band: a Z disk and the actin myofilaments that extend from either side of the Z disk to the ends of the myosin myofilaments; A band: extends the length of the myosin myofilaments, actin overlaps myosin myofilament; H zone: only myosin myofilaments, held in place by the M line
5. Actin, tropomyosin, and Ca^{2+}
6. Can bind to active site of actin to form a cross-bridge, moves on the hinge region, and has ATPase, which breaks down ATP
7. Depolarization: Na^+ move into the cell; repolarization: K^+ move out of the cell
8. If an action potential occurs, it is always of the same magnitude (all part), or an action potential does not occur (none part)
9. Action potentials cause Ca^{2+} channels in the presynaptic terminal to open. Calcium ions diffuse into the cell, causing the acetylcholine in synaptic vesicles to be secreted into the synaptic cleft. Acetylcholine diffuses across the cleft and bind to receptor molecules of the postsynaptic membrane, which becomes permeable to Na^+. Depolarization occurs, leading to an action potential in the muscle cell.
10. a. The energy from ATP is stored in myosin and later released, causing cross-bridge movement—i.e., muscle contraction.
    b. ATP is required for the cross-bridges to be released—i.e., for muscle relaxation.
    c. ATP is required for Ca^{2+} uptake into the sarcoplasmic reticulum, which initiates relaxation.
11. Lag (latent), contraction, and relaxation phase
12. For a given condition, the skeletal muscle fiber contracts maximally or not at all
13. Multiple motor unit summation: more motor units are recruited—i.e., more and more muscle fibers contract; multiple wave summation: each individual muscle fiber contracts more forcefully
14. Isometric, isotonic, concentric, and eccentric contraction
15. Psychologic, muscular, and synaptic fatigue
16. Breakdown of creatine phosphate and anaerobic respiration
17. Slow-twitch (type I) and fast-twitch (type II) skeletal muscle fibers; multiunit and visceral smooth muscle

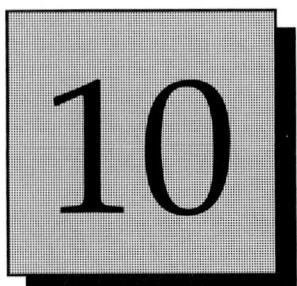

Muscular System: Gross Anatomy

CONTENT LEARNING ACTIVITY

General Principles

Match these terms with the correct statement or definition:

Agonist Insertion
Aponeurosis Origin (head)
Antagonist Prime mover
Belly Synergists
Fixators Tendon

_____ 1. Connective tissue connection between muscles and bones.

_____ 2. Very broad tendon.

_____ 3. Fixed end of the muscle; usually the most stationary, proximal end.

_____ 4. Largest part of a muscle, between its origin and insertion.

_____ 5. Muscle that accomplishes a certain movement, such as flexion.

_____ 6. Muscles that work together to cause movement.

_____ 7. Muscle working in opposition to another muscle.

_____ 8. Muscle that plays the major role in accomplishing a desired movement.

_____ 9. Muscles that stabilize the origin of the prime mover.

Muscle Shapes

Match the type of muscle with the correct description:

Multipennate Semipennate (unipennate)
Orbicular (circular) Straight
Pennate (bipennate)

_____ 1. Fasciculi arranged like the barbs of a feather along a common tendon.

_____ 2. Muscle with all fasciculi on one side of a tendon.

150 Chapter 10

_____ 3. Muscle with fasciculi arranged at many places around a central tendon.

_____ 4. Fasciculi organized parallel to the long axis of the muscle.

_____ 5. Shortens to a greater degree but shortens with less force than pennate muscles.

_____ 6. Acts as a sphincter to close an opening.

Nomenclature

Match the muscle nomenclature with the correct meaning:

Brachialis   Maximus
Brevis   Pectoralis
Deltoid   Quadriceps
Gluteus   Rectus
Masseter   Teres

_____ 1. Chest.

_____ 2. Buttock.

_____ 3. Arm.

_____ 4. Large.

_____ 5. Short.

_____ 6. Round.

_____ 7. Triangular.

_____ 8. Straight.

_____ 9. Four heads.

_____ 10. Chewer.

Movements Accomplished by Muscles

Match these terms with the correct statement or definition:

Class I lever system   Class III lever system
Class II lever system

_____ 1. The fulcrum is between the pull and the weight; a child's seesaw is an example.

_____ 2. Weight is located between the fulcrum and the pull; a wheelbarrow is an example.

_____ 3. The pull is located between the fulcrum and the weight; a person operating a shovel is an example.

_____ 4. Most common type of lever system in the body.

Chapter 10

Head and Neck Muscles

Match the neck muscle group with the correct description or example:

Anterior neck muscles   Posterior neck muscles
Lateral neck muscles

_____ 1. Flexion of the head is largely caused by gravity but is assisted by this muscle group.

_____ 2. Mainly responsible for the extension of the head.

_____ 3. Involved in rotation and lateral flexion of the head.

_____ 4. Sternocleidomastoid is an example.

Facial Expression

Match these muscles with the correct function:

Buccinator   Occipitofrontalis
Corrugator supercilii   Orbicularis oculi
Depressor anguli oris   Orbicularis oris
Levator labii superioris   Risorius
Levator palpebrae superioris   Zygomaticus major and minor

_____ 1. Raises the eyebrows and furrows the skin of the forehead.

_____ 2. Closes the eyelids.

_____ 3. Raises the upper eyelid.

_____ 4. Furrows the skin between the eyebrows, as in frowning.

_____ 5. Muscles that function in kissing.

_____ 6. Muscles that function in smiling.

_____ 7. Muscle that makes sneering possible.

_____ 8. Depresses the angle of the mouth, as in frowning or pouting.

Mastication

Match these muscles with the correct mandibular movements:

Hyoid muscles
Lateral pterygoid muscle
Masseter muscle
Medial pterygoid muscle
Temporalis muscle

_____

_____

_____ 1. Three muscles that elevate the mandible.

_____ 2. Muscles that depress the mandible.

_____

Tongue Movements

Match the type of tongue muscle with the correct description:

Extrinsic tongue muscles
Intrinsic tongue muscles

_____ 1. Muscles that make up the tongue.

_____ 2. Cause the shape of the tongue to change, as in rolling the tongue into a tube.

_____ 3. Muscles that attach the tongue to the mandible, hyoid, and skull.

_____ 4. Responsible for moving the tongue about, as in sticking out the tongue.

Swallowing and the Larynx

Match the type of hyoid muscle with the correct description:

Infrahyoid muscles
Laryngeal muscles
Palatopharyngeus
Pharyngeal constrictor muscles
Salpingopharyngeus
Soft palate muscles
Suprahyoid muscles

_____ 1. Depress the mandible when the hyoid is fixed.

_____ 2. Fix the hyoid bone when the mandible is depressed.

_____ 3. Elevate the larynx during swallowing.

_____ 4. Close the posterior opening to the nasal cavity during swallowing.

_____

_____ 5. Two muscles involved in elevation of the pharynx.

_____ 6. Muscle that opens the pharyngotympanic (auditory) tube.

_____ 7. Constrict to force food into the esophagus.

_____ 8. Prevent food from entering the larynx; shorten (relax) the vocal cords to lower the pitch of the voice.

Movements of the Eyeball

Match the eye muscle with the correct description:

Inferior oblique Medial rectus
Inferior rectus Superior oblique
Lateral rectus Superior rectus

_____ 1. Depresses and medially deviates the gaze.

_____ 2. Laterally deviates (abducts) the gaze.

_____ 3. Laterally deviates and depresses the gaze.

Muscles Moving the Vertebral Column

Match the muscle group of the back with the correct description:

Deep back muscles
Superficial back muscles

_____ 1. Also called the erector spinae group.

_____ 2. Produce the mass of muscles seen lateral to the midline of the back; in general, extend from the vertebrae to the ribs or from rib to rib.

_____ 3. Extend from one vertebra to the adjacent vertebra.

Thoracic Muscles

Match the muscle groups with their correct functions or descriptions:

Diaphragm Scalenes
External intercostals Transversus thoracis
Internal intercostals

_____ 1. Elevate the first two ribs during inspiration.

_____ 2. Elevate the ribs and sternum, increasing the diameter of the thorax.

_____ 3. Depress the ribs, forcing expiration.

_____ 4. Responsible for the majority of volume change in the thoracic cavity during quiet breathing; dome-shaped when relaxed.

Abdominal Wall

Match these terms with the correct definition or description:

External abdominal oblique Rectus abdominis
Internal abdominal oblique Tendinous intersections
Linea alba   Transversus abdominis
Linea semilunaris

_____ 1. Tendinous area that produces a vertical line from the xiphoid process through the navel to the pubis.

_____ 2. Thick muscle, on either side of the midline, with fasciculi oriented vertically in a straight line.

_____ 3. Connective tissue that transects the rectus abdominis, causing it to appear segmented.

_____ 4. Located between the rectus abdominis and the lateral abdominal wall muscles.

_____ 5. Most superficial lateral abdominal wall muscle.

_____ 6. Middle layer of lateral abdominal wall muscles.

_____ 7. Deepest lateral abdominal wall muscle.

Pelvic Floor and Perineum

Match these terms with the correct description or definition:

Pelvic diaphragm   Urogenital diaphragm
Perineum

_____ 1. Mostly formed by the coccygeus and levator ani muscles; forms the pelvic floor.

_____ 2. Diamond-shaped area superficial to the pelvic floor.

_____ 3. Consists of the deep transverse perineal muscle and the external urethral sphincter muscle.

Scapular Movements

Match these muscles with the correct function:

Levator scapulae   Serratus anterior
Pectoralis minor   Trapezius
Rhomboideus major and minor

_____ 1. Elevates, depresses, retracts, rotates, and fixes the scapula.

_____ 2. Elevates, retracts, and rotates the scapula.

_____ 3. Retracts, rotates, and fixes the scapula.

_____ 4. Rotates and protracts the scapula.

_____ 5. Depresses the scapula.

Arm Movements

Match these muscles with the correct description:

Coracobrachialis
Deltoid
Infraspinatus
Latissimus dorsi
Pectoralis major
Subscapularis
Supraspinatus
Teres major
Teres minor

_____ 1. Two muscles that abduct the arm.
_____

_____ 2. Two muscles that can flex or extend the arm.
_____

_____ 3. Two muscles that attach to the trunk and adduct the arm.
_____

_____ 4. Three muscles that attach to the scapula and adduct the arm.
_____
_____

_____ 5. Rotator cuff muscle that medially rotates the arm.

Forearm Movements

Match these muscles with the correct description:

Anconeus
Biceps brachii
Brachialis
Brachioradialis
Pronator quadratus
Pronator teres
Supinator
Triceps brachii

_____ 1. Two muscles that extend the forearm.
_____

_____ 2. Three muscles that flex the forearm.
_____
_____

_____ 3. Two muscles that supinate the forearm.
_____

_____ 4. Muscle that flexes the arm.

_____ 5. Muscle that extends the arm.

Wrist, Hand, and Finger Movements

A. Match the muscle group with the correct statement:

Anterior forearm muscles
Posterior forearm muscles

_____   1. Most originate on the medial epicondyle of the humerus; responsible for flexion of the wrist and fingers.

_____   2. Most originate on the lateral epicondyle of the humerus; responsible for extension of the wrist and fingers.

B. Match these terms with the correct description or definition:

Extrinsic hand muscles   Retinaculum
Hypothenar eminence   Thenar eminence
Intrinsic hand muscles

_____   1. Located in the forearm with tendons that extend into the hand.

_____   2. Strong band of fibrous connective tissue that covers the flexor and extensor tendons.

_____   3. Muscles located entirely with the hand.

_____   4. Fleshy prominence at the base of the thumb; responsible for the control of thumb movements.

Thigh Movements

A. Match the hip muscles with the correct function:

Deep thigh rotators   Iliacus
Gluteus maximus   Psoas major
Gluteus medius   Tensor fasciae latae
Gluteus minimus

_____   1. Two anterior hip muscles that flex the thigh.

_____

_____   2. Posterolateral hip muscle that extends, abducts, and laterally rotates the thigh.

_____   3. Two posterolateral hip muscles that abduct and medially rotate the thigh.

_____

_____   4. Muscle that inserts on the tibia through the iliotibial tract; stabilizes the knee and abducts and medially rotates the thigh.

_____   5. Group of hip muscles that laterally rotate the thigh.

Chapter 10

B. Match the thigh muscles with their function:

Anterior thigh muscles   Posterior thigh muscles
Medial thigh muscles

_____ 1. Thigh muscles that flex the thigh.

_____ 2. Thigh muscles that extend the thigh.

_____ 3. Thigh muscles that adduct the thigh.

Leg Movements

A. Match the muscle group with the correct statement:

Anterior thigh muscles   Posterior thigh muscles
Medial thigh muscles

_____ 1. Muscle group responsible for extending the leg (except for the sartorius muscle).

_____ 2. Muscle group mostly responsible for flexing the leg.

_____ 3. Muscle group that mostly adducts and flexes the thigh.

_____ 4. Rectus femoris belongs to this group.

_____ 5. Biceps femoris belongs to this group.

B. Match the thigh muscle with its function:

Adductor brevis   Sartorius
Adductor longus   Semimembranosus
Adductor magnus   Semitendinosus
Biceps femoris   Vastus intermedius
Gracilis   Vastus lateralis
Pectineus   Vastus medialis
Rectus femoris

_____ 1. Four muscles that extend the leg; constitute most of the anterior thigh muscle.

_____
_____
_____

_____ 2. Five muscles that flex the leg.

_____
_____
_____
_____

_____   3. Five muscles that adduct the thigh; constitute the medial thigh compartment.
_____
_____
_____
_____
_____   4. Two muscles of the anterior thigh compartment that flex the thigh.
_____
_____   5. Three muscles of the posterior thigh compartment that extend the thigh.
_____
_____

Ankle, Foot, and Toe Movements

A. Match the muscle group with the correct statement:

    Anterior leg muscles   Lateral leg muscles
    Intrinsic foot muscles   Posterior leg muscles

_____   1. Dorsiflexes the foot and extends the toes.

_____   2. Plantar flexes and inverts the foot and flexes the toes.

_____   3. Everts and plantar flexes the foot.

_____   4. Muscles within the foot that flex, extend, abduct, and adduct the toes.

B. Match the leg muscle with the correct description:

    Gastrocnemius, soleus, and plantaris   Tibialis anterior and posterior
    Peroneus brevis, longus, and tertius

_____   1. Muscles that invert the foot.

_____   2. Muscles that evert the foot.

_____   3. Muscles that join to form the calcaneal (Achilles) tendon; plantar flex the foot.

Location of Superficial Muscles

A. Match these muscles with the correct parts labeled in figure 10.1:

Adductors of thigh
Biceps brachii
Brachioradialis
Deltoid
External abdominal oblique
Flexors of the wrist and fingers
Pectoralis major
Quadriceps femoris
Rectus abdominis
Rectus femoris
Sartorius
Serratus anterior
Sternocleidomastoid
Tensor fasciae latae
Vastus lateralis
Vastus medialis

1. _____
2. _____
3. _____
4. _____
5. _____
6. _____
7. _____
8. _____
9. _____
10. _____
11. _____
12. _____
13. _____
14. _____
15. _____
16. _____

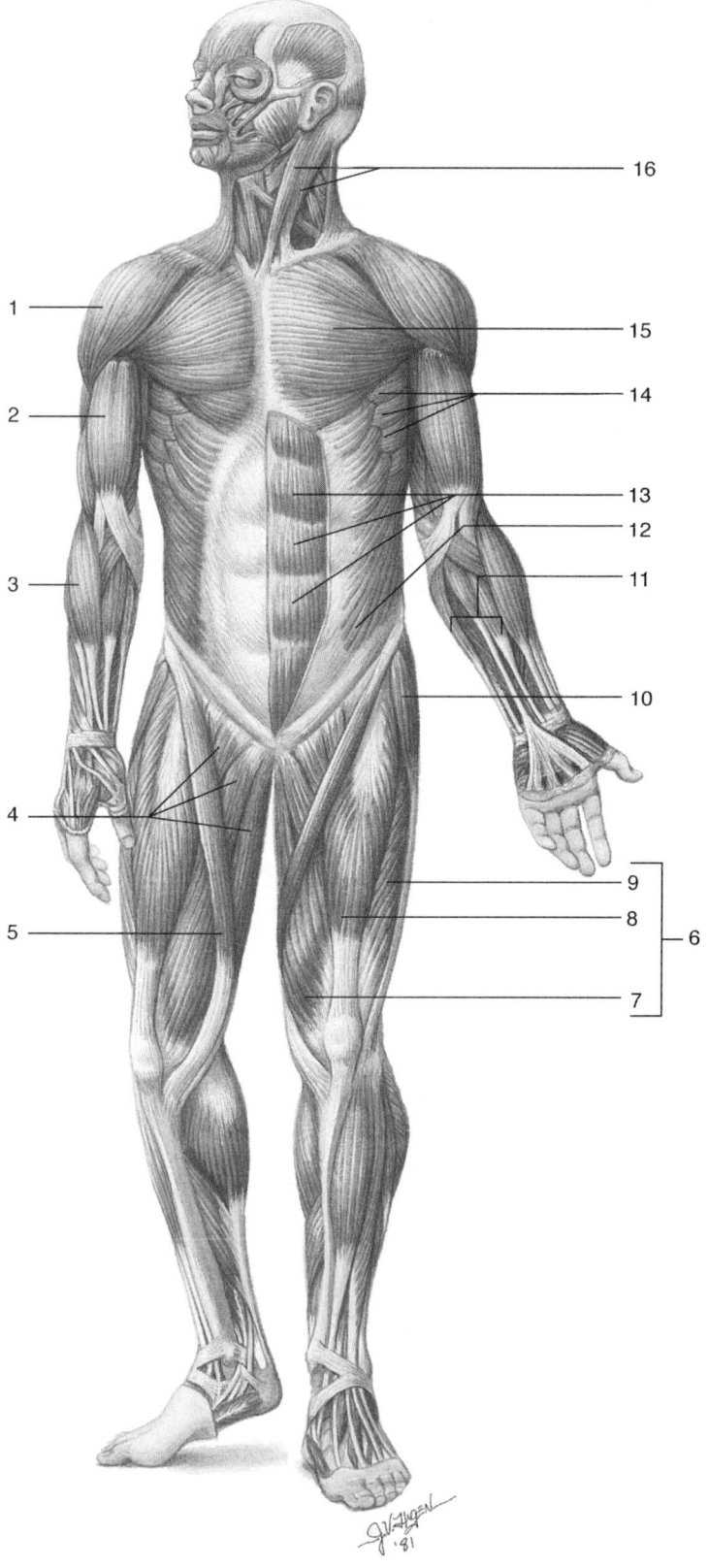

Figure 10.1

B. Match these muscles with the correct parts labeled in figure 10.2:

Biceps femoris
Extensors of the wrist and fingers
Gastrocnemius
Gluteus maximus
Gluteus medius
Hamstring muscles
Iliotibial tract
Infraspinatus
Latissimus dorsi
Semimembranosus
Semitendinosus
Soleus
Teres major
Teres minor
Trapezius
Triceps brachii

1. _____
2. _____
3. _____
4. _____
5. _____
6. _____
7. _____
8. _____
9. _____
10. _____
11. _____
12. _____
13. _____
14. _____
15. _____
16. _____

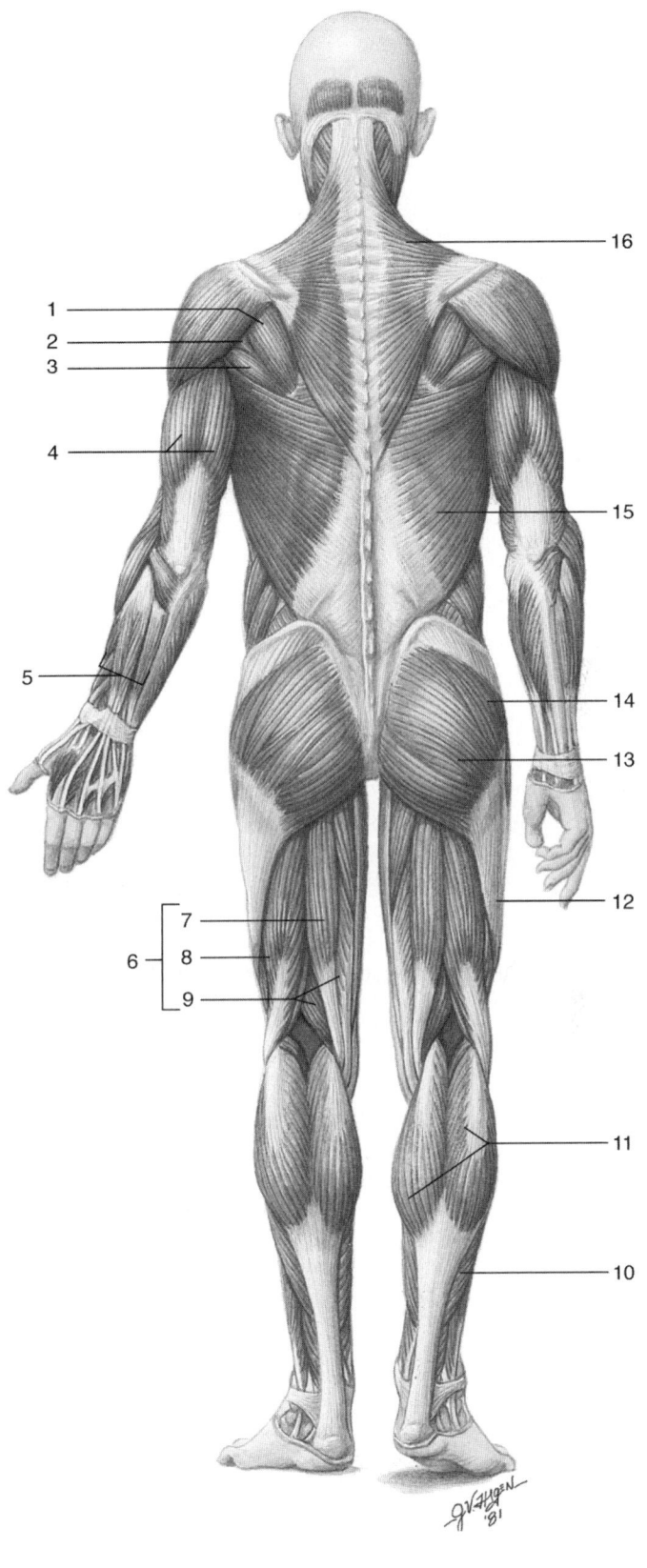

Figure 10.2

Chapter 10   161

QUICK RECALL

1. List the three basic parts of a muscle.

2. List four classes of muscle shape.

3. Name, and give an example of, the three classes of levers.

4. List the three groups of muscles that move the head.

5. Name the two muscles of facial expression that are circular muscles, and give their function.

6. List the muscles of mastication.

7. List the two major types of muscle groups that act on the vertebral column.

8. Name the three major thoracic muscles that are involved with inspiration.

9. List four muscles in the abdominal wall.

10. List two general functions for the muscles that attach the scapula to the thorax.

11. Define the rotator cuff.

12. List three muscles that flex and two muscles that extend the forearm.

13. List the two major muscle groups of the forearm, and give their major functions.

14. Define extrinsic and intrinsic hand muscles.

15. List the three groups of hip muscles and the three groups of thigh muscles that cause movement of the thigh.

16. List the three groups of thigh muscles that cause movement of the leg.

17. List the three muscle groups of the leg that act on the foot and toes.

ANSWERS TO CHAPTER 10

CONTENT LEARNING ACTIVITY

General Principles
1. Tendon; 2. Aponeurosis; 3. Origin (head);
4. Belly; 5. Agonist; 6. Synergists; 7. Antagonist;
8. Prime mover; 9. Fixators

Muscle Shapes
1. Pennate (bipennate); 2. Semipennate (unipennate); 3. Multipennate; 4. Straight;
5. Straight; 6. Orbicular (circular)

Nomenclature
1. Pectoralis; 2. Gluteus; 3. Brachialis;
4. Maximus; 5. Brevis; 6. Teres; 7. Deltoid;
8. Rectus; 9. Quadriceps; 10. Masseter

Movements Accomplished by Muscles
1. Class I lever system; 2. Class II lever system;
3. Class III lever system; 4. Class III lever system

Head and Neck Muscles
1. Anterior neck muscles; 2. Posterior neck muscles; 3. Posterior neck muscles and lateral neck muscles; 4. Lateral neck muscles

Facial Expression
1. Occipitofrontalis; 2. Orbicularis oculi;
3. Levator palpebrae superioris; 4. Corrugator supercilii; 5. Orbicularis oris and buccinator;
6. Zygomaticus major and minor and risorius;
7. Levator labii superioris; 8. Depressor anguli oris

Mastication
1. Temporalis muscle, masseter muscle, and medial pterygoid muscle; 2. Lateral pterygoid muscle and hyoid muscles

Tongue Movements
1. Intrinsic tongue muscles; 2. Intrinsic tongue muscles; 3. Extrinsic tongue muscles; 4. Extrinsic tongue muscles

Swallowing and the Larynx
1. Suprahyoid muscles; 2. Infrahyoid muscles;
3. Infrahyoid muscles; 4. Soft palate muscles;
5. Palatopharyngeus and salpingopharyngeus;
6. Salpingopharyngeus; 7. Pharyngeal constrictor muscles; 8. Laryngeal muscles

Movements of the Eyeball
1. Inferior rectus; 2. Lateral rectus; 3. Superior oblique

Muscles Moving the Vertebral Column
1. Superficial back muscles; 2. Superficial back muscles; 3. Deep back muscles

Thoracic Muscles
1. Scalenes; 2. External intercostals; 3. Internal intercostals and transversus thoracis;
4. Diaphragm

Abdominal Wall
1. Linea alba; 2. Rectus abdominis; 3. Tendinous intersections; 4. Linea semilunaris; 5. External abdominal oblique; 6. Internal abdominal oblique;
7. Transversus abdominis

Pelvic Floor and Perineum
1. Pelvic diaphragm; 2. Perineum; 3. Urogenital diaphragm

Scapular Movements
1. Trapezius; 2. Levator scapulae;
3. Rhomboideus major and minor; 4. Serratus anterior; 5. Pectoralis minor

Arm Movements
1. Deltoid and supraspinatus; 2. Deltoid and pectoralis major; 3. Latissimus dorsi and pectoralis major; 4. Coracobrachialis, teres major, and teres minor; 5. Subscapularis

Forearm Movements
1. Triceps brachii and anconeus; 2. Brachialis, biceps brachii, and brachioradialis; 3. Supinator and biceps brachii; 4. Biceps brachii; 5. Triceps brachii

Wrist, Hand, and Finger Movements
A. 1. Anterior forearm muscles; 2. Posterior forearm muscles
B. 1. Extrinsic hand muscles; 2. Retinaculum;
3. Intrinsic hand muscles; 4. Thenar eminence

Thigh Movements
A. 1. Iliacus and psoas major; 2. Gluteus maximus; 3. Gluteus medius and gluteus minimus; 4. Tensor fasciae latae; 5. Deep thigh rotators
B. 1. Anterior thigh muscles; 2. Posterior thigh muscles; 3. Medial thigh muscles

Leg Movements
A. 1. Anterior thigh muscles; 2. Posterior thigh muscles; 3. Medial thigh muscles; 4. Anterior thigh muscles; 5. Posterior thigh muscles
B. 1. Rectus femoris, vastus intermedius, vastus lateralis, and vastus medialis; 2. Biceps femoris, semimembranosus, semitendinosus, sartorius, and gracilis; 3. Adductor brevis, adductor longus, adductor magnus, gracilis, and pectineus; 4. Rectus femoris and sartorius;
5. Biceps femoris, semimembranosus, and semitendinosus

Ankle, Foot, and Toe Movements
A. 1. Anterior leg muscles; 2. Posterior leg muscles; 3. Lateral leg muscles; 4. Intrinsic foot muscles
B. 1. Tibialis anterior and posterior; 2. Peroneus brevis, longus, and tertius; 3. Gastrocnemius, soleus, and plantaris

Location of Superficial Muscles
- A. 1. Deltoid; 2. Biceps brachii; 3. Brachioradialis; 4. Adductors of thigh; 5. Sartorius; 6. Quadriceps femoris; 7. Vastus medialis; 8. Rectus femoris; 9. Vastus lateralis; 10. Tensor fasciae latae; 11. Flexors of wrist and fingers; 12. External abdominal oblique; 13. Rectus abdominis; 14. Serratus anterior; 15. Pectoralis major; 16. Sternocleidomastoid
- B. 1. Infraspinatus; 2. Teres minor; 3. Teres major; 4. Triceps brachii; 5. Extensors of wrist and fingers; 6. Hamstring muscles; 7. Semitendinosus; 8. Biceps femoris; 9. Semimembranosus; 10. Soleus; 11. Gastrocnemius; 12. Iliotibial tract; 13. Gluteus maximus; 14. Gluteus medius; 15. Latissimus dorsi; 16. Trapezius

QUICK RECALL

1. Origin (head), insertion, and belly
2. Pennate, parallel, convergent, and circular
3. Class I lever: a child's seesaw; class II lever: a wheelbarrow; class III lever: a shovel
4. Anterior, lateral, and posterior neck muscles
5. Orbicularis oculi: closing the eyelids; orbicularis oris: puckering the mouth
6. Temporalis, masseter, and medial and lateral pterygoids
7. Superficial and deep back muscles
8. Diaphragm, external intercostals, and scalenes
9. Rectus abdominis, external abdominal oblique, internal abdominal oblique, and transversus abdominis
10. Movement or fixation of the scapula
11. Four muscles that bind the humerus to the scapula and form a cuff or cap over the proximal humerus
12. Flex forearm: biceps brachii, brachialis, and brachioradialis; extend forearm: triceps brachii and anconeus
13. Anterior forearm: flex fingers, thumb, and wrist; abduct or adduct wrist; posterior forearm: extend fingers, thumb, and wrist; abduct or adduct wrist
14. Extrinsic hand muscles are in the forearm but have tendons that extend into the hand. Intrinsic hand muscles are entirely within the hand.
15. Hip muscles: anterior, posterolateral, and deep thigh rotators; thigh muscles: anterior, posterior, and medial
16. Anterior, posterior, and medial compartments
17. Anterior, posterior, and lateral compartments

11 Functional Organization of Nervous Tissue

CONTENT LEARNING ACTIVITY

Divisions of the Nervous System

A. Match these terms with the correct statement or definition:

Autonomic nervous system
Central nervous system
Enteric nervous system
Motor (efferent) division
Parasympathetic division
Peripheral nervous system
Sensory (afferent) division
Somatic nervous system
Sympathetic division

_____ 1. Consists of the brain and spinal cord.

_____ 2. Consists of nerves, ganglia, and plexuses; located outside the central nervous system (CNS).

_____ 3. PNS subdivision; action potentials go from sensory receptors to the central nervous system (CNS).

_____ 4. PNS subdivision; action potentials go from the central nervous system (CNS) to effector organs, such as muscles and glands.

_____ 5. Part of the motor division that conducts action potentials from the central nervous system (CNS) to skeletal muscle.

_____ 6. Part of the motor division that conducts action potentials from the central nervous system (CNS) to smooth muscle, cardiac muscle, and glands.

_____ 7. Autonomic division; prepares the body for physical activity.

_____ 8. Autonomic division; regulates resting or vegetative functions.

_____ 9. Consists of plexuses within the wall of the digestive tract.

_____ 10. Major site for processing information, initiating responses, and integrating mental processes.

B. Match these terms with the correct statement or definition:

Ganglia 12
Nerves 31
Plexus

_____ 1. Bundles of axons and their sheaths that connect the CNS to sensory receptors, muscles, and glands.

_____ 2. Number of cranial nerve pairs in the PNS.

_____ 3. Number of spinal nerve pairs in the PNS.

_____ 4. Collections of nerve cell bodies outside the CNS.

_____ 5. Extensive network of axons—and, in some cases neuron cell bodies—located outside the CNS.

Neurons

A. Match these terms with the correct statement or definition:

Axon Dendrite
Cell body (soma) Glial cells

_____ 1. Nonneural cells; support and protect neurons.

_____ 2. Portion of a neuron that contains the nucleus and other organelles, such as rough endoplasmic reticulum.

_____ 3. Short, often highly branched cytoplasmic process that is tapered from the neuron cell body to its tip.

_____ 4. Long, cytoplasmic process with a uniform diameter.

_____ 5. Also referred to as a nerve fiber.

B. Match these terms with the correct statement or definition:

Axolemma Nissl substance
Axon hillock Presynaptic terminals
Axoplasm Trigger zone
Dendritic spines

_____ 1. Areas of rough endoplasmic reticulum in the cytoplasm of the neuron cell body and base of dendrites; primary site of protein synthesis in a neuron.

_____ 2. Small extensions from dendrites; the location of synapses with axons of other neurons.

_____ 3. Enlarged area of each neuron cell body from which the axon arises.

_____ 4. Cytoplasm of the axon.

_____ 5. Enlarged ends of axon extensions; contain vesicles with neurotransmitters.

_____ 6. Where action potentials are generated; axon hillock and initial segment.

C. Match these terms with the correct parts labeled in figure 11.1:

Axon
Dendrites
Neuron cell body (soma)
Node of Ranvier
Nucleus
Presynaptic terminals
Schwann cell
Trigger zone

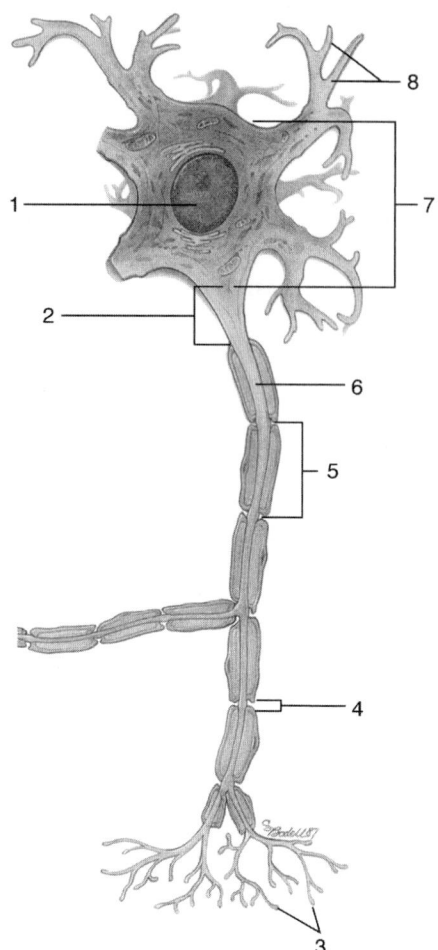

Figure 11.1

1. _____
2. _____
3. _____
4. _____
5. _____
6. _____
7. _____
8. _____

Types of Neurons

A. Match these terms with the correct statement or definition:

Interneuron (association neuron)　　Motor (efferent) neuron
　　　　　　　　　　　　　　　　Sensory (afferent) neuron

_____　1. Conducts action potentials toward the CNS.

_____　2. Conducts action potentials from the CNS to muscles or glands.

_____　3. Conducts action potentials from one neuron to another.

B. Match these terms with the correct statement or definition:

Bipolar neuron　　Unipolar neuron
Multipolar neuron

_____　1. Neuron with several dendritic processes and a single axon.

_____　2. Neuron with a single dendrite and a single axon.

_____　3. Most CNS neurons, including motor neurons, are this type.

Glial Cells of the CNS and PNS

Match these terms with the correct statement or definition:

Astrocytes
Blood–brain barrier
Ependymal cells
Microglia
Myelin sheath
Oligodendrocytes
Satellite cells
Schwann cells
  (neurolemmocytes)

_____ 1. Glial cells that are star-shaped; help regulate the composition of extracellular fluid around neurons.

_____ 2. Formed by endothelial cells that are joined by tight junctions; influenced by astrocytes.

_____ 3. Glial cells in choroid plexuses; line the ventricles of the brain; secrete and circulate cerebrospinal fluid.

_____ 4. Specialized macrophages in the CNS that become mobile and phagocytic in response to inflammation.

_____ 5. Cytoplasmic extension that wraps many times around an axon.

_____ 6. Glial cells in the CNS that form myelin sheaths around portions of several axons.

_____ 7. Glial cells in the PNS that form a myelin sheath around a portion of one axon.

_____ 8. Glial cells that surround, provide support for, and provide nutrients for neuron cell bodies in ganglia.

Myelinated and Unmyelinated Axons

Match these terms with the correct statement or definition:

Myelinated
Node of Ranvier
Unmyelinated

_____ 1. Type of axon that is surrounded by extensions from oligodendrocytes or Schwann cells that wrap repeatedly around a segment of the axon.

_____ 2. Bare area of a myelinated axon.

_____ 3. Type of axon that rests in an invagination of an oligodendrocyte or a Schwann cell.

Chapter 11

Organization of Nervous Tissue

Match these terms with the correct statement or definition:

Cortex
Ganglia
Gray matter
Nerve
Nerve tract
Nuclei
White matter

_____ 1. Bundles of parallel axons with their associated myelin sheaths that appear white in color.

_____ 2. Collection of nerve cell bodies and unmyelinated axons that appear gray in color.

_____ 3. Bundles of axons in the CNS.

_____ 4. Bundles of axons in the PNS.

_____ 5. Gray matter on the outer surface of the brain.

_____ 6. Collections of gray matter within the brain.

_____ 7. Collections of neuron cell bodies in the PNS.

Concentration Differences Across the Plasma Membrane

Match these terms with the correct statement or definition:

Cl^-
K^+
Na^+
Na^+–K^+ pump
Proteins

_____ 1. The concentration of these two ions is much greater outside the cell than inside.

_____ 2. The concentration of this ion and negatively charged proteins is much greater inside the cell than outside.

_____ 3. Movement of K^+ inward and Na^+ outward across the plasma membrane; requires active transport (ATP molecules).

_____ 4. Large, negatively charged; do not cross the plasma membrane.

_____ 5. Repelled by negative charges; pass out through membrane channels.

Leak and Gated Ion Channels

Match these terms with the correct statement or definition:

Leak channels
Ligand-gated ion channels
Other gated ion channels
Voltage-gated ion channels

_____ 1. These channels are always open; also called nongated ion channels.

_____ 2. There are more K^+ channels than Na^+ channels of this type; consequently, the membrane is more permeable to K^+ than Na^+.

_____ 3. Open and close in response to small voltage changes across the plasma membrane.

_____ 4. Open or close as a direct result of a ligand binding to a receptor.

_____ 5. Open or close as result of the activation of a G protein.

_____ 6. Examples are found in touch, temperature, and light receptors.

Establishing the Resting Membrane Potential

Using the terms provided, complete these statements:

K^+
Na^+
Na^+–K^+ pump
Negative
Positive
Resting membrane potential
Three
Two

The charge difference across the unstimulated plasma membrane is called the __(1)__. The plasma membrane is somewhat permeable to __(2)__. Thus, K^+ tend to diffuse down their concentration gradient from the inside to the outside of the plasma membrane. As K^+ leave the cell, the inside of the plasma membrane becomes more __(3)__. Because opposite charges attract, the K^+ are attracted back toward the cell and accumulate just outside the plasma membrane, making the outside of the membrane more __(4)__. Other ions, such as Na^+, have a small effect on the resting membrane potential. The concentration gradients for K^+ and Na^+ are maintained by the __(5)__. For every __(6)__ Na^+ pumped out of the cell, __(7)__ K^+ are pumped in. Consequently, the outside of the plasma membrane becomes more __(8)__ and the resting membrane potential increases.

1. _____
2. _____
3. _____
4. _____
5. _____
6. _____
7. _____
8. _____

Changing the Resting Membrane Potential

Match these terms with the correct statement or definition:

Depolarization
Hyperpolarization

_____ 1. Decrease in the resting membrane potential; the resting membrane potential is less negative.

_____ 2. Results when the K^+ concentration gradient increases.

_____ 3. Results when the membrane permeability to K^+ increases.

_____ 4. Results when the membrane permeability to Na^+ increases.

_____ 5. Results when the extracellular concentration of Ca^{2+} decreases or membrane permeability to Ca^{2+} increases.

_____ 6. Results when the membrane permeability to Cl- increases.

Chapter 11

Disorders

Match these terms with the correct description:

Hypocalcemia
Hypokalemia

_____ 1. Causes hyperpolarization of the resting membrane potential, resulting in muscular weakness and sluggish reflexes.

_____ 2. Causes action potentials to occur spontaneously, resulting in nervousness, muscular spasms, and tetany.

Graded Potentials

Using the terms provided, complete these statements:

Can
Cannot
Decrease
Graded potential
Increase
Local potential
Summate

1. _____
2. _____
3. _____
4. _____
5. _____
6. _____
7. _____

A(n) _(1)_ is a change in the resting membrane potential that can vary from small to large. This change is also called a(n) _(2)_ because it is confined to a small region of the plasma membrane. As a result of increased strength of stimulation, the magnitude of a graded potential can _(3)_. Increased frequency of stimulation can cause two or more graded potentials to add together, or _(4)_. Because graded potentials are conducted in a decremental fashion, they _(5)_ transfer information over long distances from one part of the body to another. Graded potentials _(6)_ be a depolarization and _(7)_ be a hyperpolarization.

Action Potentials

A. Match these terms with the correct statement or definition:

Afterpotential
Depolarization
Repolarization

_____ 1. The membrane potential moves away from the resting membrane potential and becomes more positive.

_____ 2. The membrane potential returns to the resting membrane potential, becoming more negative.

_____ 3. Short period of hyperpolarization observed after the action potential.

B. Match these terms with the correct statement or definition:

Action potential
Depolarization
Hyperpolarization
Less likely
More likely
Threshold

_____ 1. An action potential occurs when a graded potential reaches this level.

_____ 2. Can propagate without changing its magnitude.

_____ 3. Prevents an action potential from occurring.

_____ 4. Compared with a small depolarizing graded potential, the likelihood that a large depolarizing graded potential will produce an action potential.

_____ 5. Occurs according to the all-or-none principle.

C. Match these events with the correct statements:

Before the action potential
Depolarization
End of the action potential
Repolarization
Repolarization/afterpotential

Events	Status of Ion Gates	Ion Movement Through Voltage-Gated Ion Channels
1. _____	Na^+ activation gates are closed. Na^+ inactivation gates are open. K^+ gates are closed.	No Na^+ movement No K^+ movement
2. _____	Na^+ activation gates open. Na^+ inactivation gates are open. K^+ gates open slowly.	Na^+ move into the cell. Some K^+ move out of the cell.
3. _____	Na^+ activation gates are open. Na^+ inactivation gates close. K^+ gates are open.	Na^+ movement into the cell stops. K^+ move out of the cell.
4. _____	Na^+ activation gates close. Na^+ inactivation gates open. K^+ gates are open.	No Na^+ movement K^+ move out of the cell.
5. _____	Na^+ activation gates are closed. Na^+ inactivation gates are open. K^+ gates close.	No Na^+ movement No K^+ movement

Chapter 11

Refractory Period

A. Match these terms with the correct statement or definition:

Absolute refractory period
Relative refractory period

_____ 1. Period that occurs from the beginning of depolarization until near the end of repolarization.

_____ 2. Period during which there is complete insensitivity to another stimulus.

_____ 3. Period during which a stronger-than-threshold stimulus can initiate an action potential.

_____ 4. Period that ends when the Na^+ inactivation gates open and the Na^+ activation gates close.

_____ 5. Period that ends when the voltage-gated K^+ channels close.

B. Match these terms with the parts labeled in figure 11.2:

Absolute refractory period   Relative refractory period
Action potential   Repolarization
Afterpotential   Threshold
Depolarization

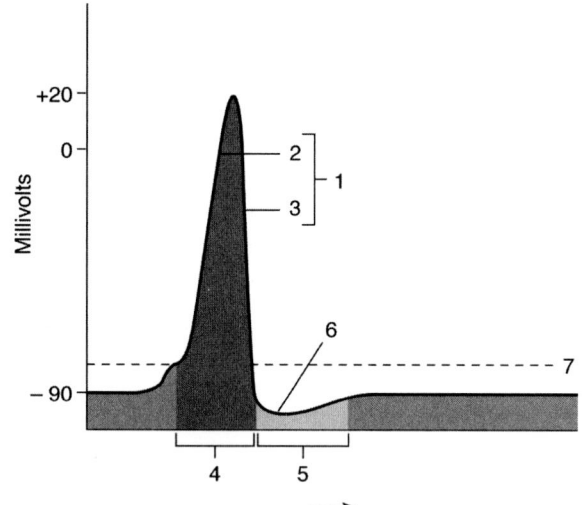

Figure 11.2

1. _____
2. _____
3. _____
4. _____
5. _____
6. _____
7. _____

Action Potential Frequency

Match these terms with the correct statement or definition:

Maximal stimulus
Submaximal stimulus
Subthreshold stimulus
Supramaximal stimulus
Threshold stimulus

_____ 1. Stimulus resulting in a graded potential so small that it does not reach threshold.

_____ 2. Stimulus just strong enough to cause a graded potential to produce a single action potential.

_____ 3. Stimulus just strong enough to produce a maximum frequency of action potentials.

_____ 4. The action potential frequency increases in proportion to the strength of the stimulus.

_____ 5. Stimulus stronger than a maximal stimulus that produces the same frequency of action potentials as a maximal stimulus.

Propagation of Action Potentials

Using the terms provided, complete these statements:

Absolute refractory period
Ionic current
Less
More
Myelinated
Nodes of Ranvier
Propagate
Saltatory conduction
Unmyelinated

Action potentials can spread, or _(1)_, across the plasma membrane. The production of an action potential results in the flow of positively charged ions, called a(n) _(2)_, which cause the plasma membrane to depolarize and produce an action potential. In a(n) _(3)_ axon, the ionic current causes the next action potential to be generated immediately adjacent to the previous action potential, whereas in a(n) _(4)_ axon the next action potential is generated at a(n) _(5)_. Conduction of action potentials from one node of Ranvier to another in myelinated axons is called _(6)_. Action potentials are propagated in one direction down an axon because the _(7)_ prevents the production of an action potential in the reverse direction. Action potentials are conducted _(8)_ rapidly in myelinated than in unmyelinated axons; large-diameter axons conduct action potentials _(9)_ rapidly than do small-diameter axons.

1. _____
2. _____
3. _____
4. _____
5. _____
6. _____
7. _____
8. _____
9. _____

Chapter 11

The Synapse

A. Match these terms with the correct statement or definition:

Ca^{2+}
Chemical
Connexons
Electrical
Neurotransmitter
Postsynaptic membrane
Presynaptic terminal
Synapse
Synaptic cleft
Synaptic vesicle

_____ 1. Junction between two cells.

_____ 2. Type of synapse that is a gap junction; allows an ionic current to flow between cells, resulting in the production of an action potential.

_____ 3. Tubular proteins in an electrical synapse.

_____ 4. Type of synapse in which an action potential causes the release of neurotransmitters.

_____ 5. Formed from the end of an axon.

_____ 6. Space between an axon ending and the cell with which it synapses.

_____ 7. Site for specific receptors that can bind to neurotransmitters.

_____ 8. Membrane-bounded organelle that contains neurotransmitters.

_____ 9. Diffuse into a presynaptic terminal when action potential occurs; cause synaptic vesicles to release their contents.

_____ 10. Chemical released from synaptic vesicles of the presynaptic terminal.

B. Match these terms with the correct statement or definition:

Acetylcholine
Acetylcholinesterase
Catechol-O-methyltransferase
Monoamine oxidase
Neuromodulator
Norepinephrine

_____ 1. Neurotransmitter broken down within the synaptic cleft.

_____ 2. Neurotransmitter taken up whole from the synaptic cleft.

_____ 3. Enzyme that breaks down acetylcholine.

_____ 4. Two enzymes that break down norepinephrine.

_____ 5. Nonneurotransmitter chemical that influences the likelihood that an action potential in the presynaptic terminal will result in the production of an action potential in the postsynaptic membrane.

Excitatory and Inhibitory Postsynaptic Potentials

A. Match these terms with the correct statement or definition:

Excitatory neuron
Excitatory postsynaptic potential (EPSP)
Inhibitory neuron
Inhibitory postsynaptic potential (IPSP)

_____ 1. Graded depolarization of the postsynaptic membrane.

_____ 2. Caused by an increase in the permeability of the cell membrane to Na^+.

_____ 3. Neuron that causes EPSPs.

_____ 4. Graded hyperpolarization of the postsynaptic membrane.

_____ 5. Caused by an increase in the permeability of the cell membrane to K^+ or Cl^-.

_____ 6. Neuron that causes IPSPs.

B. Match these terms with the correct parts labeled in figure 11.3:

EPSP
IPSP

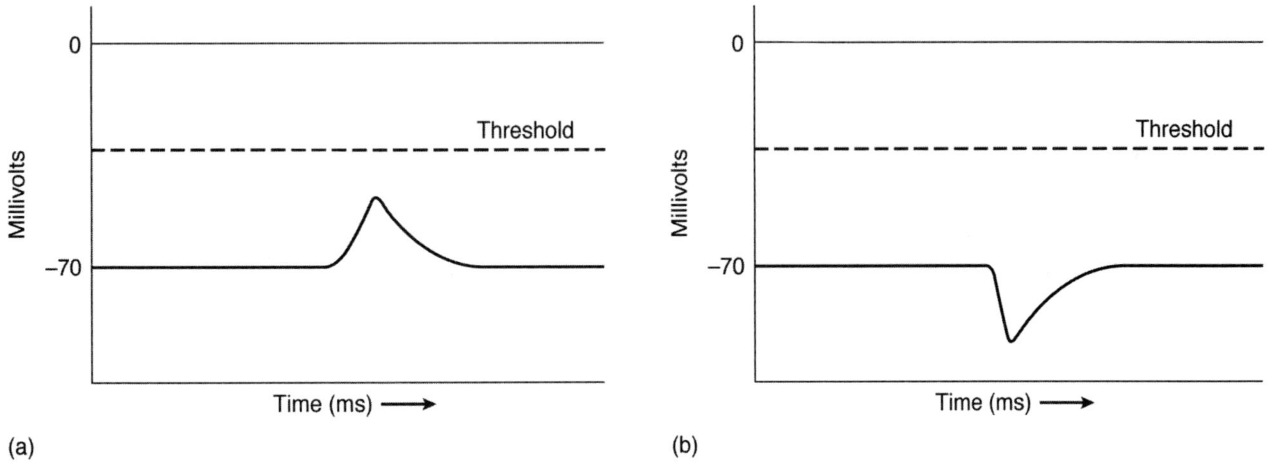

Figure 11.3

_____ 1. Change in membrane potential seen in graph A.

_____ 2. Change in membrane potential seen in graph B.

Chapter 11

Presynaptic Inhibition and Facilitation

Using the terms provided, complete these statements:

Axoaxonic
Neuromodulators
Presynaptic facilitation
Presynaptic inhibition

Many of the synapses of the CNS are _(1)_ synapses, in which the axon of one neuron synapses with the presynaptic terminal (axon) of another neuron. When an action potential reaches the presynaptic terminal, _(2)_ released in the axoaxonic synapse can alter the amount of neurotransmitter released from the presynaptic terminal. In _(3)_, there is decreased neurotransmitter release from the presynaptic terminal—e.g., when endorphins and enkephalins are released in the brain and spinal cord. In _(4)_, there is increased neurotransmitter release from the presynaptic terminal—e.g., when serotonin stimulates the release of neurotransmitter from presynaptic neurons.

1. _____
2. _____
3. _____
4. _____

Spatial and Temporal Summation

Using the terms provided, complete these statements:

Graded potentials
IPSPs
Spatial
Summation
Temporal
Threshold

Presynaptic action potentials produce _(1)_ in the postsynaptic neuron. These combine in a process called _(2)_ at the trigger zone of the postsynaptic neuron. If this combination of graded potentials exceeds _(3)_ at the trigger zone, an action potential is produced. _(4)_ summation results when two action potentials arrive simultaneously at two different presynaptic terminals that synapse with the same postsynaptic neuron. _(5)_ summation occurs when two action potentials arrive in very close succession at a single presynaptic terminal. Excitatory and inhibitory neurons may synapse with a single postsynaptic neuron. Summation occurs in the postsynaptic neuron, and the _(6)_ tend to cancel the EPSPs.

1. _____
2. _____
3. _____
4. _____
5. _____
6. _____

Neuronal Pathways and Circuits

Match these terms with the correct statement or definition:

Afterdischarge Divergent pathway
Convergent pathway Oscillating circuit

_____ 1. Pathway in which many neurons synapse with fewer neurons; spatial summation can occur.

_____ 2. Pathway in which a smaller number of presynaptic neurons synapse with a larger number of postsynaptic neurons.

_____ 3. Circuits with neurons arranged in a circular fashion.

_____ 4. A single action potential entering the circuit results in the production of more than one action potential further along the circuit.

_____ 5. Type of circuit responsible for periodic activity, such as the cycle of wakefulness and sleep.

QUICK RECALL

1. List three types of neurons based on their shape. Give an example of each type.

2. List five types of glial cells. Give a function for each type.

3. Contrast the structural and functional characteristics of myelinated and unmyelinated axons.

4. List two differences between white matter and gray matter. Give an example of each in the CNS and PNS.

5. List three components of a chemical synapse.

6. Distinguish between an EPSP and an IPSP.

7. List two types of summation and distinguish between them.

8. Name the types of neuronal pathways and circuits.

ANSWERS TO CHAPTER 11

CONTENT LEARNING ACTIVITY

Divisions of the Nervous System
A. 1. Central nervous system; 2. Peripheral nervous system; 3. Sensory (afferent) division; 4. Motor (efferent) division; 5. Somatic nervous system; 6. Autonomic nervous system; 7. Sympathetic division; 8. Parasympathetic division; 9. Enteric nervous system; 10. Central nervous system
B. 1. Nerves; 2. 12; 3. 31; 4. Ganglia; 5. Plexus

Neurons
A. 1. Glial cells; 2. Cell body (soma); 3. Dendrite; 4. Axon; 5. Axon
B. 1. Nissl substance; 2. Dendritic spines; 3. Axon hillock; 4. Axoplasm; 5. Presynaptic terminals; 6. Trigger zone
C. 1. Nucleus; 2. Trigger zone; 3. Presynaptic terminal; 4. Node of Ranvier; 5. Schwann cell; 6. Axon; 7. Neuron cell body (soma); 8. Dendrite

Types of Neurons
A. 1. Sensory (afferent) neuron; 2. Motor (efferent) neuron; 3. Interneuron (association neuron)
B. 1. Multipolar neuron; 2. Bipolar neuron; 3. Multipolar neuron

Glial Cells of the CNS and PNS
1. Astrocytes; 2. Blood–brain barrier; 3. Ependymal cells; 4. Microglia; 5. Myelin sheath; 6. Oligodendrocytes; 7. Schwann cells (neurolemmocytes); 8. Satellite cells

Myelinated and Unmyelinated Axons
1. Myelinated; 2. Node of Ranvier; 3. Unmyelinated

Organization of Nervous Tissue
1. White matter; 2. Gray matter; 3. Nerve tract; 4. Nerve; 5. Cortex; 6. Nuclei; 7. Ganglia

Concentration Differences Across the Plasma Membrane
1. Na^+ and Cl^-; 2. K^+; 3. Na^+–K^+ pump; 4. Proteins; 5. Cl^-

Leak and Gated Ion Channels
1. Leak channels; 2. Leak channels; 3. Voltage-gated ion channels; 4. Ligand-gated ion channels; 5. Ligand-gated ion channels; 6. Other gated ion channels

Establishing the Resting Membrane Potential
1. Resting membrane potential; 2. K^+; 3. Negative; 4. Positive; 5. Na^+–K^+ pump; 6. Three; 7. Two; 8. Positive

Changing the Resting Membrane Potential
1. Depolarization; 2. Hyperpolarization; 3. Hyperpolarization; 4. Depolarization; 5. Depolarization; 6. Hyperpolarization

Disorders
1. Hypokalemia; 2. Hypocalcemia

Graded Potentials
1. Graded potential; 2. Local potential; 3. Increase; 4. Summate; 5. Cannot; 6. Can; 7. Can

Action Potentials
A. 1. Depolarization; 2. Repolarization; 3. Afterpotential
B. 1. Threshold; 2. Action potential; 3. Hyperpolarization; 4. More likely; 5. Action potential
C. 1. Before the action potential; 2. Depolarization; 3. Repolarization; 4. Repolarization/afterpotential; 5. End of the action potential

Refractory Period
   A. 1. Absolute refractory period; 2. Absolute refractory period; 3. Relative refractory period; 4. Absolute refractory period; 5. Relative refractory period
   B. 1. Action potential; 2. Depolarization; 3. Repolarization; 4. Absolute refractory period; 5. Relative refractory period; 6. Afterpotential; 7. Threshold

Action Potential Frequency
   1. Subthreshold stimulus; 2. Threshold stimulus; 3. Maximal stimulus; 4. Submaximal stimulus; 5. Supramaximal stimulus

Propagation of Action Potentials
   1. Propagate; 2. Ionic current; 3. Unmyelinated; 4. Myelinated; 5. Node of Ranvier; 6. Saltatory conduction; 7. Absolute refractory period; 8. More; 9. More

The Synapse
   A. 1. Synapse; 2. Electrical; 3. Connexons; 4. Chemical; 5. Presynaptic terminal; 6. Synaptic cleft; 7. Postsynaptic membrane; 8. Synaptic vesicle; 9. Ca^{2+}; 10. Neurotransmitter
   B. 1. Acetylcholine; 2. Norepinephrine; 3. Acetylcholinesterase; 4. Monoamine oxidase and catechol-O-methyltransferase; 5. Neuromodulator

Excitatory and Inhibitory Postsynaptic Potentials
   A. 1. Excitatory postsynaptic potential (EPSP); 2. Excitatory postsynaptic potential (EPSP); 3. Excitatory neuron; 4. Inhibitory postsynaptic potential (IPSP); 5. Inhibitory postsynaptic potential (IPSP); 6. Inhibitory neuron
   B. 1. EPSP; 2. IPSP

Presynaptic Inhibition and Facilitation
   1. Axoaxonic; 2. Neuromodulators; 3. Presynaptic inhibition; 4. Presynaptic facilitation

Spatial and Temporal Summation
   1. Graded potentials; 2. Summation; 3. Threshold; 4. Spatial; 5. Temporal; 6. IPSPs

Neuronal Pathways and Circuits
   1. Convergent circuit; 2. Divergent circuit; 3. Oscillating circuit; 4. Afterdischarge; 5. Oscillating circuit

QUICK RECALL

1. Multipolar neurons: most CNS neurons and motor neurons; bipolar neurons: sensory organs in the eye (retina) and nasal cavity; unipolar neurons: most sensory neurons
2. Astrocytes: form a supporting matrix and help regulate the composition of extracellular fluid around neurons in the CNS; ependymal cells: produce and circulate cerebrospinal fluid; microglia: macrophages that phagocytize foreign or necrotic tissue; Schwann cells: form myelin sheaths around axons in the PNS; oligodendrocytes: form myelin sheaths around axons in the CNS
3. Myelinated axon: wrapped by several layers of oligodendrocytes or Schwann cells, saltatory conduction (rapid); unmyelinated axon: rests in invagination of oligodendrocytes or Schwann cell, conducts action potentials more slowly
4. White matter: myelinated axons, propagates action potentials, forms nerve tracts in the CNS and nerves in the PNS; gray matter: neuron cell bodies and dendrites, site of integration (synapses), forms nuclei (in CNS) and ganglia (in PNS)
5. Presynaptic terminal, synaptic cleft, and postsynaptic membrane
6. EPSP: depolarization of postsynaptic membrane caused by increase in membrane permeability to Na^+; IPSP: hyperpolarization of postsynaptic membrane caused by an increase in membrane permeability to Cl^- or K^+
7. Spatial summation occurs when two or more presynaptic terminals simultaneously stimulate a postsynaptic membrane; temporal summation occurs when two or more action potentials arrive in succession at a single presynaptic neuron.
8. Convergent pathways, divergent pathways, and oscillating circuits

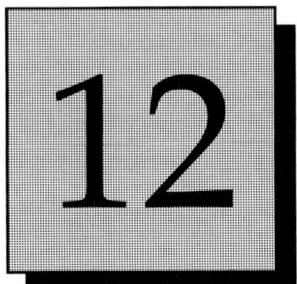

Spinal Cord and Spinal Nerves

CONTENT LEARNING ACTIVITY

Spinal Cord

Match these terms with the correct statement or definition:
    Cauda equina
    Cervical enlargement
    Conus medullaris
    Lumbar enlargement

_____ 1. Region where nerves that supply the upper limbs enter and exit the spinal cord.

_____ 2. Region where nerves that supply the lower limbs enter and exit the spinal cord.

_____ 3. The spinal cord tapers to form this conelike region immediately inferior to the lumbar enlargement.

_____ 4. Conus medullaris and numerous nerves extending inferiorly from it.

Meninges of the Spinal Cord

Match these terms with the correct statement or definition:
    Arachnoid mater
    Denticulate ligaments
    Dura mater
    Epidural space
    Filum terminale
    Pia mater
    Subarachnoid space
    Subdural space

_____ 1. Most superficial and thickest meningeal layer; continuous with the epineurium of the spinal nerves.

_____ 2. Space that separates the dura mater from the periosteum of the vertebral canal; contains blood vessels, connective tissue, and fat.

_____ 3. Middle, wispy meningeal layer.

_____ 4. Space between the dura mater and the arachnoid mater; contains a small amount of serous fluid.

_____ 5. Innermost meningeal layer; bound very tightly to the surface of the spinal cord.

_____ 6. Connective tissue strands that hold the spinal cord in place.

_____ 7. Connective tissue filament that anchors the inferior end of the spinal cord to the coccyx.

_____ 8. Space between the arachnoid mater and the pia mater; contains CSF.

Cross Section of the Spinal Cord

A. Match these terms with the correct statement or definition:

Anterior median fissure   Funiculi
Fasciculi   Posterior median sulcus

_____ 1. Deep clefts that partially separate the two halves of the spinal cord.

_____ 2. Ventral, lateral, and dorsal columns of white matter located in each half of the spinal cord.

_____ 3. Nerve tracts, or pathways, in the funiculi.

B. Match these terms with the correct statement or definition:

Anterior (ventral) horn   Lateral horn
Dorsal root   Posterior (dorsal) horn
Dorsal root (spinal) ganglion   Spinal nerve
Gray and white commissures   Ventral root

_____ 1. Many sensory neurons synapse with interneurons in this structure.

_____ 2. Contains the cell bodies of somatic motor neurons.

_____ 3. Contains the cell bodies of autonomic neurons.

_____ 4. Allows communication between the halves of the spinal cord.

_____ 5. Contains only motor axons.

_____ 6. Contains only sensory axons.

_____ 7. Contains the cell bodies of sensory neurons.

_____ 8. Formed by the union of the dorsal and ventral roots.

C. Match these terms with the correct parts of the diagram labeled in figure 12.1:

Anterior median fissure
Anterior (ventral) horn
Central canal
Dorsal column
Dorsal root
Dorsal root (spinal) ganglion
Gray commissure
Gray matter
Lateral column
Lateral horn
Posterior (dorsal) horn
Posterior median sulcus
Rootlets
Spinal nerve
Ventral column
Ventral root
White commissure
White matter

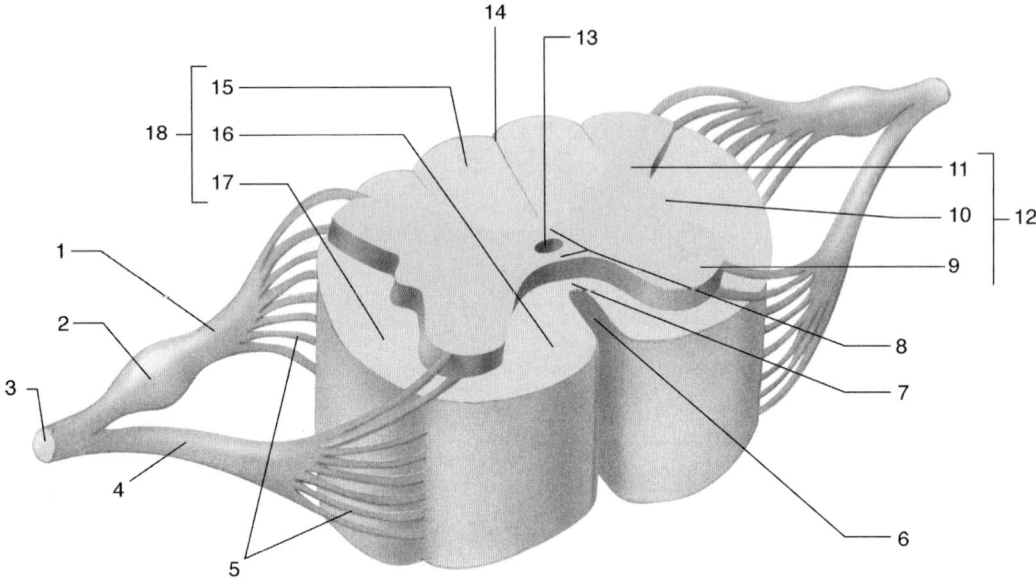

Figure 12.1

1. _____ 7. _____ 13. _____
2. _____ 8. _____ 14. _____
3. _____ 9. _____ 15. _____
4. _____ 10. _____ 16. _____
5. _____ 11. _____ 17. _____
6. _____ 12. _____ 18. _____

Reflexes

A. Using the terms provided, complete these statements:

Effector organs
Interneurons
neurons
Reflex
Reflex arc
Sensory neurons Motor
Sensory receptors

The (1) is the smallest, simplest portion of the nervous system capable of receiving a stimulus and producing a response. Action potentials initiated in (2) are propagated along (3) within the PNS to the CNS, where they usually synapse with (4) . These neurons synapse with (5) , which send axons out of the spinal cord through the PNS to muscles or glands. Action potentials of the motor neurons cause (6) to respond. This automatic response produced by a reflex arc is called a(n) (7) .

1. _____
2. _____
3. _____
4. _____
5. _____
6. _____
7. _____

B. Match these terms with the correct parts labeled in figure 12.2:

Effector organ
Interneuron
Motor neuron
Sensory neuron
Sensory receptor

1. _____
2. _____
3. _____
4. _____
5. _____

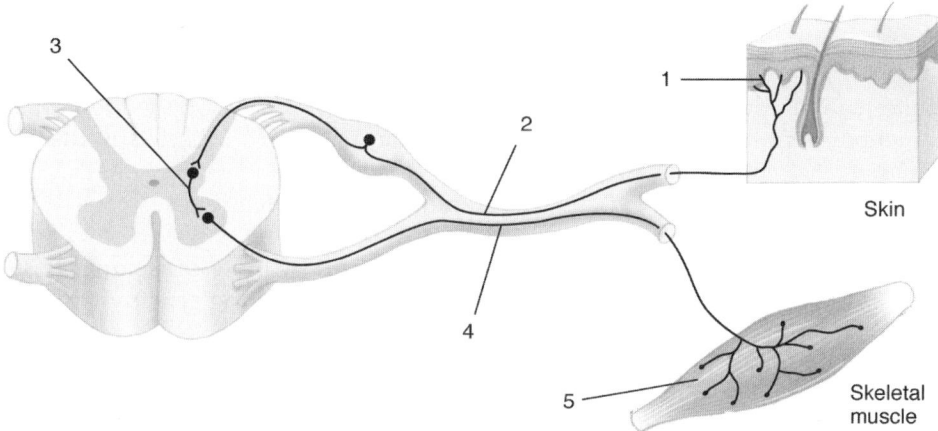

Figure 12.2

C. Using the terms provided, complete these statements:

Complexity
Conscious
Excitatory
Homeostasis
Inhibitory
Monosynaptic
Polysynaptic
Suppressing

Reflexes do not require _(1)_ thought but they do vary in _(2)_. For example, _(3)_ reflexes involve simple neuronal pathways in which sensory neurons synapse directly with motor neurons, whereas _(4)_ reflexes involve more complex pathways with one or more interneurons between the sensory neuron and motor neuron. Reflexes maintain _(5)_—for example, keeping the body from suddenly falling or maintaining relatively constant blood pressure. Some reflexes involve _(6)_ neurons and result in responses such as when a muscle contracts. Other reflexes involve _(7)_ neurons and result in responses such as when a muscle is inhibited and relaxes. In addition, higher brain centers influence reflexes by _(8)_ or exaggerating them.

1. _____
2. _____
3. _____
4. _____
5. _____
6. _____
7. _____
8. _____

D. Match these terms with the correct description or function:

Crossed extensor reflex Stretch reflex
Golgi tendon reflex Withdrawal reflex
Reciprocal innervation

_____ 1. Stretch of a muscle results in contraction of the muscle.

_____ 2. Excess tension causes the muscle to relax.

_____ 3. Removes a limb from a painful stimulus.

_____ 4. Causes the antagonist of a muscle to relax.

_____ 5. Prevents falling when the withdrawal reflex occurs in one leg.

E. Match these terms with the correct statement or definition:

Alpha motor neuron Golgi tendon organ
Gamma motor neuron Muscle spindle

_____ 1. Sensory receptor for the stretch reflex.

_____ 2. Action potentials carried by these cause muscle spindles to contract; important for regulating sensitivity of muscle spindles.

_____ 3. Innervate the muscle in which a muscle spindle is embedded.

_____ 4. Stimulation of this causes inhibition of alpha motor neurons.

Interactions with Spinal Cord Reflexes

Using the terms provided, complete these statements:

Anterior Inhibit
Ascending Posterior
Descending Reflexes

_(1)_ do not operate as isolated entities within the nervous system because of divergent and convergent pathways. Diverging branches of sensory neurons or interneurons in a reflex arc send action potentials along _(2)_ nerve tracts to the brain. Axons within _(3)_ nerve tracts from the brain carry action potentials to motor neurons in the _(4)_ horn of the spinal cord, converging with neurons of reflex arcs. Neurotransmitters released from the axons of these tracts either stimulate or _(5)_ the motor neurons. Various ascending and _(6)_ tracts occupy specific areas of the spinal cord.

1. _____
2. _____
3. _____
4. _____
5. _____
6. _____

Structure of Peripheral Nerves

A. Match these terms with the correct statement or definition:

Endoneurium Perineurium
Epineurium

_____ 1. Delicate connective tissue layer that surrounds each axon and its Schwann cell sheath.

_____ 2. Heavier connective tissue layer that surrounds groups of axons to form nerve fascicles.

_____ 3. Connective tissue layer that binds nerve fascicles together to form a nerve.

B. Match these terms with the correct parts labeled in figure 12.3:

Axon Nerve
Endoneurium Perineurium
Epineurium Schwann cell
Fascicle

1. _____
2. _____
3. _____
4. _____
5. _____
6. _____
7. _____

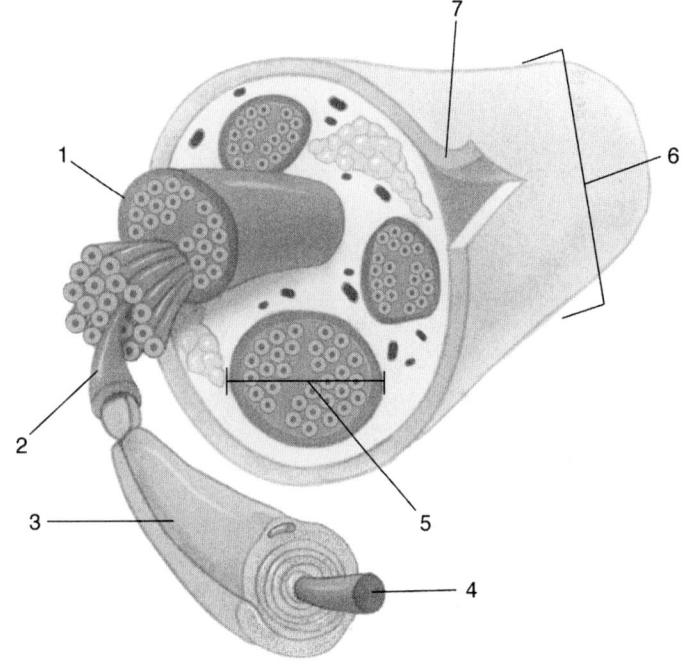

Figure 12.3

Chapter 12 187

Spinal Nerves

A. Match these terms with the correct statement or definition:

Dermatome   Roots of the plexus
Dorsal rami   Ventral rami
Plexuses

_____ 1. Area of skin supplied with sensory innervation by a pair of spinal nerves.

_____ 2. Innervate most of the deep muscles of the dorsal trunk responsible for movement of the vertebral column.

_____ 3. Spinal nerve branches that become intercostal nerves or plexuses.

_____ 4. Organization produced by intermingling of nerves.

_____ 5. Ventral rami of different spinal nerves that join with each other to form a plexus.

B. Match the spinal nerve region with the number of pairs of spinal nerves in each:

Cervical   Sacral
Coccygeal   Thoracic
Lumbar

_____ 1. One pair.

_____ 2. Five pairs (two regions).

_____ 3. Eight pairs.

_____ 4. Twelve pairs.

C. Match these terms with the correct parts labeled in figure 12.4:

Dorsal ramus
Dorsal root
Dorsal root ganglion
Rootlets
Spinal nerve
Ventral ramus
Ventral root

1. _____
2. _____
3. _____
4. _____
5. _____
6. _____
7. _____

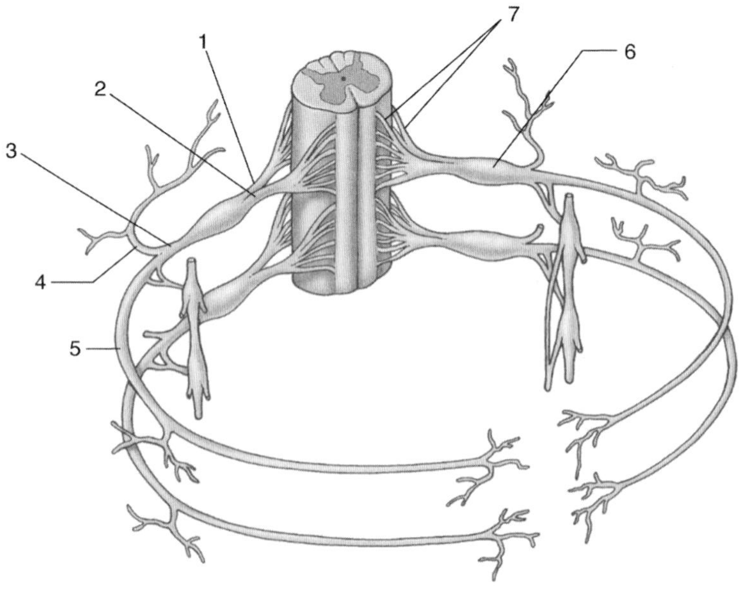

Figure 12.4

188 Chapter 12

Cervical Plexus

Match these terms with the
correct statement or definition:

Diaphragm
Hyoid muscles
Neck and posterior head

_____ 1. Skin of these areas is innervated by the cervical plexus.

_____ 2. Motor innervation from the cervical plexus (except phrenic nerve).

_____ 3. Phrenic nerve innervates this structure.

Brachial Plexus

A. Match these nerves with their correct motor function or innervation:

Axillary nerve
Median nerve
Musculocutaneous nerve
Radial nerve
Ulnar nerve

_____ 1. Innervates the deltoid and teres minor muscles.

_____ 2. Innervates all the extensor muscles of the upper limb.

_____ 3. Innervates the anterior muscles of the arm.

_____ 4. Innervates most of the intrinsic hand muscles.

_____ 5. Two nerves that innervate the flexor muscles of the forearm.

_____

B. Match these terms with the correct statement or definition:

Axillary nerve
Median nerve
Musculocutaneous nerve
Radial nerve
Ulnar nerve

_____ 1. Sensory innervation to the inferior lateral shoulder.

_____ 2. Sensory innervation to the posterior arm, forearm, and lateral two-thirds of the dorsum of the hand.

_____ 3. Sensory innervation to the lateral surface of the forearm.

_____ 4. Sensory innervation to the medial third of the hand, little finger, and medial half of the ring finger.

_____ 5. Sensory innervation to the lateral two-thirds of the palm of the hand, thumb, and index and middle fingers.

Chapter 12

Lumbar, Sacral, and Coccygeal Plexuses

A. Match these nerves with their correct motor function or innervation:

    Coccygeal nerve   Femoral nerve
    Common fibular   Obturator nerve
      (peroneal) nerve   Tibial nerve

_____ 1. Innervates the muscles that adduct the thigh.

_____ 2. Innervates the iliopsoas, sartorius, and quadriceps femoris muscles.

_____ 3. Innervates most of the posterior thigh and leg muscles.

_____ 4. Innervates the anterior and lateral muscles of the leg and foot.

_____ 5. Supplies motor innervation to the pelvic floor.

B. Match these nerves with the correct description:

    Common fibular (peroneal) nerve   Tibial nerve
    Ischiadic (sciatic) nerve

_____ 1. Tibial and common fibular (peroneal) nerves bound together in the same connective tissue sheath.

_____ 2. Branches to form medial plantar, lateral plantar, and sural nerves.

_____ 3. Branches to form deep and superficial fibular (peroneal) nerves.

C. Match these nerves with their correct sensory innervation:

    Femoral nerve   Plantar nerves and
    Obturator nerve   sural nerve

_____ 1. Sensory innervation to the medial side of the thigh.

_____ 2. Sensory innervation to the anterior and lateral thigh and medial leg and foot.

_____ 3. Sensory innervation to the lateral and posterior one-third of the leg and the sole of the foot.

Quick Recall

1. List the three meninges and the three spaces associated with them.

2. List the five components of a reflex arc.

3. List five major reflexes.

4. Name two distribution patterns for the ventral rami of spinal nerves.

5. Name the five major plexuses formed by the spinal nerves, and list the level of the spinal cord from which each plexus arises.

6. List the five major nerves arising from the brachial plexus.

7. List the four major nerves arising from the lumbosacral plexus.

ANSWERS TO CHAPTER 12

CONTENT LEARNING ACTIVITY

Spinal Cord
1. Cervical enlargement; 2. Lumbar enlargement; 3. Conus medullaris; 4. Cauda equina

Meninges of the Spinal Cord
1. Dura mater; 2. Epidural space; 3. Arachnoid mater; 4. Subdural space; 5. Pia mater; 6. Denticulate ligaments; 7. Filum terminale; 8. Subarachnoid space

Cross Section of the Spinal Cord
A. 1. Anterior median fissure and posterior median sulcus; 2. Funiculi; 3. Fasciculi
B. 1. Posterior (dorsal) horn; 2. Anterior (ventral) horn; 3. Lateral horn; 4. Gray and white commissures; 5. Ventral root; 6. Dorsal root; 7. Dorsal root (spinal) ganglion; 8. Spinal nerve
C. 1. Dorsal root; 2. Dorsal root (spinal) ganglion; 3. Spinal nerve; 4. Ventral root; 5. Rootlets; 6. Anterior median fissure; 7. White commissure; 8. Gray commissure; 9. Anterior (ventral) horn; 10. Lateral horn; 11. Posterior (dorsal) horn; 12. Gray matter; 13. Central canal; 14. Dorsal median sulcus; 15. Dorsal column; 16. Ventral column; 17. Lateral column; 18. White matter

Reflexes
A. 1. Reflex arc; 2. Sensory receptors; 3. Sensory neurons; 4. Interneurons; 5. Motor neurons; 6. Effector organs; 7. Reflex
B. 1. Sensory receptor; 2. Sensory neuron; 3. Interneuron; 4. Motor neuron; 5. Effector organ
C. 1. Conscious; 2. Complexity; 3. Monosynaptic; 4. Polysynaptic; 5. Homeostasis; 6. Excitatory; 7. Inhibitory; 8. Suppressing
D. 1. Stretch reflex; 2. Golgi tendon reflex; 3. Withdrawal reflex; 4. Reciprocal innervation; 5. Crossed extensor reflex
E. 1. Muscle spindle; 2. Gamma motor neuron; 3. Alpha motor neuron; 4. Golgi tendon organ

Interactions with Spinal Cord Reflexes
1. Reflexes; 2. Ascending; 3. Descending; 4. Anterior; 5. Inhibit; 6. Descending

Structure of Peripheral Nerves
A. 1. Endoneurium; 2. Perineurium; 3. Epineurium
B. 1. Perineurium; 2. Endoneurium; 3. Schwann cell; 4. Axon; 5. Fascicle; 6. Nerve; 7. Epineurium

Spinal Nerves
A. 1. Dermatome; 2. Dorsal rami; 3. Ventral rami; 4. Plexuses; 5. Roots of the plexus
B. 1. Coccygeal; 2. Lumbar and sacral; 3. Cervical; 4. Thoracic
C. 1. Dorsal root; 2. Ventral root; 3. Spinal nerve; 4. Dorsal ramus; 5. Ventral ramus; 6. Dorsal root ganglion; 7. Rootlets

Cervical Plexus
1. Neck and posterior head; 2. Hyoid muscles; 3. Diaphragm

Brachial Plexus
A. 1. Axillary nerve; 2. Radial nerve; 3. Musculocutaneous nerve; 4. Ulnar nerve; 5. Median nerve and ulnar nerve
B. 1. Axillary nerve; 2. Radial nerve; 3. Musculocutaneous nerve; 4. Ulnar nerve; 5. Median nerve

Lumbar, Sacral, and Coccygeal Plexuses
A. 1. Obturator nerve; 2. Femoral nerve; 3. Tibial nerve; 4. Common fibular (peroneal) nerve; 5. Coccygeal nerve
B. 1. Ischiadic (sciatic) nerve; 2. Tibial nerve; 3. Common fibular (peroneal) nerve
C. 1. Obturator nerve; 2. Femoral nerve; 3. Plantar nerves and sural nerve

QUICK RECALL

1. Dura mater, arachnoid mater, and pia mater. The subdural space separates the dura mater and the arachnoid mater, and the subarachnoid space separates the arachnoid mater and the pia mater. In the spinal cord, the dura mater is separated from the periosteum of the vertebral canal by the epidural space.
2. Sensory receptors, sensory neurons, interneurons, motor neurons, and effector organs
3. Stretch reflex, Golgi tendon reflex, withdrawal reflex, reciprocal innervation, and crossed extensor reflex
4. In the thoracic region, the ventral rami form intercostal nerves. All other ventral rami form plexuses.
5. Cervical plexus: C1 to C4; brachial plexus: C5 to T1; lumbar plexus: L1 to L4; sacral plexus: L4 to S4; Coccygeal plexus: S4, S5, and coccygeal nerve
6. Axillary, radial, musculocutaneous, median, and ulnar nerves
7. Obturator, femoral, tibial, and common fibular (peroneal)

13 Brain and Cranial Nerves

CONTENT LEARNING ACTIVITY

Development

A. Match these terms with the correct statement or definition:

Neural crest cells Neural tube
Neural folds Neural tube cavities
Neural plate Neural tube wall

_____ 1. Flat plate of tissue on the upper surface of the embryo.

_____ 2. Elevated edges of the neural plate.

_____ 3. Structure formed when the neural folds move toward each other and fuse.

_____ 4. Separate from the neural folds; give rise to part of the PNS.

_____ 5. Gives rise to the tissues of the brain and spinal cord.

_____ 6. Give rise to the ventricles and central canal of the spinal cord.

B. Match these terms with the correct statement or definition:

Diencephalon Myelencephalon
Mesencephalon Telencephalon
Metencephalon

_____ 1. Becomes the cerebrum in the adult.

_____ 2. Becomes the thalamus, subthalamus, epithalamus, and hypothalamus in the adult.

_____ 3. Becomes the midbrain.

_____ 4. Becomes the pons and cerebellum in the adult.

_____ 5. Becomes the medulla oblongata in the adult.

Brainstem

A. Match the structure with the part of the brainstem in which it is located:

Medulla oblongata
Midbrain
Pons

_____ 1. Pyramids.

_____ 2. Heart rate centers.

_____ 3. Swallowing center.

_____ 4. Respiratory centers (two locations).

_____

_____ 5. Colliculi.

_____ 6. Cerebral peduncles.

B. Match these parts of the brain with the correct function:

Cerebral peduncles   Red nucleus
Colliculi   Reticular formation
Olives   Substantia nigra
Pyramids

_____ 1. Descending tracts in the medulla oblongata that decussate; involved in the conscious control of skeletal muscles.

_____ 2. Nuclei in the medulla oblongata; involved with balance, coordination, and modulation of sound.

_____ 3. Involved with visual reflexes and auditory pathways; form the corpora quadrigemina.

_____ 4. Nucleus located in the midbrain; aids in the unconscious regulation and coordination of motor activities.

_____ 5. Descending tracts in the midbrain; carry motor information from the cerebrum to the spinal cord.

_____ 6. Nucleus located in the midbrain; involved in maintaining muscle tone and in coordinating movements.

_____ 7. Nuclei scattered throughout the brainstem; maintains consciousness and arousal.

C. Match these terms with the correct parts labeled in figure 13.1:

Brainstem
Cerebral peduncle
Inferior colliculus
Medulla oblongata
Midbrain
Pineal body
Pons
Spinal cord
Superior colliculus
Thalamus

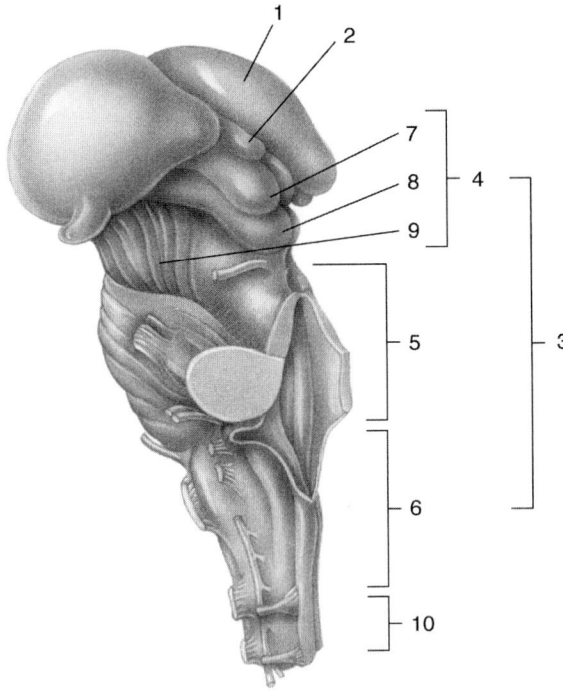

Figure 13.1

1. _____
2. _____
3. _____
4. _____
5. _____
6. _____
7. _____
8. _____
9. _____
10. _____

Cerebellum

Match these terms with the correct statement or definition:

Cerebellar peduncles
Flocculonodular lobe
Lateral hemispheres (lateral portion)
Lateral hemispheres (medial portion)
Vermis

_____

_____

_____

_____

1. Tracts that communicate with other areas of the CNS.

2. Small, inferior portion of the cerebellum involved in balance.

3. Two parts of the cerebellum involved in the control of posture, locomotion, and fine motor coordination.

4. Major parts of the cerebellum involved with planning, practicing, and learning complex movements.

Chapter 13

Diencephalon

A. Match these terms with the correct statement or definition:

Epithalamus　　　Subthalamus
Hypothalamus　　Thalamus

_____ 1. Consists of the thalamus, subthalamus, epithalamus, and hypothalamus.

_____ 2. Consists of two large, lateral portions connected by the interthalamic adhesion.

_____ 3. Consists of many nuclei; receives most sensory input and projects to the cerebral cortex.

_____ 4. Has motor nuclei that connect to the basal nuclei, cerebellum, and motor cortex of the cerebrum.

_____ 5. Small area immediately inferior to the thalamus; contains nuclei involved with motor functions.

_____ 6. Contains the pineal body, which plays a role in the onset of puberty and the sleep–wake cycle.

_____ 7. Contains the mammillary bodies, which are involved in olfactory reflexes.

_____ 8. Regulates the secretion of hormones by the pituitary gland.

_____ 9. Involved in the control of many autonomic functions, such as movement of food through the digestive tract.

_____ 10. Involved with temperature regulation.

_____ 11. Contains nuclei involved with hunger, thirst, and sex drive.

_____ 12. Three parts of the diencephalon involved with mood and emotions.

_____

_____

B. Match these terms with the correct parts of the diagram labeled in figure 13.2:

Cerebellum
Corpus callosum
Hypothalamus
Interthalamic adhesion
Midbrain
Optic chiasm
Pineal body
Pituitary gland
Pons
Subthalamus
Thalamus

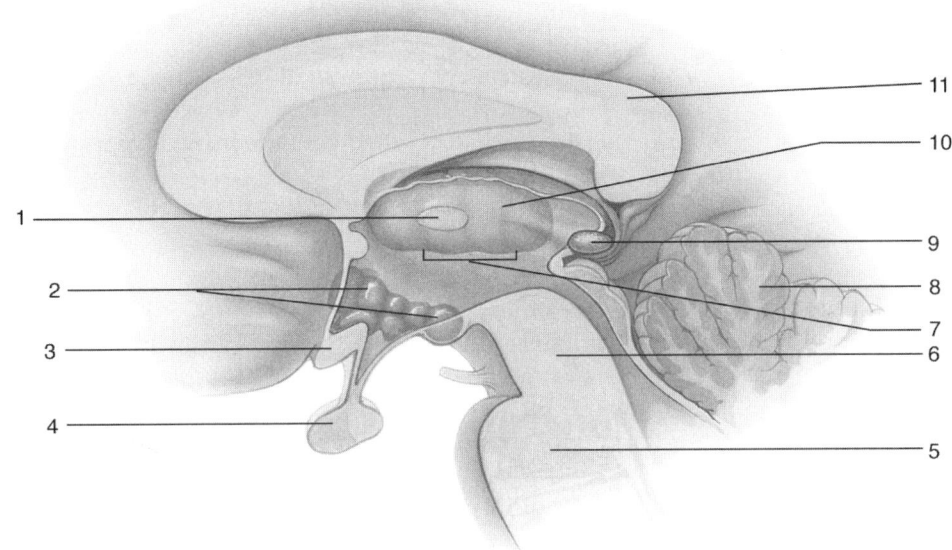

Figure 13.2

1. _____
2. _____
3. _____
4. _____
5. _____
6. _____
7. _____
8. _____
9. _____
10. _____
11. _____

Cerebrum

A. Match these terms with the correct statement or definition:

Central sulcus
Gyri
Lateral fissure
Longitudinal fissure
Postcentral gyrus
Precentral gyrus
Sulci

_____ 1. Deep groove that separates the right and left cerebral hemispheres.

_____ 2. Raised folds or ridges on the surface of the cerebrum.

_____ 3. Grooves between gyri.

_____ 4. Groove that separates the frontal and parietal lobes.

_____ 5. Deep groove that separates the temporal lobe from the rest of the cerebrum.

_____ 6. Primary motor cortex.

_____ 7. Primary somatic sensory cortex.

B. Match these lobes of the cerebrum with the correct primary function:

Frontal lobe
Occipital lobe
Parietal lobe
Temporal lobe

_____ 1. Voluntary motor function, motivation, aggression, the sense of smell, and mood.

_____ 2. Evaluation of most sensory input, except for smell, hearing, and vision.

_____ 3. Reception and integration of visual input.

_____ 4. Receives input for smell and hearing; important in memory.

_____ 5. Abstract thought and judgment ("psychic cortex").

C. Match these terms with the correct statement or definition:

Association fibers
Commissural fibers
Gray matter
Projection fibers
White matter

_____ 1. The cortex and nuclei of the cerebrum are composed of this.

_____ 2. Tracts of the cerebrum between the cortex and nuclei (the cerebral medulla) are composed of this.

_____ 3. Tracts that connect areas of the cerebral cortex within the same hemisphere.

_____ 4. Tracts that connect one cerebral hemisphere to the other cerebral hemisphere.

_____ 5. Tracts that connect the cerebrum with other parts of the brain and spinal cord.

Basal Nuclei

Match these terms with the correct statement or definition:

Basal nuclei
Caudate nucleus
Corpus striatum
Lentiform nucleus
Substantia nigra
Subthalamic nucleus

_____ 1. Group of nuclei located in the inferior cerebrum, diencephalon, and midbrain; involved in the control of motor function.

_____ 2. Collectively, the nuclei in the cerebrum.

_____ 3. Two nuclei that are part of the corpus striatum.

_____ 4. Located in the diencephalon.

_____ 5. Located in the midbrain.

Limbic System

Match these terms with the correct statement or definition:

Cingulate gyrus   Hippocampus
Fornix   Limbic system

_____ 1. Deep parts of the cerebrum that form a ring around the diencephalon; involved with emotions, memory, and basic survival functions, such as reproduction and nutrition.

_____ 2. Cerebral parts of the limbic system.

_____

_____ 3. Connects the hippocampus to the thalamus and mammillary bodies.

_____ 4. Includes parts of the thalamus, hypothalamus (especially mammillary bodies), parts of the basal nuclei, and the olfactory cortex.

Meninges

A. Match these terms with the correct statement or definition:

Falx cerebelli   Tentorium cerebelli
Falx cerebri

_____ 1. Dural fold in the longitudinal fissure between the cerebral hemispheres.

_____ 2. Dural fold oriented horizontally between the cerebrum and the cerebellum.

_____ 3. Dural fold between the cerebellar hemispheres.

B. Match these terms with the correct statement or definition:

Arachnoid mater   Pia mater
Dura mater   Subarachnoid space
Dural venous sinus Subdural space

_____ 1. Most superficial and thickest meningeal layer; consists of an outer, periosteal dura and an inner, meningeal dura.

_____ 2. Middle, wispy meningeal layer.

_____ 3. Deepest meningeal layer; attached to the surface of the brain.

_____ 4. Space formed by the separation of the periosteal dura and meningeal dura.

_____ 5. Space between the dura mater and the arachnoid mater.

_____ 6. Space between the arachnoid mater and the pia mater.

C. Match these terms with the correct part of the diagram labeled in figure 13.3:

Arachnoid granulation
Arachnoid mater
Meningeal dura
Periosteal dura
Pia mater
Subarachnoid space
Superior sagittal sinus (dural sinus)

1. _____
2. _____
3. _____
4. _____
5. _____
6. _____
7. _____

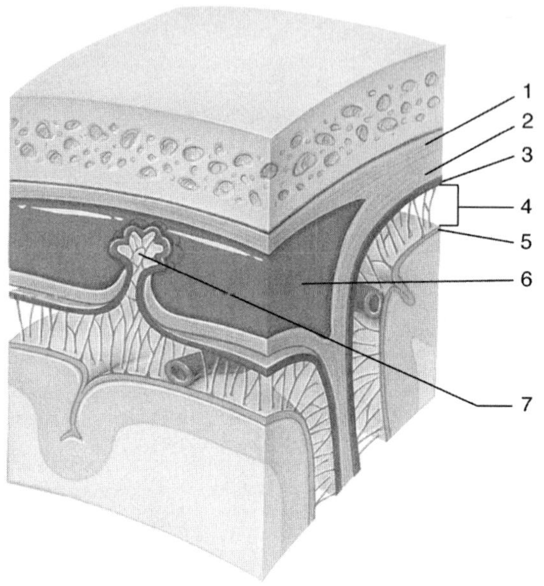

Figure 13.3

Ventricles

Match these terms with the correct statement or definition:

Blood–cerebrospinal fluid barrier
Central canal
Cerebral aqueduct
Ependymal
Fourth ventricle
Interventricular foramina
Lateral ventricles
Septa pellucida
Third ventricle

_____ 1. Cells lining the ventricles.

_____ 2. Formed by tight junctions between ependymal cells.

_____ 3. Large cavities in each cerebral hemisphere.

_____ 4. Separate the lateral ventricles from each other.

_____ 5. Midline cavity located between the lobes of the thalamus.

_____ 6. Connect the lateral ventricles to the third ventricle.

_____ 7. Cavity located in the inferior pontine and superior medullary regions.

_____ 8. Connects the third and fourth ventricles.

_____ 9. Continuation of the fourth ventricle into the spinal cord.

Cerebrospinal Fluid

A. Match these terms with the correct statement or definition:

Arachnoid granulations
Cerebrospinal fluid
Choroid plexuses
Dural venous sinuses
Lateral and median apertures
Subarachnoid space

_____ 1. Fills ventricles and subarachnoid spaces.

_____ 2. Site of the production of cerebrospinal fluid.

_____ 3. Cerebrospinal fluid from the fourth ventricle enters this.

_____ 4. Passageways from the fourth ventricle into the subarachnoid space.

_____ 5. Vessels into which the cerebrospinal fluid enters the blood.

_____ 6. Cerebrospinal fluid enters the dural venous sinuses through these.

B. Match these terms with the correct parts of the diagram labeled in figure 13.4:

Arachnoid granulation
Central canal
Cerebral aqueduct
Choroid plexuses
Fourth ventricle
Interventricular foramen
Lateral foramen
Lateral ventricle
Median aperture
Subarachnoid space
Superior sagittal sinus
Third ventricle

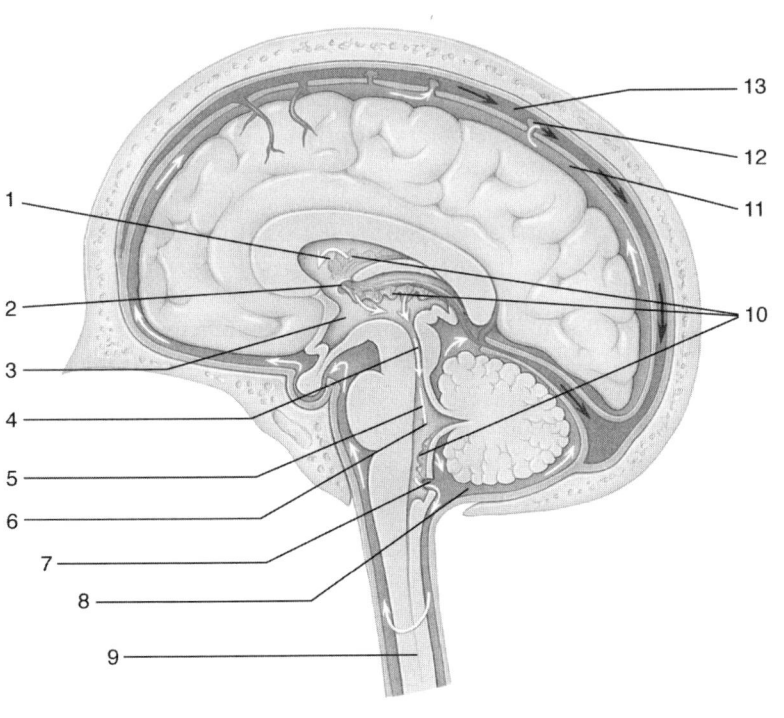

Figure 13.4

1. _____
2. _____
3. _____
4. _____
5. _____
6. _____
7. _____
8. _____
9. _____
10. _____
11. _____
12. _____
13. _____

Chapter 13

Blood Supply to the Brain

Match these terms with the correct statement or definition:

Basilar artery
Blood–brain barrier
Cerebral arterial circle
Internal carotid artery
Vertebral artery

_____ 1. Two arteries from the neck that supply the brain.

_____ 2. The vertebral arteries join to form this artery.

_____ 3. The internal carotid arteries and the basilar artery contribute to this collection of arteries, from which cerebral arteries supply the brain.

_____ 4. Produced by the tight junctions of the capillaries supplying the brain; lipid-soluble substances pass through it by diffusion, but water-soluble substances are moved by mediated transport.

Cranial Nerves

A. Match these terms with the correct statement or definition:

Parasympathetic
Proprioception
Sensory
Somatic motor

_____ 1. Includes vision, touch, and pain.

_____ 2. Involves the control of skeletal muscles through motor neurons.

_____ 3. Informs the brain about the position of various body parts, including joints and muscles.

_____ 4. Involves the regulation of glands, smooth muscle, and cardiac muscle.

_____ 5 Included with somatic motor function because motor nerves to muscles also convey impulses from those muscles to the CNS.

_____ 6. Two types of ganglia associated with cranial nerves.

B. Match the name of the cranial nerve with its number:

Abducent
Accessory
Facial
Glossopharyngeal
Hypoglossal
Oculomotor
Olfactory
Optic
Trigeminal
Trochlear
Vagus
Vestibulocochlear

_____ I. _____ VII.

_____ II. _____ VIII.

_____ III. _____ IX.

_____ IV. _____ X.

_____ V. _____ XI.

_____ VI. _____ XII.

A mnemonic is a formula or another memory aid. In these mnemonics, the first letter(s) of each word in the mnemonic is the same letter(s) as the name of a cranial nerve.

On Occasion, Our Trusty Truck Acts Funny; Very Good Vehicle, AnyHow!
Oh, Oh, Oh, To Touch And Feel Very Good Velvet, AH!
Only Old Oranges Taste Terrible And Feel Very Grainy, Veiny, And Hard.
Old Opie Occasionally Tries Trigonometry And Feels Very Gloomy, Vague, And Hypoactive.

C. Match these cranial nerves with their correct sensory function:

Facial (VII)
Glossopharyngeal (IX)
Olfactory (I)
Optic (II)
Trigeminal (V)
Vagus (X)
Vestibulocochlear (VIII)

_____ 1. Sensory from face, teeth, upper and lower jaw, and oral cavity.

_____ 2. Alveolar nerves arise from maxillary and mandibular branches.

_____ 3. Sense of taste from anterior two-thirds of tongue.

_____ 4. Sense of hearing and balance.

_____ 5. Sense of taste from posterior one-third of tongue; sensory from receptors in the carotid arteries and aortic arch.

_____ 6. Sensory from the thoracic and abdominal organs; sense of taste from posterior tongue.

Chapter 13

D. Match these cranial nerves with their correct motor function:

Abducent (VI)
Accessory (XI)
Facial (VII)
Glossopharyngeal (IX)
Hypoglossal (XII)
Oculomotor (III)
Trigeminal (V)
Trochlear (IV)
Vagus (X)

_____ 1. Motor to four extrinsic eye muscles.

_____ 2. Two nerves, each motor to an extrinsic eye muscle.

_____ 3. Motor to muscles of mastication.

_____ 4. Motor to muscles of facial expression.

_____ 5. Motor to soft palate, pharynx, and laryngeal muscles (voice).

_____ 6. Motor to sternocleidomastoid and trapezius.

_____ 7. Motor to intrinsic and extrinsic tongue muscles.

E. Match these terms with the correct statement or definition:

Facial (VII)
Glossopharyngeal (IX)
Oculomotor (III)
Vagus (X)

_____ 1. Parasympathetic to pupil of eye and ciliary muscle of lens.

_____ 2. Parasympathetic to two salivary glands and lacrimal glands.

_____ 3. Parasympathetic to the parotid salivary gland.

_____ 4. Parasympathetic to the thoracic and abdominal viscera.

Quick Recall

1. Complete the following table:

Structure	Function
Medulla oblongata	
Pons	
Midbrain	
Cerebellum	
Thalamus	
Hypothalamus	
Cerebrum	
Cerebral medulla	
Basal nuclei	
Limbic system	

2. Name the three parts of the brainstem.

3. List four major parts of the diencephalon.

4. List the four largest lobes of the cerebrum, and list their major functions.

5. List the three types of tracts found in the cerebral medulla.

6. List the three meninges and three spaces associated with them.

7. List the four ventricles of the brain and the openings that connect them to each other.

8. List the three basic functions of the cranial nerves.

9. List the 12 cranial nerves.

ANSWERS TO CHAPTER 13

CONTENT LEARNING ACTIVITY

Development
- A. 1. Neural plate; 2. Neural folds; 3. Neural tube; 4. Neural crest cells; 5. Neural tube walls; 6. Neural tube cavities
- B. 1. Telencephalon; 2. Diencephalon; 3. Mesencephalon; 4. Metencephalon; 5. Myelencephalon

Brainstem
- A. 1. Medulla oblongata; 2. Medulla oblongata; 3. Medulla oblongata; 4. Medulla oblongata and pons; 5. Midbrain; 6. Midbrain
- B. 1. Pyramids; 2. Olives; 3. Colliculi; 4. Red nucleus; 5. Cerebral peduncles; 6. Substantia nigra; 7. Reticular formation
- C. 1. Thalamus; 2. Pineal body; 3. Brainstem; 4. Midbrain; 5. Pons; 6. Medulla oblongata; 7. Superior colliculus; 8. Inferior colliculus; 9. Cerebral peduncle; 10. Spinal cord

Cerebellum
1. Cerebellar peduncles; 2. Flocculonodular lobe; 3. Vermis and lateral hemispheres (medial portion); 4. Lateral hemispheres (lateral portion)

Diencephalon
- A. 1. Diencephalon; 2. Thalamus; 3. Thalamus; 4. Thalamus; 5. Subthalamus; 6. Epithalamus; 7. Hypothalamus; 8. Hypothalamus; 9. Hypothalamus; 10. Hypothalamus; 11. Hypothalamus; 12. Thalamus, hypothalamus, and epithalamus
- B. 1. Intermediate adhesion; 2. Hypothalamus; 3. Optic chiasm; 4. Pituitary gland; 5. Pons; 6. Midbrain; 7. Subthalamus; 8. Cerebellum; 9. Pineal body; 10. Thalamus; 11. Corpus callosum

Cerebrum
- A. 1. Longitudinal fissure; 2. Gyri; 3. Sulci; 4. Central sulcus; 5. Lateral fissure; 6. Precentral gyrus; 7. Postcentral gyrus
- B. 1. Frontal lobe; 2. Parietal lobe; 3. Occipital lobe; 4. Temporal lobe; 5. Temporal lobe
- C. 1. Gray matter; 2. White matter; 3. Association fibers; 4. Commissural fibers; 5. Projection fibers

Basal Nuclei
1. Basal nuclei; 2. Corpus striatum; 3. Caudate nucleus and lentiform nucleus; 4. Subthalamic nucleus; 5. Substantia nigra

Limbic System
1. Limbic system; 2. Hippocampus and cingulate gyrus; 3. Fornix; 4. Limbic system

Meninges
- A. 1. Falx cerebri; 2. Tentorium cerebelli; 3. Falx cerebelli
- B. 1. Dura mater; 2. Arachnoid mater; 3. Pia mater; 4. Dural venous sinus; 5. Subdural space; 6. Subarachnoid space
- C. 1. Periosteal dura; 2. Meningeal dura; 3. Arachnoid mater; 4. Subarachnoid space; 5. Pia mater; 6. Superior sagittal sinus; 7. Arachnoid granulations

Ventricles
- A. 1. Ependymal; 2. Blood–cerebrospinal fluid barrier; 3. Lateral ventricles; 4. Septa pellucida; 5. Third ventricle; 6. Interventricular foramina; 7. Fourth ventricle; 8. Cerebral aqueduct; 9. Central canal

Cerebrospinal Fluid
- A. 1. Cerebrospinal fluid; 2. Choroid plexuses; 3. Subarachnoid space; 4. Lateral and median apertures; 5. Dural venous sinuses; 6. Arachnoid granulations
- B. 1. Lateral ventricle; 2. Interventricular foramen; 3. Third ventricle; 4. Cerebral aqueduct; 5. Fourth ventricle; 6. Lateral aperture; 7. Median aperture; 8. Subarachnoid space; 9. Central canal; 10. Choroid plexuses; 11. Subarachnoid space; 12. Arachnoid granulation; 13. Superior sagittal sinus

Blood Supply to the Brain
1. Internal carotid artery and vertebral artery; 2. Basilar artery; 3. Cerebral arterial circle; 4. Blood–brain barrier

Cranial Nerves
- A. 1. Sensory; 2. Somatic motor; 3. Proprioception; 4. Parasympathetic; 5. Proprioception; 6. Parasympathetic and sensory
- B. I. Olfactory; II. Optic; III. Oculomotor; IV. Trochlear; V. Trigeminal; VI. Abducent; VII. Facial; VIII. Vestibulocochlear; IX. Glossopharyngeal; X. Vagus; XI. Accessory; XII. Hypoglossal
- C. 1. Trigeminal (V); 2. Trigeminal (V); 3. Facial (VII); 4. Vestibulocochlear (VIII); 5. Glossopharyngeal (IX); 6. Vagus (X)
- D. 1. Oculomotor (III); 2. Trochlear (IV), abducent (VI); 3. Trigeminal (V); 4. Facial (VII); 5. Vagus (X); 6. Accessory (XI); 7. Hypoglossal (XII)
- E. 1. Oculomotor (III); 2. Facial (VII); 3. Glossopharyngeal (IX); 4. Vagus (X)

QUICK RECALL

1. | | |
    |---|---|
    | Medulla oblongata | Motor pathways (pyramids), balance (olives), autonomic reflexes (e.g., heart rate, breathing, swallowing), ascending and descending tracts, consciousness (reticular activating system) |
    | Pons | Connects cerebellum to brain (cerebellar peduncles), regulates respiration, and contains ascending and descending tracts |
    | Midbrain | Visual and auditory reflexes (colliculi), motor pathways (cerebral peduncles), and motor nuclei (red nuclei and substantia nigra) |
    | Cerebellum | Controls balance and eye movements; controls posture, locomotion, and fine motor coordination; and is involved in the planning, practicing, and learning of complex movements |
    | Thalamus | Receives most sensory input that projects to the cerebrum; connects motor functions of the basal nuclei, cerebellum, and motor cortex; and influences mood and strong emotions |
    | Hypothalamus | Endocrine control, autonomic control, temperature regulation, hunger, thirst, sex drive, emotions, and sleep–wake cycle |
    | Cerebrum | Interprets sensations and initiates motor activities, thoughts, reasoning, speech, and other higher brain functions |
    | Cerebral medulla | Tract system that connects parts of a cerebral hemisphere with other parts of the same hemisphere, with the opposite cerebral hemisphere, or with other parts of the brain and spinal cord |
    | Basal nuclei | Involved in the control of motor functions |
    | Limbic system | Emotions and memory |

2. Medulla oblongata, pons, and midbrain
3. Thalamus, subthalamus, hypothalamus, and epithalamus
4. Frontal lobe: voluntary motor function, motivation, aggression, the sense of smell, and mood; parietal lobe: major center for reception and evaluation of most sensory information; occipital lobe: reception and integration of visual input; temporal lobe: abstract thought, judgment, and memory and receives and evaluates input for smell and hearing
5. Association, commissural, and projection tracts
6. Dura mater, arachnoid mater, and pia mater. The subdural space separates the dura mater and arachnoid mater and the subarachnoid space separates the arachnoid mater and the pia mater. The periosteal dura and the meningeal dura separate to form dural venous sinuses.
7. The two lateral ventricles are connected to the third ventricle by the interventricular foramina; the third ventricle is connected to the fourth ventricle by the cerebral aqueduct.
8. Olfactory (I), optic (II), oculomotor (III), trochlear (IV), trigeminal (V), abducent (VI), facial (VII), vestibulocochlear (VIII), glossopharyngeal (IX), vagus (X), accessory (XI), and hypoglossal (XII)
9. Sensory, somatic motor and proprioceptive, and parasympathetic

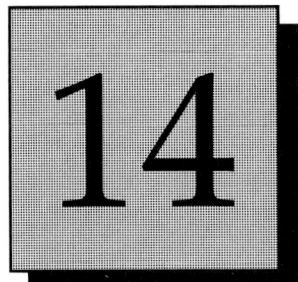

Integration of Nervous System Functions

CONTENT LEARNING ACTIVITY

Sensation

A. Match these terms with the correct statement or definition:

    Sensation (perception)   Special senses
    Somatic senses   Visceral senses

_____ 1. Conscious awareness of stimuli received by sensory receptors.

_____ 2. General senses that provide sensory information about the body and the environment—e.g., touch and temperature.

_____ 3. General senses that provide pain and pressure information from various internal organs.

_____ 4. Senses that are more specialized in structure and are localized in specific parts of the body—e.g., sight, smell, and hearing.

B. Using the terms provided, complete these statements:

    Action potentials   Sensory receptors
    Cerebral cortex   Translated

In order for sensation to occur, stimuli either inside or outside the body must be detected by _(1)_ and converted into _(2)_, which are propagated to the CNS by nerves. Within the CNS, tracts convey the action potentials to the _(3)_ and to other areas of the CNS. Action potentials reaching the cerebral cortex must be _(4)_ so that the person becomes aware of the stimulus.

1. _____
2. _____
3. _____
4. _____

Types of Sensory Receptors

A. Match these terms with the correct statement or definition:

    Chemoreceptors   Photoreceptors
    Mechanoreceptors   Thermoreceptors
    Nociceptors (pain receptors)

_____ 1. Sensory receptors that respond to compression, bending, or stretching.

_____ 2. Sensory receptors that respond to chemicals that become attached to receptors on their membranes.

_____ 3. Sensory receptors that respond to temperature.

_____ 4. Sensory receptors that respond to light striking the receptor cells.

_____ 5. Receptors that respond to painful mechanical, chemical, or thermal stimuli.

B. Match these terms with the correct statement or definition:

    Exteroreceptors   Visceroreceptors
    Proprioceptors

_____ 1. Sensory nerve endings associated with the skin; cutaneous receptors.

_____ 2. Sensory nerve endings associated with the viscera or organs.

_____ 3. Sensory nerve endings associated with joints, tendons, and other connective tissue.

_____ 4. Sensory nerve endings that provide information about the internal environment of the body.

_____ 5. Sensory nerve endings that provide information about body position and movement.

C. Match these types of nerve ending with their correct function:

    Free nerve endings   Merkel's disks
    Golgi tendon apparatus   Muscle spindle
    Hair follicle receptors   Pacinian corpuscles
    Meissner corpuscles   Ruffini end organs

_____ 1. Simplest and most common nerve ending; responds to pain, warm, cold, itch, tickle, and movement.

_____ 2. Tactile disks; respond to light touch and superficial pressure.

_____ 3. Respond to very slight bending of the hair; involved in light touch.

_____ 4. Lamellated corpuscles; complex nerve endings resembling an onion; respond to deep cutaneous pressure and vibration.

_____ 5. Tactile corpuscles; distributed throughout the dermal papillae; involved in two-point discrimination (fine touch).

_____ 6. Located in the dermis, primarily in the fingers; respond to skin displacement in continuous touch or pressure.

_____ 7. Consist of specialized skeletal muscle fibers; important to the control and muscle tone of postural muscles.

_____ 8. Proprioception associated with tendon movement.

Responses of Sensory Receptors

Match these terms with the correct statement or definition:

Accommodation (adaptation) Receptor (generator) potential
Phasic   Secondary receptors
Primary receptors   Tonic

_____ 1. Graded potential produced by the interaction of a stimulus with a sensory receptor.

_____ 2. Sensory receptor cells that have axons that conduct action potentials in response to the receptor potential.

_____ 3. Sensory receptor cells that have no axons; receptor potentials produced in these cells cause the release of a neurotransmitter.

_____ 4. Decreased sensitivity to a continued stimulus.

_____ 5. Receptors that accommodate very slowly and generate action potentials as long as a stimulus is applied.

_____ 6. Receptors that accommodate rapidly and are most sensitive to changes in stimuli.

Sensory Tracts

A. Match these spinal pathways with the correct function:

Dorsal-column/medial-   Spinoolivary
  lemniscal system   Spinoreticular
Spinocerebellar   Spinothalamic
Spinomesencephalic   Trigeminothalamic

_____ 1. Three tracts that are part of the anterolateral system.

_____
_____

_____ 2. Tracts that are involved in the *conscious* perception of external stimuli.

_____ 3. Pain and temperature from the face and teeth.

Chapter 14

_____ 4. Two-point discrimination (fine touch), proprioception, pressure, and vibration.

_____ 5. Proprioception to the cerebellum (two).

_____

B. Match the tract with the location where it crosses over from one side of the body to the other:

Cerebellum   Spinal cord
Medulla oblongata   Uncrossed

_____ 1. Spinothalamic tract.

_____ 2. Dorsal-column/medial-lemniscal system.

_____ 3. Posterior spinocerebellar tract.

_____ 4. Spinoolivary (crosses twice).

_____

_____ 5. Spinomesencephalic.

C. Match the type of neuron with the correct location of the neuron cell body:

Primary neuron   Tertiary neuron
Secondary neuron

_____ 1. Dorsal root ganglion.

_____ 2. Spinal cord.

_____ 3. Medulla oblongata.

_____ 4. Thalamus.

D. Match the pathway with the correct description or definition:

Fasciculus cuneatus   Spinoolivary tract
Fasciculus gracilis   Spinotectal tract

_____ 1. Tract that is part of the dorsal-column/medial-lemniscal system; conveys sensation from below the midthoracic level.

_____ 2. Tract that is part of the dorsal-column/medial-lemniscal system; conveys sensation from above the midthorax.

_____ 3. Projects to the accessory olivary nucleus and cerebellum; associated primarily with proprioception relating to balance.

_____ 4. Projects to the superior colliculi; involved in reflexes that turn the head and eye toward a point of cutaneous stimulation.

Sensory Areas of the Cerebral Cortex

A. Match these terms with the correct statement or definition:

    Association areas   Primary somatic sensory cortex
    Primary sensory areas   Projection

_____ 1. Regions of the cerebral cortex where sensory sensations are perceived.

_____ 2. Area that receives pain, pressure, and temperature input; located in the postcentral gyrus.

_____ 3. The brain refers a cutaneous sensation to the superficial site at which the stimulus interacts with sensory receptors.

_____ 4. Cortical areas involved in the recognition of sensations; immediately adjacent to the primary sensory areas.

_____ 5. Present sensory input is compared with past sensory input in these areas.

B. Match these sensory areas with the correct location in the cerebral cortex:

    Olfactory cortex   Taste area
    Primary auditory cortex   Visual cortex

_____ 1. Located at the inferior end of the postcentral gyrus.

_____ 2. Located on the inferior surface of the frontal lobe.

_____ 3. Located in the superior part of the temporal lobe.

_____ 4. Located in the occipital lobe.

Control of Skeletal Muscles

Match the type of neuron with the correct location of the neuron cell body:

    Lower motor neuron
    Upper motor neuron

_____ 1. Neuron cell body in the cerebral cortex.

_____ 2. Neuron cell body in the anterior horns of spinal cord gray matter.

_____ 3. Neuron cell body in the cranial nerve nuclei.

Motor Areas of the Cerebral Cortex

Match these areas of the cerebral cortex with the correct description:

    Prefrontal area   Primary motor cortex
    Premotor area

_____ 1. Located in the precentral gyrus.

_____ 2. Organized topographically similar to the primary somatic sensory cortex.

_____ 3. Staging area where motor functions are organized before they are actually initiated.

_____ 4. Motivation to plan and initiate movements occur here.

Motor Tracts

A. Match these terms with the correct statement or definition:

    Direct (pyramidal) system
    Indirect (extrapyramidal) system

_____ 1. Involved in maintaining muscle tone and fine movements involved in dexterity.

_____ 2. Involved in less precise control of motor functions and with overall body coordination and cerebellar function.

_____ 3. Includes the corticospinal and corticobulbar tracts.

_____ 4. Includes the rubrospinal, vestibulospinal, and reticulospinal tracts.

B. Match the tract with the correct function:

    Corticobulbar tract   Rubrospinal tract
    Corticospinal tract   Vestibulospinal tract
    Reticulospinal tract

_____ 1. Primarily involved with the control of movements below the head.

_____ 2. Controls head and neck movement.

_____ 3. Cerebellar-like control of the distal part of the upper limbs.

_____ 4. Two tracts involved with the maintenance of upright posture.

_____

C. Match the tract with the location where it crosses over from one side to the other:

    Medulla oblongata
    Spinal cord

_____ 1. Lateral corticospinal tract.

_____ 2. Anterior corticospinal tract.

Modifying and Refining Motor Activities

Match these terms with the correct statement or definition:

Basal nuclei Spinocerebellum
Cerebrocerebellum Vestibulocerebellum

_____ 1. Located in the cerebrum; important in planning, organizing, and coordinating motor movements and posture.

_____ 2. Flocculonodular lobe of the cerebellum; receives input from the semicircular canals.

_____ 3. Vermis and medial portion of the lateral hemispheres of the cerebellum; fine motor coordination by use of its comparator function.

_____ 4. Lateral two-thirds of the lateral hemispheres of the cerebellum; helps plan and practice rapid, complex motor actions.

Brainstem Functions

Match true or false with these statements:

True
False

_____ 1. The major ascending and descending pathways project through the brainstem.

_____ 2. Cranial nerves II, V, VII, VIII, IX, and X have sensory nuclei in the brainstem.

_____ 3. The reticular activating system (RAS) is formed from collateral branches of the trigeminothalamic tract neurons that project to the reticular formation.

_____ 4. Visual and acoustic stimuli, as well as mental activities, stimulate the RAS to maintain alertness and attention.

_____ 5. Several important reflexes, including the gag reflex, are integrated by nuclei in the brainstem.

_____ 6. Critical functions, such as heart rate, blood pressure, respiration, sleep, swallowing, vomiting, coughing, and sneezing, are regulated by brainstem nuclei.

_____ 7. Direct pathways originate in the cerebral cortex and pass directly through the brainstem, whereas indirect pathways synapse with brainstem nuclei and send descending fibers to the spinal cord.

_____ 8. Fibers from the reticular formation are critical for controlling vital functions, such as respiratory movements and cardiac rhythms.

_____ 9. Nuclei in the reticular formation coordinate reflexes for visual tracking and pupil constriction.

Speech

Match these areas of the cerebral cortex with their correct role in speech production:

Broca's area Primary motor cortex
Premotor area Wernicke's area

_____ 1. Association area necessary for understanding and formulating coherent speech.

_____ 2. Receives input from Wernicke's area and initiates the movements necessary for speech; also known as the motor speech area.

_____ 3. Broca's area sends action potentials here.

_____ 4. Activation of this area is the last step in speech production.

Right and Left Cerebral Cortex

Match these terms with the correct statement or definition:

Corpus callosum Right cerebral hemisphere
Left cerebral hemisphere

_____ 1. Controls muscular activity in and receives sensory input from the right half of the body.

_____ 2. In most people, the analytical hemisphere involved in mathematics and speech.

_____ 3. In most people, the hemisphere involved in spatial perception, recognition of faces, and musical ability.

_____ 4. Largest commissure between the right and left hemispheres.

Brain Waves and Sleep

Match these brain waves with the activity that produces them:

Alpha waves Delta waves
Beta waves Theta waves

_____ 1. Quiet resting state with eyes closed.

_____ 2. Intense mental activity.

_____ 3. In children and frustrated adults.

_____ 4. Deep sleep, in infancy, or in severe brain disorders.

Memory

A. Match these terms with the correct description or definition:

Declarative (explicit) memory
Procedural (implicit) memory
Sensory memory
Short-term memory

_____ 1. Very short-term retention of sensory input received by the brain while something is scanned, evaluated, and acted upon.

_____ 2. Type of memory in which information is retained for a few seconds to a few minutes.

_____ 3. Type of long-term memory that involves the retention of facts, such as dates, names, and places.

_____ 4. Type of long-term memory that is accessed by the hippocampus and amygdaloid nucleus in the temporal lobe.

_____ 5. Type of long-term memory that involves the development of skills, such as riding a bicycle or playing the piano.

_____ 6. Type of long-term memory that is stored primarily in the cerebellum and the premotor area of the cerebrum.

B. Match these terms with the correct description or definition:

Calmodulin
Glutamate
Long-term potentiation
Memory engram

_____ 1. Long-term changes in neurons that facilitate future transmission of action potentials.

_____ 2. Production of this neurotransmitter is increased in the presynaptic neuron as part of long-term potentiation.

_____ 3. As part of long-term potentiation, this intracellular molecule is activated by Ca^{2+} and in turn activates cAMP, which stimulates protein synthesis.

_____ 4. Whole series of neurons and their pattern of activity; a memory trace.

Chapter 14

Limbic System

Using the terms provided, complete these statements:

Docility
Memory
Olfactory
Pheromones
Pleasure
Satisfaction center
Species
Survival

The limbic system influences emotions, the visceral response to emotions, motivation, mood, and the sensations of pain and (1) . This system is associated with basic (2) instincts, including the acquisition of food and water and reproduction. One of the major sources of sensory input into the limbic system is the (3) nerves. In animals such as dogs or cats, olfactory detection of (4) is important in reproduction. Pheromones are chemicals released by one animal that attract another animal of the same (5) , usually of the opposite sex. Apparently, the cingulate gyrus is a(n) (6) for the brain and is associated with the feeling of satisfaction after a meal or after sexual intercourse. The relationship of the hippocampus with the limbic system and with (7) is probably important to survival. Lesions in the limbic system can result in a voracious appetite, often inappropriate sexual activity, loss of memory formation, and (8) , including the loss of normal fear and anger response.

1. _____
2. _____
3. _____
4. _____
5. _____
6. _____
7. _____
8. _____

Effects of Aging on the Nervous System

Match these terms as they apply to conditions that change with aging:

Decrease(s)
Increase(s)
No change

_____   1. Sensory function and the number of sensory neurons.

_____   2. Balance, coordination, and the ability to detect length, position, and tension of muscles and tendons.

_____   3. Likelihood of high blood pressure, dehydration, constipation, or incontinence.

_____   4. Number of motor neurons and the muscle fibers they innervate.

_____   5. Number of neurotransmitters and receptors.

_____   6. Size and weight of the brain.

_____   7. Neurofibrillar tangles in the cell and amyloid plaques in synapses.

_____   8. Function of short-term memory, voluntary movement, conscious sensations, reflexes, and sleep.

_____   9. Long-term memory.

QUICK RECALL

1. Name five types of sensory receptors based on the type of stimulus to which they respond.

2. Complete this table for sensory nerve tracts:

SPINAL PATHWAY	FUNCTION
Spinothalamic system	
Dorsal-column/medial-lemniscal system	
Spinocerebellar system	

3. Distinguish between the direct (pyramidal) and indirect (extrapyramidal) systems.

4. List the three areas of the cerebral cortex, in order, that are required to plan, organize, and control most voluntary movements.

5. List the critical functions regulated by brainstem nuclei.

6. Name and list the function of the two speech areas in the cerebral cortex.

7. Compare the general functions of the right and left cerebral hemispheres.

Chapter 14

8. List the three major types of memory, and compare their retention time.

9. Name two types of long-term memory, and list the type of information retained in each.

ANSWERS TO CHAPTER 14

CONTENT LEARNING ACTIVITY

Sensation
- A. 1. Sensation (perception); 2. Somatic senses; 3. Visceral senses; 4. Special senses
- B. 1. Sensory receptors; 2. Action potentials; 3. Cerebral cortex; 4. Translated

Types of Sensory Receptors
- A. 1. Mechanoreceptors; 2. Chemoreceptors; 3. Thermoreceptors; 4. Photoreceptors; 5. Nociceptors (pain receptors)
- B. 1. Exteroceptors; 2. Visceroreceptors; 3. Proprioceptors; 4. Visceroreceptors; 5. Proprioceptors
- C. 1. Free nerve endings; 2. Merkel's disks; 3. Hair follicle receptors; 4. Pacinian corpuscles; 5. Meissner corpuscles; 6. Ruffini end organs; 7. Muscle spindle; 8. Golgi tendon organ

Responses of Sensory Receptors
1. Receptor (generator) potential; 2. Primary receptors; 3. Secondary receptors; 4. Accommodation (adaptation); 5. Tonic; 6. Phasic

Sensory Tracts
- A. 1. Spinothalamic, spinoreticular, and spinomesencephalic; 2. Spinothalamic and dorsal-column/medial-lemniscal system; 3. Trigeminothalamic; 4. Dorsal-column/medial-lemniscal system; 5. Spinocerebellar and spinoolivary
- B. 1. Spinal cord; 2. Medulla oblongata; 3. Uncrossed; 4. Spinal cord and cerebellum; 5. Spinal cord
- C. 1. Primary neuron; 2. Secondary neuron; 3. Secondary neuron; 4. Tertiary neuron
- D. 1. Fasciculus gracilis; 2. Fasciculus cuneatus; 3. Spinoolivary tract; 4. Spinotectal tract

Sensory Areas of the Cerebral Cortex
- A. 1. Primary sensory areas; 2. Primary somatic sensory cortex; 3. Projection; 4. Association areas; 5. Association areas
- B. 1. Taste area; 2. Olfactory cortex; 3. Primary auditory cortex; 4. Visual cortex

Control of Skeletal Muscles
1. Upper motor neuron; 2. Lower motor neuron; 3. Lower motor neuron

Motor Areas of the Cerebral Cortex
1. Primary motor cortex; 2. Primary motor cortex; 3. Premotor area; 4. Prefrontal area

Motor Tracts
- A. 1. Direct (pyramidal) system; 2. Indirect (extrapyramidal) system; 3. Direct (pyramidal) system; 4. Indirect (extrapyramidal) system
- B. 1. Corticospinal tract; 2. Corticobulbar tract; 3. Rubrospinal tract; 4. Vestibulospinal tract and reticulospinal tract
- C. 1. Medulla oblongata; 2. Spinal cord

Modifying and Refining Motor Activities
1. Basal nuclei; 2. Vestibulocerebellum; 3. Spinocerebellum; 4. Cerebrocerebellum

Brainstem Functions
1. True; 2. False; 3. True; 4. True; 5. True; 6. True; 7. True; 8. True; 9. True

Speech
1. Wernicke's area; 2. Broca's area; 3. Premotor area; 4. Primary motor cortex

Right and Left Cerebral Cortex
1. Left cerebral hemisphere; 2. Left cerebral hemisphere; 3. Right cerebral hemisphere; 4. Corpus callosum

Brain Waves and Sleep
1. Alpha waves; 2. Beta waves; 3. Theta waves; 4. Delta waves

Memory
- A. 1. Sensory memory; 2. Short-term memory; 3. Declarative (explicit) memory; 4. Declarative (explicit) memory; 5. Procedural (implicit) memory; 6. Procedural (implicit) memory
- B. 1. Long-term potentiation; 2. Glutamate; 3. Calmodulin; 4. Memory engram

Limbic System
1. Pleasure; 2. Survival; 3. Olfactory;
4. Pheromones; 5. Species; 6. Satisfaction center;
7. Memory; 8. Docility

Effects of Aging on the Nervous System
1. Decrease; 2. Decreases; 3. Increases;
4. Decreases; 5. Decrease; 6. Decrease;
7. Increase; 8. Decrease; 9. No change

QUICK RECALL

1. Mechanoreceptors, chemoreceptors, thermoreceptors, photoreceptors, and nociceptors, (pain receptors)
2. Spinothalamic system: ascending pathway carrying pain, temperature, light touch, pressure, tickle, and itch sensations; dorsal-column/medial-lemniscal system: ascending pathways carrying senses of two-point discrimination, proprioception, vibration, and pressure; spinocerebellar system: ascending pathways carrying unconscious proprioception
3. Direct (pyramidal) system: conscious control of skeletal muscles, especially concerned with fine movements and speed, increases muscle tone; indirect (extrapyramidal) system: conscious and unconscious control of skeletal muscle, involved with error-correcting function of the cerebellum, maintenance of posture and balance, control of large movements (e.g., in the trunk and limbs), and inhibition of muscle activity
4. Prefrontal area: motivation and foresight to plan movements; premotor area: organization of motor movements before they are initiated; primary motor cortex: initiation of action potentials that control voluntary movement of skeletal muscle
5. Heart rate, respiration, blood pressure, sleep, swallowing, vomiting, coughing, and sneezing
6. Wernicke's area: necessary for understanding and formulating coherent speech; Broca's area: initiates the complex series of movements necessary for speech
7. Left hemisphere: mathematics and speech; right hemisphere: three-dimensional or spatial perception, recognition of faces, and musical ability
8. Sensory memory: less than a second; short-term memory: a few seconds to a few minutes; long-term memory: a lifetime
9. Explicit (declarative) memory: names, dates, and places; implicit (procedural) memory: development of skills

15 The Special Senses

CONTENT LEARNING ACTIVITY

Olfaction

A. Match these terms with the correct statement or definition:

Basal cells
Olfactory epithelium
Olfactory hairs
Olfactory tracts
Olfactory vesicles

_____ 1. Specialized cells lining the olfactory recess.

_____ 2. Axons of olfactory neurons project through the cribriform plate to these structures.

_____ 3. Bulbous enlargements of the dendrites of olfactory neurons.

_____ 4. Have chemoreceptors that bind to odorants, resulting in action potential production.

_____ 5. Olfactory neurons lost from the olfactory epithelium are replaced by these cells.

B. Match these terms with the correct statement or definition:

Intermediate olfactory area
Lateral olfactory area
Medial olfactory area
Olfactory cortex
Olfactory bulb
Olfactory nerve
Olfactory tract

_____ 1. Formed by axons from the olfactory neurons.

_____ 2. Where the olfactory nerves synapse with mitral and tufted cells.

_____ 3. Structures that project from the olfactory bulbs to the cerebral cortex.

_____ 4. Where the olfactory tracts terminate.

_____ 5. Part of the olfactory cortex involved with the conscious perception of smell.

_____ 6. Part of the olfactory cortex responsible for visceral and emotional reactions to odors.

_____ 7. Part of the olfactory cortex that modifies input from the olfactory bulb.

Taste

A. Match these terms with the correct statement or definition:

Filiform papillae   Fungiform papillae
Foliate papillae   Vallate papillae

_____ 1. Largest but least numerous papillae; surrounded by a groove or valley.

_____ 2. Mushroom-shaped papillae; appear as small, red dots scattered irregularly over the tongue.

_____ 3. Leaf-shaped papillae; distributed over the sides of the tongue and containing the most sensitive taste buds.

_____ 4. Filament-shaped papillae; most numerous papillae, but with no taste buds; provide a rough surface on the tongue.

B. Match these terms with the correct statement or definition:

Gustatory hairs   Taste (gustatory) cells
Support cells   Taste (gustatory) pore
Taste buds

_____ 1. Oval structures, consisting of two types of cells, embedded in the epithelium of the tongue and mouth.

_____ 2. Forms the exterior supporting capsule of the taste bud.

_____ 3. About 50 of these cells are found internally in each taste bud.

_____ 4. Microvilli located on a gustatory cell.

_____ 5. Opening in a taste bud containing gustatory hairs.

C. Match these terms with the correct statement or definition:

Acid   Sweet
Bitter   Tastants
Salt   Umami (savory)

_____ 1. Substances dissolved in saliva that cause taste cells to depolarize.

_____ 2. Taste that results from the diffusion of Na^+ into taste cells.

_____ 3. Taste that results from H^+ diffusing into taste cells, binding to ligand-gated K^+ channels, or binding to ligand-gated channels that let positive ions into taste cells.

_____ 4. Three tastes that result from the activation of G protein mechanisms.

Chapter 15

D. Match these nerves with their function or description:

Facial nerve (VII)   Vagus nerve (X)
Glossopharyngeal nerve (IX)

_____ 1. Branch of this nerve (chorda tympani) crosses the tympanic membrane.

_____ 2. Carries taste from the anterior two-thirds of the tongue.

_____ 3. Carries taste from the posterior one-third of the tongue.

_____ 4. Carries a few fibers for taste sensation from the epiglottis.

Accessory Structures of the Visual System

A. Match these terms with the correct statement or definition:

Bulbar conjunctiva   Meibomian glands
Canthi   Palpebrae
Caruncle   Palpebral conjunctiva
Chalazion   Sty
Ciliary glands   Tarsal plate

_____ 1. Eyelids.

_____ 2. Angles where the eyelids join.

_____ 3. Reddish-pink mound in the medial canthus.

_____ 4. Dense connective tissue; maintains the shape of the eyelid.

_____ 5. Modified sweat glands that open into the eyelash follicles.

_____ 6. Inflamed ciliary gland.

_____ 7. Eyelid glands that produce sebum.

_____ 8. Infection or blockage of a meibomian gland; a meibomian cyst.

_____ 9. Mucous membrane covering the inner surface of the eyelids.

_____ 10. Mucous membrane covering the anterior surface of the eye.

B. Match these terms with the correct statement or definition:

Lacrimal apparatus Lacrimal gland
Lacrimal canaliculus Lacrimal sac
Lacrimal duct Nasolacrimal duct

_____ 1. Produces and conducts tears over the surface of the eye and into the nasal cavity.

_____ 2. Gland that produces tears.

_____ 3. Passageway through which tears exit the lacrimal gland.

_____ 4. Passageway in the medial corner of the eye into which excess tears flow; its opening is the punctum, which is located on the lacrimal papilla.

_____ 5. Structure located between the lacrimal canaliculus and the nasolacrimal duct.

_____ 6. Passageway that opens into the nasal cavity.

C. Match these terms with the correct parts of the diagram labeled in figure 15.1:

Lacrimal canaliculi
Lacrimal ducts
Lacrimal gland
Lacrimal sac
Nasolacrimal duct

Puncta

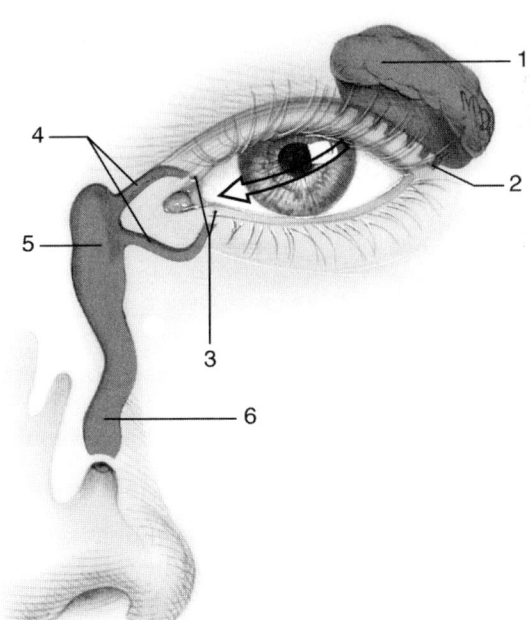

Figure 15.1

1. _____
2. _____
3. _____
4. _____
5. _____
6. _____

Chapter 15   225

D. Match these extrinsic eye muscles
with the correct description:

Oblique muscles
Rectus muscles

_____ 1. Extrinsic eye muscles that run more or less straight anteroposteriorly.

_____ 2. Extrinsic eye muscles that are placed at an angle to the eye.

_____ 3. There are four of these muscles.

Anatomy of the Eye

A. Match these terms with the
correct statement or definition:

Cornea   Sclera
Fibrous layer   Vascular layer
Nervous layer

_____ 1. Outer layer of the eye, consisting of the sclera and cornea.

_____ 2. Middle eye layer; the choroid, ciliary body, and iris.

_____ 3. Inner layer of the eye, consisting of the retina.

_____ 4. Firm, opaque, white, outer layer of the posterior five-sixths of the eye.

_____ 5. Avascular, transparent structure; permits light entry.

B. Match these terms with the
correct statement or definition:

Choroid   Dilator pupillae
Ciliary body   Iris
Ciliary processes Pupil
Ciliary ring   Sphincter pupillae

_____ 1. Vascular layer associated with the scleral portion of the eye.

_____ 2. Continuous with the choroid; the iris attaches to its lateral margin.

_____ 3. Part of the ciliary body that contains smooth muscles (intrinsic eye muscles); attaches to the lens by the suspensory ligaments and changes the shape of the lens.

_____ 4. Capillaries and epithelium that produce aqueous humor.

_____ 5. Smooth muscle structure surrounding the pupil.

_____ 6. Radial group of iris muscles; increases the size of the pupil.

C. Match these terms with the correct statement or definition:

Fovea centralis   Optic disc
Macula   Pigmented layer
Nervous layer   Rods and cones

_____ 1. Outer part of the retina, consisting of simple cuboidal epithelium.

_____ 2. Photoreceptor cells in the retina.

_____ 3. Small, yellow spot near the center of the posterior retina.

_____ 4. Small pit in the retina with the greatest visual acuity.

_____ 5. Blind spot containing no photoreceptor cells; where blood vessels enter and nerve processes exit the eye.

D. Match these terms with the correct statement or definition:

Anterior chamber   Scleral venous sinus
Aqueous humor   Vitreous chamber
Posterior chamber   Vitreous humor

_____ 1. Chamber between the cornea and the iris.

_____ 2. Chamber between the iris and lens.

_____ 3. Fluid that fills the anterior and posterior chambers of the eye; maintains intraocular pressure.

_____ 4. Venous ring that returns aqueous humor to the circulatory system; also called the canal of Schlemm.

_____ 5. Transparent, jellylike substance that fills the vitreous chamber.

Chapter 15

E. Match these terms with the correct parts of the diagram labeled in figure 15.2:

Anterior chamber
Choroid
Ciliary body
Conjunctiva
Cornea
Iris
Lens
Optic nerve
Posterior chamber
Pupil
Retina
Sclera
Suspensory ligaments
Vitreous humor

1. _____
2. _____
3. _____
4. _____
5. _____
6. _____
7. _____
8. _____
9. _____
10. _____
11. _____
12. _____
13. _____
14. _____

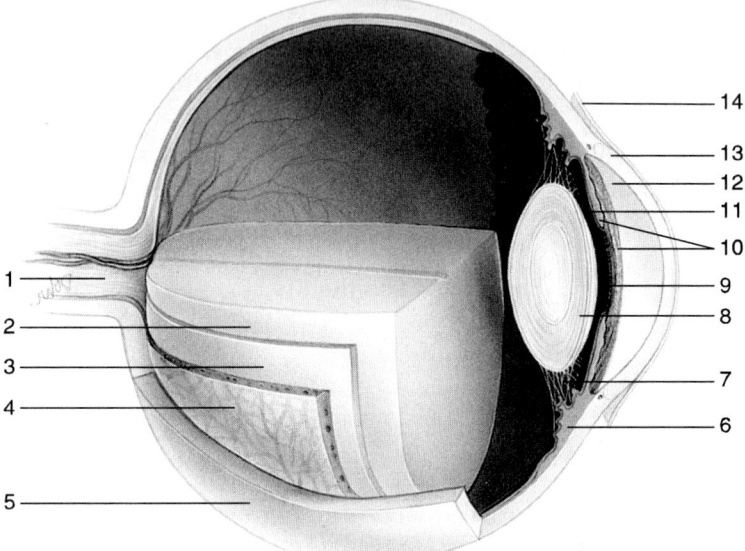

Figure 15.2

F. Using the terms provided, complete these statements:

Capsule
Crystallines
Cuboidal epithelial
Lens fibers
Suspensory ligaments

The lens consists of a layer of __(1)__ cells on its anterior surface and a posterior portion of very long columnar epithelial cells called __(2)__. Lens fibers lose their nuclei and other cellular organelles and accumulate a special set of proteins called __(3)__. The lens is covered by a highly elastic, transparent __(4)__. The lens is suspended between the two eye compartments by the __(5)__, which are connected to the lens capsule and to the ciliary body.

1. _____
2. _____
3. _____
4. _____
5. _____

Functions of the Complete Eye

A. Match these terms with the correct statement or definition:

    Concave   Reflection
    Convex   Refraction
    Focal point   Visible light
    Focusing

_____ 1. Portion of the electromagnetic spectrum that can be detected by the human eye.

_____ 2. Bending of light rays as they pass into a new medium, such as light passing from air into water.

_____ 3. Type of lens surface that causes light rays to converge.

_____ 4. Type of lens surface that causes light rays to diverge.

_____ 5. Point where convergent light rays cross.

_____ 6. Act of causing light rays to converge.

_____ 7. Light rays that bounce off a nontransparent object.

B. Match these terms with the correct statement or definition:

    Accommodation   Emmetropia
    Ciliary muscles   Far point of vision
    Convergence   Near point of vision
    Depth of focus   Suspensory ligaments

_____ 1. Structures that maintain elastic pressure on the lens, keeping it relatively flat.

_____ 2. Normal resting, flattened condition of the lens.

_____ 3. Structures that contract and reduce the tension on the lens, allowing it to assume a more spherical shape.

_____ 4. Point beyond which accommodation is not required; usually 20 feet or more from the eye.

_____ 5. Process of allowing the lens to assume a more spherical (convex) shape; results in greater refraction of light.

_____ 6. Point at which an object is close to the eye that accommodation is no longer possible.

_____ 7. Greatest distance through which an object can be moved and still remain in focus.

_____ 8. Medial movement of the eyeballs that keeps objects in view as they move closer to the eye.

Structure and Function of the Retina

A. Match these parts of the retina with the correct description:

    Neural layer
    Pigmented layer

1. This portion of the retina has three layers of neurons.

2. This portion of the retina is a layer of cells filled with melanin; enhances visual acuity by isolating individual photoreceptors in a black matrix.

B. Match these terms with the correct statement or definition:

    11-*cis*   Rhodopsin
    All-*trans*   Rods
    False   Transducin
    Retinal   True

1. Bipolar photoreceptor cells that cannot detect color; most important for vision under conditions of reduced light.

2. Molecule containing opsin in loose chemical combination with retinal; located in rod cells.

3. Pigment molecule derived from vitamin A.

4. G protein in rods.

5. The configuration of retinal in inactive rhodopsin.

6. True or false: Light causes the shape of opsin and retinol to change.

7. Configuration of retinal in activated rhodopsin.

8. True or false: Activated rhodopsin activates transducin.

9. True or false: All-*trans* retinol is released from opsin.

10. True or false: Energy from ATP converts all-*trans* retinol to 11-*cis* retinol.

11. True or false: 11-*cis* retinol attaches to opsin to form inactive rhodopsin.

C. Using the terms provided, complete these statements:

Activated
Close
Depolarize
Glutamate
Hyperpolarize
Inactivated
Into
Less
More
Open
Out of
Transducin

In the dark, rhodopsin is (1), transducin is (2), gated Na⁺ channels are (3), and Na⁺ move (4) the rods. As a result, the rods release (5), which binds to bipolar cells, causing them to (6). Thus, no action potentials are propagated toward the CNS. In the light, rhodopsin is activated, which activates transducin, causing gated Na⁺ channels to (7) and block Na⁺ movement into the rods, which causes them to (8). As a result, the rods release (9) glutamate and bipolar cells are less inhibited. The bipolar cells release neurotransmitters, causing the production of action potentials in ganglionic cells, and action potentials are propagated toward the CNS.

1. _____
2. _____
3. _____
4. _____
5. _____
6. _____
7. _____
8. _____
9. _____

D. Match the type of adaptation to changing light conditions with the correct statement:

Dark adaptation
Light adaptation

_____ 1. Amount of rhodopsin increases.

_____ 2. Pupil constricts.

_____ 3. Rod function increases; cone function decreases.

_____ 4. More rapid of the two processes.

E. Match these terms with the correct statement or definition:

Cones
Fovea centralis
Iodopsin

_____ 1. Bipolar photoreceptor cells that are sensitive to blue, red, or green light; most important in visual acuity.

_____ 2. Visual pigment in cones.

_____ 3. Part of the retina that has only cone cells.

Chapter 15

F. Match these terms with the correct statement or definition:

Bipolar cells
Ganglion cells
Interneurons neurons
Optic nerve
Photoreceptor cells

_____ 1. Neurons in the outermost nuclear layer of the neural layer of the retina; next to the pigmented layer of the retina.

_____ 2. Neurons in the middle nuclear layer of the neural layer of the retina.

_____ 3. Innermost nuclear layer of neurons in the neural layer of the retina; closest to the vitreous humor.

_____ 4. Collection of all the ganglion cells as they exit the eye.

_____ 5. Neurons that modify the signal from photoreceptor cells before they leave the retina.

G. Using the terms provided, complete these statements:

Bipolar
Decreases
Ganglion
Increases
Spatial summation

1. _____
2. _____
3. _____
4. _____
5. _____

In the retina, numerous rods usually synapse with one _(1)_ cell. Further, many of these cells then synapse with one _(2)_ cell. This arrangement results in _(3)_ that _(4)_ light sensitivity and _(5)_ visual acuity. In comparison, cones exhibit little convergence. This decreases light sensitivity and increases visual acuity.

Neuronal Pathways for Vision

A. Using the terms provided, complete these statements:

Optic chiasm
Optic nerve
Optic radiations
Optic tract
Superior colliculi
Thalamus
Visual cortex

1. _____
2. _____
3. _____
4. _____
5. _____
6. _____
7. _____

Ganglion cells from the sensory retina converge and exit the eye in the _(1)_, then pass into the cranial cavity through the optic canal. Just inside the cranial cavity, the optic nerves are connected at the _(2)_, where some of the axons cross to the opposite side of the brain. Beyond the optic chiasm, the route of the ganglionic axons is called the _(3)_. Most of the optic tract axons end in the _(4)_, although some separate from the optic tract and terminate in the _(5)_, the center for reflexes initiated by visual stimuli. Neurons of the thalamus form the _(6)_, projecting to the _(7)_ in the occipital lobe.

B. Match these terms with the correct statement or definition:

Binocular vision Opposite side
Depth perception Same side
Nasal retina Temporal retina

_____ 1. Image from the temporal part of a visual field falls on this part of the retina.

_____ 2. Image from this part of the retina crosses to the opposite side of the brain.

_____ 3. Side of the brain to which images from the temporal visual field projects.

_____ 4. The image from the nasal part of a visual field falls on this part of the retina.

_____ 5. Area of overlap of the visual fields.

_____ 6. Ability to distinguish between near and far objects.

Clinical Applications

A. Match these terms with the correct statement or definition:

Astigmatism Myopia
Hyperopia Presbyopia

_____ 1. Occurs when the lens and cornea are optically too strong or when the eyeball is too long.

_____ 2. The eye becomes less able to accommodate; a result of aging.

_____ 3. Occurs when the cornea or lens is not uniformly curved.

_____ 4. Corrected by a concave lens.

_____ 5. Corrected by reading glasses or bifocals.

_____ 6. Corrected by lenses with curvature opposite to the defect.

B. Match these terms with the correct statement or definition:

Cataract Glaucoma
Color blindness Strabismus
Diabetes

_____ 1. Lack of parallelism of light paths through the eye.

_____ 2. Disease resulting from increased intraocular pressure.

_____ 3. Recessive X-linked trait producing a deficiency of cone pigments.

_____ 4. Clouding of the lens resulting from a buildup of proteins.

_____ 5. Results in defective blood circulation to the eye; a leading cause of blindness.

Auditory Structures and Their Functions

A. Match these terms with the correct description or definition:
   Auricle (pinna)   External acoustic meatus
   Cerumen   Tympanic membrane

_____ 1. Fleshy part of the external ear on the outside of the head.

_____ 2. Passageway from the outside to the eardrum.

_____ 3. Modified sebum, commonly called earwax, that helps prevent foreign objects from reaching the tympanic membrane.

_____ 4. Thin, semitransparent, nearly oval, three-layered membrane that vibrates in response to sound waves.

B. Match these terms with the correct parts of the diagram labeled in figure 15.3:

Auditory ossicles
Auditory tube
Auricle (pinna)
Cochlea
External acoustic meatus
External ear
Incus
Inner ear
Malleus
Oval window
Round window
Semicircular canals
Stapes
Tympanic membrane
Vestibule

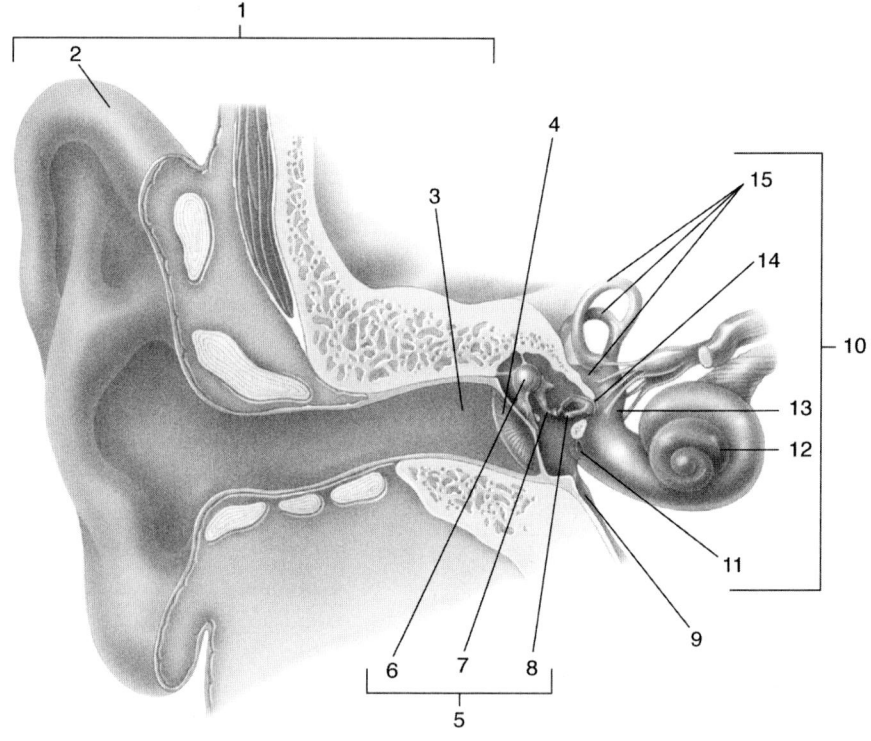

Figure 15.3

1. _____   6. _____   11. _____
2. _____   7. _____   12. _____
3. _____   8. _____   13. _____
4. _____   9. _____   14. _____
5. _____   10. _____   15. _____

C. Match these terms with the correct description or definition:

Auditory tube
Incus
Malleus
Mastoid air cells
Oval window
Round window
Stapes

_____ 1. Spaces in the mastoid processes of the temporal bones.

_____ 2. Structure that allows air pressure to equalize between the middle ear and the outside air.

_____ 3. Auditory ossicle attached to the tympanic membrane.

_____ 4. Middle auditory ossicle.

_____ 5. Opening in which the foot plate of the stapes is held by the annular ligament.

_____ 6. Membrane-covered opening on the medial side of the middle ear with nothing attached.

D. Match these terms with the correct statement or definition:

Basilar membrane
Bony labyrinth
Cochlea
Cochlear duct
Helicotrema
Membranous labyrinth
Modiolus
Scala tympani
Scala vestibuli
Vestibular membrane

_____ 1. Interconnecting bony tunnels and chambers in the petrous portion of the temporal bone; subdivided into the cochlea, vestibule, and semicircular canals.

_____ 2. Part of the inner ear that can be divided into scala tympani, scala vestibuli, and cochlear duct.

_____ 3. Cochlear chamber that contains perilymph and extends from the oval window to the helicotrema.

_____ 4. Cochlear chamber that contains perilymph and extends from the helicotrema to the round window.

_____ 5. Opening between the scala vestibuli and the scala tympani.

_____ 6. Wall of the membranous labyrinth bordering the scala vestibuli.

_____ 7. Wall of the membranous labyrinth bordering the scala tympani.

_____ 8. Interior of the membranous labyrinth containing endolymph; space between the vestibular and basilar membranes.

_____ 9. Bony core of the cochlea, which has a projection called the spiral lamina.

_____ 10. Attached to the spiral lamina and the spiral ligament.

Chapter 15

E. Using the terms provided, complete these statements:

Cochlear ganglion
Cochlear nerve
Gating spring
Hair bundles
Hair cells
Spiral organ
Tectorial membrane
Tip links
Vestibulocochlear nerve

The cells inside the cochlear duct are highly modified to form a structure called the __(1)__, or organ of Corti. This structure contains supporting epithelial cells and specialized sensory cells called __(2)__. These cells have stereocilia organized to form a __(3)__. The ends of the stereocilia are connected by __(4)__. Each tip link is a __(5)__ connected to the gate of a gated K⁺ channel. When a stereocilia bends, the gating protein pulls the K⁺ gate open. The apical ends of the longest stereocilia are embedded within an acellular, gelatinous shelf called the __(6)__, which is attached to the spiral lamina. Hair cells have no axons, but the basilar region of each hair cell is covered by synaptic terminals of sensory neurons, the cell bodies of which are grouped into a __(7)__ within the cochlear modiolus. The proximal, afferent fibers of these neurons join to form the __(8)__, which then joins the vestibular nerve to become the __(9)__.

1. _____
2. _____
3. _____
4. _____
5. _____
6. _____
7. _____
8. _____
9. _____

F. Using the terms provided, complete these statements:

Amplitude
Attenuation reflex
Auditory ossicles
Basilar membrane
Endolymph
Frequency
Round window
Scala vestibuli
Tectorial membrane
Timbre
Tympanic membrane

Vibrations are propagated through the air as sound waves. Volume (loudness) of sound is a function of the __(1)__ (height) of the waves, and pitch is a function of __(2)__ (how far the waves are apart). The resonance quality, or overtones, of a sound is __(3)__. Sound waves are collected by the auricle and are conducted through the external acoustic meatus toward the __(4)__. Sound waves strike the tympanic membrane and cause it to vibrate, which in turn causes the __(5)__ to vibrate. Two small skeletal muscles attached to the ossicles reflexively dampen excessively loud sounds. The contraction of these muscles, called the __(6)__, protects the delicate ear structures from being damaged by loud noises. The vibration of the auditory ossicles is transferred to the oval window and causes vibration in the perilymph of the __(7)__. This produces waves in the perilymph, which causes the vestibular membrane and __(8)__ of the cochlear duct to vibrate. Consequently, the __(9)__ vibrates, causing the hairs embedded in the __(10)__ to bend, inducing action potentials in cochlear neurons. Vibration of the basilar membrane also causes vibration of the perilymph within the scala tympani. These vibrations are dissipated by vibration of the membrane of the __(11)__.

1. _____
2. _____
3. _____
4. _____
5. _____
6. _____
7. _____
8. _____
9. _____
10. _____
11. _____

G. Match these terms with the correct parts of the diagram labeled in figure 15.4:

Basilar membrane
Cochlear duct
Helicotrema
Membranous labyrinth
Oval window
Round window
Scala tympani
Scala vestibuli
Spiral organ
Tectorial membrane
Vestibular membrane

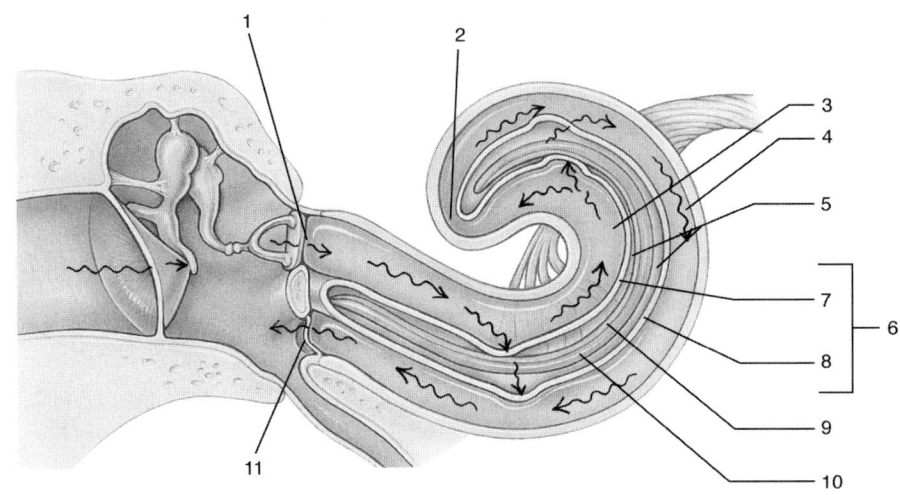

Figure 15.4

1. _____
2. _____
3. _____
4. _____
5. _____
6. _____
7. _____
8. _____
9. _____
10. _____
11. _____

Neuronal Pathways for Hearing

Using the terms provided, complete these statements:

Auditory cortex
Inferior colliculi
Medulla oblongata
Superior colliculus
Superior olivary nucleus
Thalamus

The neurons from the cochlear ganglion synapse with the central nervous system neurons in the dorsal or ventral cochlear nucleus in the (1) . These neurons, in turn, either synapse in or pass through the (2) . Ascending neurons from this point travel in the lateral lemniscus. All ascending fibers synapse in the (3) , and neurons from there project to the medial geniculate nucleus of the (4) , where they synapse with neurons that terminate in the (5) in the dorsal portion of the temporal lobe. Neurons from the inferior colliculus also project to the (6) , where reflexes that turn the head and eyes in response to a loud sound are initiated.

1. _____
2. _____
3. _____
4. _____
5. _____
6. _____

Chapter 15   237

Balance

A. Match these terms with the correct statement or definition:

Ampulla Macula
Crista ampullaris Otoliths
Cupula Semicircular canals
Kinetic labyrinth Static labyrinth

_____ 1. Consists of the utricle and saccule; involved in evaluating the position of the head relative to gravity or linear acceleration.

_____ 2. Specialized epithelium in the utricle and saccule.

_____ 3. Add weight to the gelatinous mass that embeds hair cells.

_____ 4. Arranged in three planes; enable a person to sense movement in all directions.

_____ 5. Specialized sensory epithelium located in each ampulla.

_____ 6. Curved, gelatinous mass suspended over the crista ampullaris.

B. Using the terms provided, complete these statements:

Endolymph Opposite
Gelatinous mass Same
Hair cells

When the position of the head relative to the ground changes, the (1) within a macula moves, causing the stimulation of (2). As a result, action potentials are produced that are interpreted by the brain as a change in head position. As the head begins to move in a given direction, (3) in the semicircular canals does not move at the same rate as the surrounding bone. Consequently, the cupula moves in the (4) direction as the endolymph, resulting in the stimulation of (5) in the crista ampullaris. As a result, action potentials are produced that are interpreted by the brain as head movement.

1. _____
2. _____
3. _____
4. _____
5. _____

Neuronal Pathways for Balance

Using the terms provided, complete these statements:

Nystagmus Vestibular ganglion
Proprioceptive neurons Vestibular nucleus

Neurons synapsing on the hair cells of the maculae and cristae ampullares converge into the (1), where their cell bodies are located. Afferent fibers from these neurons terminate in the (2) within the medulla. The vestibular nuclear complex also receives input from (3) throughout the body. Reflex pathways exist between the kinetic portion of the vestibular system and the nuclei controlling the extrinsic eye muscles. Spinning the head causes the eyes to track slowly in the direction of motion and return with a rapid recovery movement. This is called (4).

1. _____
2. _____
3. _____
4. _____

Clinical Applications

Match these terms with the correct statement or definition:

Otitis media
Otosclerosis
Presbyacusis
Tinnitus

_____ 1. Infection of the middle ear.

_____ 2. Caused by spongy bone growth immobilizing the stapes.

_____ 3. Disorder characterized by noises in the ears.

_____ 4. Age-related hearing loss.

Effects of Aging on the Special Senses

Match these terms with the correct statement or definition:

Cones
Hair cells

_____ 1. An age-related decrease in the number of these can cause decreased visual acuity and decreased color perception.

_____ 2. An age-related decrease in the number of these can cause decreased color perception.

QUICK RECALL

1. Name the three functional areas of the olfactory cortex and the function of each.

2. List the five basic tastes detected by the taste buds.

3. List the three layers of the eye.

4. Name the two types of photoreceptor cells in the retina, and list two important differences between them.

5. List the three chambers of the eye and the substances that fill each.

6. List the three nuclear layers of neurons in the neural layer of the retina.

7. List the visual pigments located in rod and cone cells.

8. List the three types of cone cells.

9. Name the three auditory ossicles in the middle ear.

10. List the three subdivisions of the bony labyrinth, and give their function.

11. List the three cochlear chambers and the fluid found in each.

12. List the two functional parts of the organs of balance.

13. Name the structures that relieve pressure in the middle ear and the inner ear.

ANSWERS TO CHAPTER 15

CONTENT LEARNING ACTIVITY

Olfaction
A. 1. Olfactory epithelium; 2. Olfactory bulbs; 3. Olfactory vesicles; 4. Olfactory hairs; 5. Basal cells
B. 1. Olfactory nerves; 2. Olfactory bulb; 3. Olfactory tracts; 4. Olfactory cortex; 5. Lateral olfactory area; 6. Medial olfactory area; 7. Intermediate olfactory area

Taste
A. 1. Vallate papillae; 2. Fungiform papillae; 3. Foliate papillae; 4. Filiform papillae
B. 1. Taste buds; 2. Support cells; 3. Taste (gustatory) cells; 4. Gustatory hairs; 5. Taste (gustatory) pores
C. 1. Tastants; 2. Salt; 3. Acid; 4. Sweet; 5. Bitter; 6. Umami (savory)
D. 1. Facial nerve (VII); 2. Facial nerve (VII); 3. Glossopharyngeal nerve (IX); 4. Vagus nerve (X)

Accessory Structures of the Visual System
A. 1. Palpebrae; 2. Canthi; 3. Caruncle; 4. Tarsal plate; 5. Ciliary glands; 6. Sty; 7. Meibomian glands; 8. Chalazion; 9. Palpebral conjunctiva; 10. Bulbar conjunctiva
B. 1. Lacrimal apparatus; 2. Lacrimal gland; 3. Lacrimal duct; 4. Lacrimal canaliculus; 5. Lacrimal sac; 6. Nasolacrimal duct
C. 1. Lacrimal gland; 2. Lacrimal ducts; 3. Puncta; 4. Lacrimal canaliculi; 5. Lacrimal sac; 6. Nasolacrimal duct
D. 1. Rectus muscles; 2. Oblique muscles; 3. Rectus muscles

Anatomy of the Eye
A. 1. Fibrous layer; 2. Vascular layer; 3. Nervous layer; 4. Sclera; 5. Cornea
B. 1. Choroid; 2. Ciliary body; 3. Ciliary ring; 4. Ciliary processes; 5. Iris; 6. Dilator pupillae
C. 1. Pigmented layer; 2. Rods and cones; 3. Macula; 4. Fovea centralis; 5. Optic disc
D. 1. Anterior chamber; 2. Posterior chamber; 3. Aqueous humor; 4. Scleral venous sinus; 5. Vitreous humor
E. 1. Optic nerve; 2. Vitreous humor; 3. Retina; 4. Choroid; 5. Sclera; 6. Ciliary body; 7. Suspensory ligaments; 8. Lens; 9. Pupil; 10. Iris; 11. Posterior chamber; 12. Anterior chamber; 13. Cornea; 14. Conjunctiva
F. 1. Cuboidal epithelial; 2. Lens fibers; 3. Crystallines; 4. Capsule; 5. Suspensory ligaments

Functions of the Complete Eye
A. 1. Visible light; 2. Refraction; 3. Convex; 4. Concave; 5. Focal point; 6. Focusing; 7. Reflection
B. 1. Suspensory ligaments; 2. Emmetropia; 3. Ciliary muscles; 4. Far point of vision; 5. Accommodation; 6. Near point of vision; 7. Depth of focus; 8. Convergence

Structure and Function of the Retina
A. 1. Neural layer; 2. Pigmented layer
B. 1. Rods; 2. Rhodopsin; 3. Retinal; 4. Transducin; 5. 11-*cis*; 6. True; 7. All-*trans*; 8. True; 9. True; 10. True; 11. True
C. 1. Inactivated; 2. Inactivated; 3. Open; 4. Into; 5. Glutamate; 6. Hyperpolarize; 7. Close; 8. Hyperpolarize; 9. Less
D. 1. Dark adaptation; 2. Light adaptation; 3. Dark adaptation; 4. Light adaptation
E. 1. Cones; 2. Iodopsin; 3. Fovea centralis
F. 1. Photoreceptor cells; 2. Bipolar cells; 3. Ganglion cells; 4. Optic nerve; 5. Interneurons neurons
F. 1. Bipolar; 2. Ganglion; 3. Spatial summation; 4. Increases; 5. Decreases

Neuronal Pathways for Vision
A. 1. Optic nerve; 2. Optic chiasm; 3. Optic tract; 4. Thalamus; 5. Superior colliculi; 6. Optic radiations; 7. Visual cortex
B. 1. Nasal retina; 2. Nasal retina; 3. Opposite side; 4. Temporal retina; 5. Binocular vision; 6. Depth perception

Clinical Applications
A. 1. Myopia; 2. Presbyopia; 3. Astigmatism; 4. Myopia; 5. Presbyopia; 6. Astigmatism
B. 1. Strabismus; 2. Glaucoma; 3. Color blindness; 4. Cataract; 5. Diabetes

Auditory Structures and Their Functions
A. 1. Auricle (pinna); 2. External acoustic meatus; 3. Cerumen; 4. Tympanic membrane
B. 1. External ear; 2. Auricle (pinna); 3. External acoustic meatus; 4. Tympanic membrane; 5. Auditory ossicles; 6. Malleus; 7. Incus 8. Stapes; 9. Auditory tube; 10. Inner ear; 11. Round window; 12. Cochlea; 13. Vestibule; 14. Oval window; 15. Semicircular canals
C. 1. Mastoid air cells; 2. Auditory tube; 3. Malleus; 4. Incus; 5. Oval window; 6. Round window
D. 1. Bony labyrinth; 2. Cochlea; 3. Scala vestibuli; 4. Scala tympani; 5. Helicotrema; 6. Vestibular membrane; 7. Basilar membrane; 8. Cochlear duct; 9. Modiolus; 10. Basilar membrane
E. 1. Spiral organ; 2. Hair cells; 3. Hair bundle; 4. Tip links; 5. Gating spring; 6. Tectorial membrane; 7. Cochlear ganglion; 8. Cochlear nerve; 9. Vestibulocochlear nerve
F. 1. Amplitude; 2. Frequency; 3. Timbre; 4. Tympanic membrane; 5. Auditory ossicles; 6. Attenuation reflex; 7. Scala vestibuli; 8. Endolymph; 9. Basilar membrane; 10. Tectorial membrane; 11. Round window
G. 1. Oval window; 2. Helicotrema; 3. Scala vestibuli; 4. Scala tympani; 5. Cochlear duct; 6. Membranous labyrinth; 7. Vestibular membrane; 8. Basilar membrane; 9. Tectorial membrane; 10. Spiral organ; 11. Round window

Neuronal Pathways for Hearing
1. Medulla oblongata; 2. Superior olivary nucleus; 3. Inferior colliculi; 4. Thalamus; 5. Auditory cortex; 6. Superior colliculus

Balance
A. 1. Static labyrinth; 2. Macula; 3. Otoliths; 4. Semicircular canals; 5. Crista ampullaris; 6. Cupula
B. 1. Gelatinous mass; 2. Hair cells; 3. Endolymph; 4. Opposite; 5. Hair cells

Neuronal Pathways for Balance
1. Vestibular ganglion; 2. Vestibular nucleus; 3. Proprioceptive neurons; 4. Nystagmus

Clinical Applications
1. Otitis media; 2. Otosclerosis; 3. Tinnitus; 4. Presbyacusis

Effects of Aging on the Special Senses
1. Cones; 2. Hair cells

Quick Recall

1. Lateral olfactory area: conscious perception of smell; intermediate olfactory area: modulating smell: medial olfactory area: visceral and emotional responses to smell
2. Sour, salty, bitter, sweet, and umami
3. Fibrous, vascular, and nervous layers
4. Rods: high sensitivity to light, but lower visual acuity; shades of gray; cones: lower sensitivity to light, but greater visual acuity; color vision
5. Anterior and posterior chambers filled with aqueous humor and vitreous chamber filled with vitreous humor
6. Photoreceptors (rods and cones), bipolar cells, and ganglion cells
7. Rods: rhodopsin; cones: iodopsin
8. Red-sensitive, green-sensitive, and blue-sensitive cone cells
9. Malleus, incus, and stapes
10. Vestibule, cochlea, and semicircular canals. The vestibule and semicircular canals are involved primarily in balance, and the cochlea is involved in hearing.
11. Scala vestibuli: perilymph; scala tympani: perilymph; cochlear duct: endolymph
12. Static labyrinth and kinetic labyrinth
13. Middle ear: auditory tube; inner ear: round window

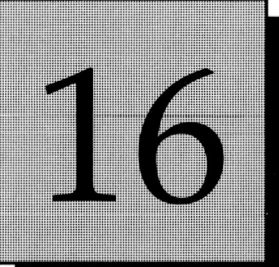

Autonomic Nervous System

CONTENT LEARNING ACTIVITY

Contrasting the Somatic and Autonomic Nervous Systems

A. Match the type of neuron with the correct description:

Autonomic motor neurons   Sensory neurons
Motor neurons   Somatic motor neurons

_____ 1. Neurons that carry action potentials from the periphery to the CNS.

_____ 2. Extend from the CNS to skeletal muscle.

_____ 3. Two neurons in series between the CNS and innervated organs.

_____ 4. Effect on target tissues is usually consciously controlled.

_____ 5. Effect on target tissues can be excitatory or inhibitory.

B. Match these terms from the autonomic division with the correct statement or definition:

Autonomic ganglion   Postganglionic neuron
Effector organ   Preganglionic neuron

_____ 1. First ANS neuron between CNS and the organs innervated.

_____ 2. Cell bodies are located in the brainstem or spinal cord.

_____ 3. Second ANS neuron between CNS and the organs innervated.

_____ 4. Contains the cell bodies of the postganglionic neurons.

_____ 5. Postganglionic neuron extends to and synapses with this; smooth muscle, cardiac muscle, or glands.

Chapter 16

C. Match the terms from part B with the parts labeled in figure 16.1:

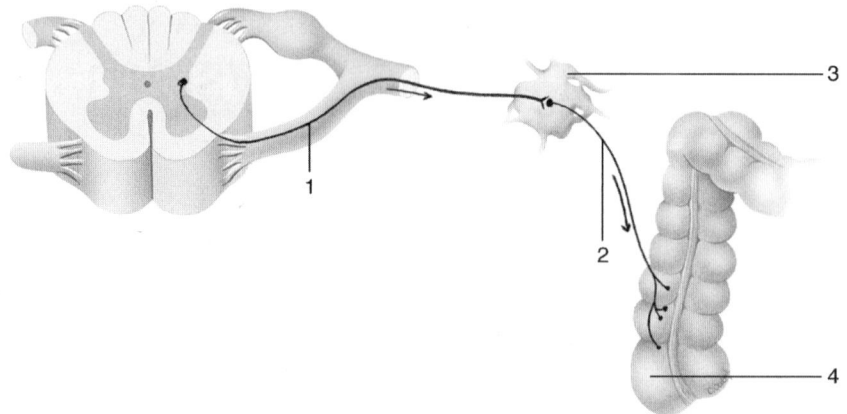

Figure 16.1

1. _____   3. _____

2. _____   4. _____

Sympathetic Division

A. Match these terms with the correct parts labeled in figure 16.2:

Collateral (prevertebral) ganglion
Gray ramus communicans
Splanchnic nerve
Sympathetic chain ganglion
White ramus communicans

1. _____

2. _____

3. _____

4. _____

5. _____

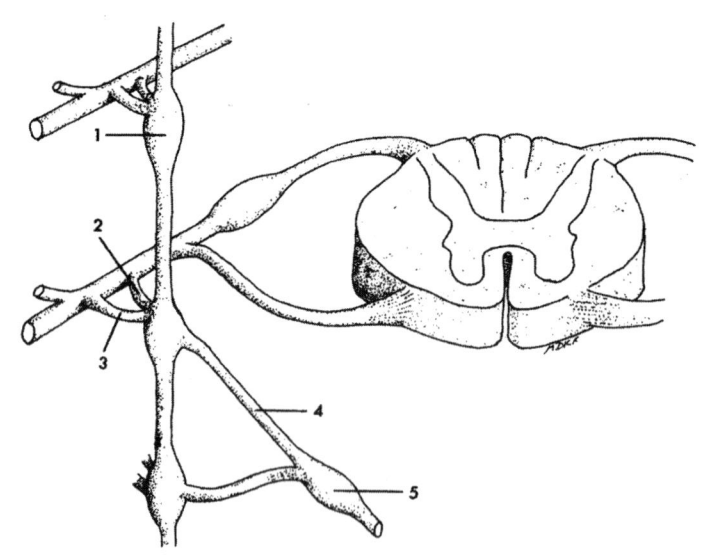

Figure 16.2

B. Using the terms provided, complete these statements:

- Abdominopelvic cavity
- Cervical and sacral
- Collateral ganglia
- Lateral horns
- Paravertebral ganglia
- Prevertebral ganglia
- Sympathetic ganglia
- Thoracolumbar

The _(1)_ of the spinal cord gray matter from T1 to L2 contain the cell bodies of the sympathetic preganglionic neurons. Because of the location of the preganglionic cell bodies, this division is sometimes called the _(2)_ division. Axons of sympathetic preganglionic neurons run through ventral roots of spinal nerves and project to _(3)_. There are two types of sympathetic ganglia: sympathetic chain ganglia and _(4)_. Sympathetic chain ganglia are also called _(5)_ because they are located on the left and right sides of the vertebral column. Although the sympathetic division originates in the thoracic and lumbar vertebral regions, the sympathetic chain ganglia extend into the _(6)_ regions. Collateral ganglia are unpaired ganglia located in the _(7)_. They are also called _(8)_ because they are anterior to the vertebral column.

1. _____
2. _____
3. _____
4. _____
5. _____
6. _____
7. _____
8. _____

C. Using the terms provided, complete these statements:

- Adrenal medulla
- Collateral ganglion
- Epinephrine
- Gray ramus communicans
- Postganglionic
- Preganglionic
- Splanchnic nerve
- Sympathetic nerve
- White ramus communicans

Because sympathetic _(1)_ axons are myelinated, the short connection between the spinal nerve and the sympathetic chain ganglion is called a(n) _(2)_. After entering the sympathetic chain ganglion, preganglionic neurons may synapse with _(3)_ neurons at the same level, or above or below the ganglion first entered. The postganglionic neurons then pass through a(n) _(4)_ (unmyelinated axons) to reenter a spinal nerve or project through a(n) _(5)_ to the organs they innervate. Preganglionic axons from T5 to T12 of the spinal cord exit the chain ganglia without synapsing and pass through a(n) _(6)_ to a(n) _(7)_, where they synapse with postganglionic neurons. Preganglionic axons of splanchnic nerves also extend to the _(8)_, where they synapse with specialized postganglionic neurons that contain _(9)_, and norepinephrine, which can be released into the blood.

1. _____
2. _____
3. _____
4. _____
5. _____
6. _____
7. _____
8. _____
9. _____

Chapter 16

Parasympathetic Division

Using the terms provided, complete these statements:

Brainstem
Craniosacral
Gray matter
Postganglionic
Preganglionic
Terminal ganglia

Preganglionic cell bodies of the parasympathetic division are either within cranial nerve nuclei in the __(1)__ or within the lateral parts of the __(2)__ from S2 to S4 in the spinal cord. Therefore, this division is sometimes called the __(3)__ division. Axons of the __(4)__ neurons pass through cranial nerves and pelvic nerves to __(5)__ located either near to or embedded in the wall of the organ innervated. The axons of the __(6)__ neurons extend from the terminal ganglia to the target organ.

1. _____
2. _____
3. _____
4. _____
5. _____
6. _____

Enteric Nervous System

A. Match the neuron types with the correct description:

ANS motor neurons   Sensory neurons
Enteric neurons

_____ 1. Neurons that connect the digestive tract to the CNS.

_____ 2. Neurons that connect the CNS to the digestive tract.

_____ 3. Neurons that are confined to the enteric plexuses.

B. Match the neuron types with the correct description:

Enteric interneurons   Enteric sensory neurons
Enteric motor neurons

_____ 1. Neurons that detect changes in the chemical composition of the digestive tract or stretch of the digestive tract wall.

_____ 2. Neurons that stimulate or inhibit smooth muscle contraction or gland secretion.

_____ 3. Neurons that connect enteric sensory and enteric motor neurons with each other.

Distribution of Sympathetic Nerve Fibers

A. Match the parts of the body with the sympathetic fibers that supply them:

    Abdominopelvic organs   Limbs and body
    Head and neck   Thoracic organs

_____ 1. Spinal nerves.

_____ 2. Sympathetic nerves from the superior cervical chain ganglia.

_____ 3. Sympathetic nerves from the cervical and thoracic chain ganglia (to T5) supplying thoracic nerve plexuses.

_____ 4. Splanchnic nerves from chain ganglia below T5.

B. Match the nerves or plexuses with the specific structures they supply:

    Cardiac and pulmonary plexuses   Hypogastric plexuses
    Celiac plexus   Inferior mesenteric plexus
    Head and neck nerve plexuses   Spinal nerves
        Superior mesenteric plexus

_____ 1. Sweat glands, skeletal and skin blood vessels, and arrector pili.

_____ 2. Sweat glands, blood vessels, arrector pili, salivary glands, and iris and ciliary muscles of the eye.

_____ 3. Heart and lungs.

_____ 4. Diaphragm, stomach, spleen, liver, gallbladder, adrenal glands, kidneys, testes, and ovaries.

_____ 5. Pancreas, small intestine, ascending colon, and transverse colon.

_____ 6. Transverse colon to the rectum.

_____ 7. Descending colon to the rectum, urinary bladder, and reproductive organs.

Distribution of Parasympathetic Nerve Fibers

A. Match the parts of the body with the parasympathetic fibers that supply them:

    Abdominal organs   Head and neck
    Abdominopelvic organs   Thoracic organs

_____ 1. Oculomotor (III) nerve, facial (VII) nerve, and glossopharyngeal (IX) nerve.

_____ 2. Vagus (X) nerve to the cardiac, pulmonary, and esophageal plexuses.

_____ 3. Vagus (X) nerve to the stomach and the celiac and superior mesenteric plexuses.

_____ 4. Pelvic nerves and hypogastric plexus

Chapter 16

B. Match the ganglia with the cranial nerve and structures innervated:

Ciliary ganglion   Pterygopalatine ganglion
Otic ganglion   Submandibular ganglion

_____ 1. The oculomotor (III) nerve, through this ganglion, supplies the ciliary muscles and iris of the eye.

_____ 2. The facial (VII) nerve, through this ganglion, supplies the lacrimal gland and mucosal glands of the nasal cavity and palate.

_____ 3. The facial (VII) nerve, through this ganglion, supplies the submandibular and sublingual salivary glands.

_____ 4. The glossopharyngeal (IX) nerve, through this ganglion, supplies the parotid salivary gland.

Physiology of the Autonomic Nervous System

Match these terms with the correct statement or definition:

Adrenergic neuron   Cholinergic neuron
Adrenergic receptors   Muscarinic receptors
 (α or β)   Nicotinic receptors

_____ 1. Neuron that secretes acetylcholine.

_____ 2. All preganglionic neurons of the autonomic division are of this type.

_____ 3. All postganglionic neurons of the parasympathetic division are this type.

_____ 4. Most postganglionic neurons of the sympathetic division are of this type.

_____ 5. Located in postganglionic neurons and the membranes of skeletal muscle cells, these cholinergic receptors produce an excitatory response to acetylcholine.

_____ 6. Located in effector organs, these cholinergic receptors may produce either an excitatory or inhibitory response to acetylcholine.

_____ 7. Norepinephrine binds to and activates these receptors, which are located in effector organs; the response can be excitatory or inhibitory.

Regulation of the Autonomic Nervous System

Using the terms provided, complete these statements:

Autonomic reflexes
Brainstem and spinal cord
Effector cells
Enteric nervous system
Hypothalamus
Local reflexes
Unconscious

Much of the regulation of structures by the ANS occurs through _(1)_, but other CNS activities also influence autonomic function. Autonomic reflexes, like other reflexes, involve sensory receptors, sensory neurons, interneurons, motor neurons, and _(2)_. The _(3)_ is in overall control of the ANS, but the _(4)_ contain important autonomic reflex centers responsible for maintaining homeostasis. The _(5)_ is involved with autonomic and local reflexes that regulate the activity of the digestive tract. The neurons of the enteric nervous system also operate independently of the CNS to produce _(6)_. A local reflex does not involve the CNS, but it does produce an involuntary, _(7)_, stereotypic response to a stimulus.

1. _____
2. _____
3. _____
4. _____
5. _____
6. _____
7. _____

Functional Generalizations About the Autonomic Nervous System

A. Using the terms provided, complete these statements:

Coordinate
Diverges
Exceptions
General
Identical
Innervated
Opposite
Physical activity or stress
Resting

Generalizations can be made about the effects of the ANS on effector organs, but most of the generalizations have _(1)_. Both divisions of the ANS produce stimulatory and _(2)_ effects. Most organs are _(3)_ by both the sympathetic and the parasympathetic divisions. When a single structure is innervated by both autonomic divisions, the two divisions usually produce _(4)_ effects. One or both autonomic divisions can _(5)_ the activities of different structures. The sympathetic division has a more _(6)_ effect than the parasympathetic division because the sympathetic division _(7)_ more, and activation of the sympathetic division often causes secretion of epinephrine and norepinephrine from the adrenal medulla. The sympathetic division has a major influence under conditions of _(8)_, whereas the parasympathetic division tends to have a greater role during _(9)_ conditions.

1. _____
2. _____
3. _____
4. _____
5. _____
6. _____
7. _____
8. _____
9. _____

B. Match these terms with the correct statement or definition:

Parasympathetic division
Sympathetic division

_____ 1. Sweat glands and blood vessels are almost exclusively innervated by this division.

_____ 2. Salivation, lacrimation, urination, digestion, and defecation increase when this division is stimulated.

_____ 3. Has the greatest activity during physical activity or stress.

_____ 4. Increases metabolism and causes breakdown of glycogen from liver and skeletal muscle.

_____ 5. Causes an increase in heart rate and force of contraction.

_____ 6. Has a more localized effect.

QUICK RECALL

1. Complete this table:

CHARACTERISTIC	SOMATIC MOTOR	AUTONOMIC
Number of neurons		
Effector organs		
Neurotransmitter		
Conscious vs. unconscious control		

2. Complete this table by indicating which ANS division is involved and what kind of cell body is found in each location:

LOCATION	PARASYMPATHETIC OR SYMPATHETIC DIVISION	PREGANGLIONIC OR POSTGANGLIONIC CELL BODY
Cranial nuclei		
Lateral horns T1–L2		
Lateral horns S2–S4		
Chain ganglia		
Collateral ganglia		
Terminal ganglia		

3. Name two types of neurotransmitters released by the ANS, and indicate which type is released in sympathetic and parasympathetic preganglionic and postganglionic neurons.

4. List four types of receptors in the ANS and give their location.

5. List six functional generalizations concerning the ANS.

ANSWERS TO CHAPTER 16

CONTENT LEARNING ACTIVITY

Contrasting the Somatic and Autonomic Nervous Systems
- A. 1. Sensory neurons; 2. Somatic motor neurons; 3. Autonomic motor neurons; 4. Somatic motor neurons; 5. Autonomic motor neurons
- B. 1. Preganglionic neuron; 2. Preganglionic neuron; 3. Postganglionic neuron; 4. Autonomic ganglion; 5. Effector organ
- C. 1. Preganglionic neuron; 2. Postganglionic neuron; 3. Autonomic ganglion; 4. Effector organ

Sympathetic Division
- A. 1. Sympathetic chain ganglion; 2. Gray ramus communicans; 3. White ramus communicans; 4. Splanchnic nerve; 5. Collateral (prevertebral) ganglion
- B. 1. Lateral horns; 2. Thoracolumbar; 3. Sympathetic ganglia; 4. Collateral ganglia; 5. Paravertebral ganglia; 6. Cervical and sacral; 7. Abdominopelvic cavity; 8. Prevertebral ganglia
- C. 1. Preganglionic; 2. White ramus communicans; 3. Postganglionic; 4. Gray ramus communicans; 5. Sympathetic nerve; 6. Splanchnic nerve; 7. Collateral ganglion; 8. Adrenal medulla; 9. Epinephrine

Parasympathetic Division
- 1. Brainstem; 2. Gray matter; 3. Craniosacral; 4. Preganglionic; 5. Terminal ganglia; 6. Postganglionic

Enteric Nervous System
- A. 1. Sensory neurons; 2. ANS motor neurons; 3. Enteric neurons
- B. 1. Enteric sensory neurons; 2. Enteric motor neurons; 3. Enteric interneurons

Distribution of Sympathetic Nerve Fibers
- A. 1. Limbs and body; 2. Head and neck; 3. Thoracic organs; 4. Abdominopelvic organs
- B. 1. Spinal nerves; 2. Head and neck nerve plexuses; 3. Cardiac and pulmonary plexuses; 4. Celiac plexus; 5. Superior mesenteric plexus; 6. Inferior mesenteric plexus; 7. Hypogastric plexus

Distribution of Parasympathetic Nerve Fibers
- A. 1. Head and neck; 2. Thoracic organs; 3. Abdominal organs; 4. Abdominopelvic organs
- B. 1. Ciliary ganglion; 2. Pterygopalatine ganglion; 3. Submandibular ganglion; 4. Otic ganglion

Physiology of the Autonomic Nervous System
 1. Cholinergic neuron; 2. Cholinergic neuron; 3. Cholinergic neuron; 4. Adrenergic neuron; 5. Nicotinic receptors; 6. Muscarinic receptors; 7. Adrenergic receptors (α or β)

Regulation of the Autonomic Nervous System
 1. Autonomic reflexes; 2. Effector cells; 3. Hypothalamus; 4. Brainstem and spinal cord; 5. Enteric nervous system; 6. Local reflexes; 7. Unconscious

Functional Generalizations About the Autonomic Nervous System
 A. 1. Exceptions; 2. Inhibitory; 3. Innervated; 4. Opposite; 5. Coordinate; 6. General; 7. Diverges; 8. Physical activity or stress; 9. Resting
 B. 1. Sympathetic division; 2. Parasympathetic division; 3. Sympathetic division; 4. Sympathetic division; 5. Sympathetic division; 6. Parasympathetic division

QUICK RECALL

1. Number of neurons: somatic motor, one; autonomic, two; effector organs: somatic motor, skeletal muscle; autonomic, smooth muscle, cardiac muscle, and glands; neurotransmitter: somatic motor, acetylcholine; autonomic, acetylcholine or norepinephrine; conscious vs. unconscious control: somatic motor, conscious; autonomic, unconscious

2. Cranial nuclei: parasympathetic division, preganglionic cell body; lateral horns T1–L2: sympathetic division, preganglionic cell body; lateral horns S2–S4: parasympathetic, preganglionic cell body; chain ganglia: sympathetic, postganglionic cell body; collateral ganglia: sympathetic, postganglionic cell body; terminal ganglia: parasympathetic, postganglionic cell body

3. All preganglionic fibers: acetylcholine; postganglionic fibers of parasympathetic: acetylcholine; postganglionic fibers of sympathetic: mostly norepinephrine, some release acetylcholine

4. Alpha and beta adrenergic receptors: effector organs of sympathetic division; nicotinic cholinergic receptors: postganglionic receptors for both sympathetic and parasympathetic divisions; muscarinic cholinergic receptors: effector organ receptors of parasympathetic and some sympathetic divisions

5. In most cases, the influence of the two autonomic divisions is opposite on structures that receive dual innervation; each division can produce inhibitory or excitatory effects; and the parasympathetic division is consistent with resting conditions, whereas the sympathetic division is consistent with physical activity or stress; most organs that receive autonomic neurons are innervated by both sympathetic and parasympathetic divisions; one division alone or both divisions acting together can coordinate the activities of different structures; and the sympathetic division has a more general effect when activated.

17 Functional Organization of the Endocrine System

CONTENT LEARNING ACTIVITY

General Characteristics of the Endocrine System

A. Match these terms with the correct statement or definition:

Amplitude-modulated
Endocrine system
Frequency-modulated
Hormones

_____ 1. Composed of glands that secrete their products into the circulatory system.

_____ 2. Secretions of endocrine glands; acts on target tissues.

_____ 3. Communication between cells that involves increases or decreases in the concentration of hormones in the body fluids.

_____ 4. Method of communication between neurons and their effectors; an all-or-none response.

B. Match these chemical signals with their correct location in the following table:

Autocrine
Hormone
Neurohormone
Neurotransmitter
Paracrine
Pheromone

INTERCELLULAR CHEMICAL SIGNALS	PRODUCED BY	FUNCTION	EXAMPLE
1. _____	Wide variety of tissue	Secreted into tissue space; localized effect on SAME cell type from which the chemical signal is released	Prostaglandins
2. _____	Wide variety of tissue	Secreted into tissue space; localized effect on DIFFERENT cell type from which the chemical signal is released	Prostaglandins, histamine

Chapter 17

INTERCELLULAR CHEMICAL SIGNALS	PRODUCED BY	FUNCTION	EXAMPLE
3. _____	Specialized cells	Travels in blood; influences specific activities	Thyroxine, insulin
4. _____	Neurons	Like hormones	Oxytocin, antidiuretic hormone
5. _____	Neurons	Released from pre-synaptic terminals; affects postsynaptic cells	Acetylcholine, norepinephrine
6. _____	Specialized cells	Secreted into the external environment; modifies the physiology and behavior of other individuals of the same species	Found in urine

C. Match these terms with the correct parts of the diagram labeled in figure 17.1:

Adrenals
Hypothalamus
Ovaries
Pancreas
Parathyroid glands
Pineal body
Pituitary
Testes
Thymus
Thyroid

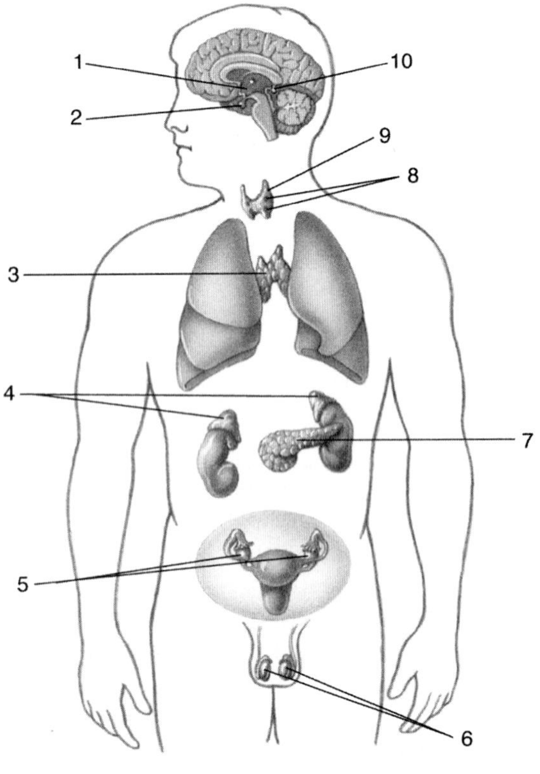

1. _____
2. _____
3. _____
4. _____
5. _____
6. _____
7. _____
8. _____
9. _____
10. _____

Figure 17.1

Control of Secretion Rate

Using the terms provided, complete these statements:

Hormone
Negative-feedback
Neural
Nonhormone
Positive-feedback

There are three major patterns of regulation for hormones. An example is the effect glucose has on the secretion of insulin. This is a case of a _(1)_ influencing an endocrine gland. The release of epinephrine from the adrenal glands is a different situation. In this pattern, the release of a hormone is regulated by _(2)_ control. A third pattern occurs when thyroid-stimulating hormone causes the thyroid to release thyroid hormones. In this case, a _(3)_ controls the secretory activity of an endocrine gland. When hormones control the release of other hormones, it usually involves a _(4)_ mechanism.

1. _____
2. _____
3. _____
4. _____

Transport and Distribution in the Body

Using the terms provided, complete these statements:

Binding proteins
Equilibrium
Free hormone
Target tissue

Hormones are dissolved in blood plasma and are transported either in a free form or bound to plasma proteins called _(1)_. A(n) _(2)_ exists between the unbound hormone and those bound to plasma proteins. A large increase or decrease in plasma protein concentration can influence the concentration of _(3)_ in the blood. In general, the amount of free hormone that reaches the _(4)_ is directly correlated with the concentration of hormone in the blood.

1. _____
2. _____
3. _____
4. _____

Metabolism and Excretion

A. The length of time it takes for half a dose of a substance to be eliminated from the circulatory system is its half-life. Match the length of the half-life with the correct hormone characteristic:

Long half-life
Short half-life

_____ 1. Typical of water-soluble hormones (proteins, glycoproteins, and epinephrine and norepinephrine).

_____ 2. These hormones generally regulate activities that have a rapid onset and a short duration.

_____ 3. The concentration of these hormones is maintained at a relatively constant level through time.

B. Match these factors with the correct statement or definition:

Active transport Metabolism
Conjugation Reversible binding
Excretion Structural protection

_____ 1. Elimination of hormones from the blood into the urine or bile.

_____ 2. Process in which hormones are enzymatically degraded.

_____ 3. Hormones are made less active and eliminated by attaching compounds, such as sulfates or glucuronic acid, to them.

_____ 4. Method of prolonging the half-life of hormones by reacting with binding proteins.

_____ 5. Half-life of hormones is prolonged by the carbohydrate components of glycoprotein hormones.

Interaction of Hormones with Their Target Tissues

Match these terms with the correct statement or definition:

Agonist Receptor site
Antagonist Specificity
Down-regulation Up-regulation
Receptor

_____ 1. Protein or glycoprotein to which hormones bind.

_____ 2. Portion of a protein or glycoprotein where chemical signals bind.

_____ 3. Tendency for a chemical signal to bind to a specific type of receptor.

_____ 4. The presence or absence of these determines whether or not a specific hormone produces a response from a specific tissue.

_____ 5. Drug that binds to a hormone receptor and activates it.

_____ 6. Decrease in the number of hormone receptors after exposure to certain hormones.

_____ 7. Increase in the number of hormone receptors after exposure to certain hormones.

Classes of Hormone Receptors

A. Match these terms with the correct statement or definition:

Intracellular receptors
Membrane-bound receptors

_____ 1. Binds to water-soluble or large molecular weight hormones.

_____ 2. Binds to lipid-soluble hormones (e.g., steroids, thyroid hormones).

_____ 3. When activated, can cause changes in G protein activity or intracellular enzyme activity.

B. Using the terms provided, complete these statements:

Binds to
Enzymes
G proteins
Ion channels
Splits apart

Many receptors have regulatory proteins, called _(1)_. When a hormone _(2)_ the receptor, the G protein splits apart and the α subunit releases GDP and _(3)_ guanine triphosphate (GTP). The GTP-bound α subunit binds to _(4)_ or activates _(5)_. Consequently, other activities result that produce a response from the cell. The activity of the α subunit is terminated when the GTP _(6)_ to form GDP. The α subunit then recombines with the β and γ subunits.

1. _____
2. _____
3. _____
4. _____
5. _____
6. _____

C. Match these terms with the correct statement or definition:

Adenylyl cyclase
Cascade effect
Intracellular mediator
Phosphodiesterase
Phosphorylation

_____ 1. G proteins can alter the activity of this enzyme, which produces cyclic AMP

_____ 2. Cyclic AMP and Ca^{2+} are examples; alter enzyme activity, which produces a response by cells.

_____ 3. Addition of a phosphate group to a protein, which then influences the activities of enzymes, producing a response by cells.

_____ 4. Enzyme that limits the activity of cyclic AMP and cyclic GMP.

_____ 5. A few intracellular mediators activate several enzymes, which activate many more enzymes, and so on.

D. Using the terms provided, complete these statements:

DNA
Enzymes
mRNA
Protein synthesis

Hormones can bind to intracellular receptors in the cytoplasm or the nucleus of the cell. Once in the nucleus, the hormone-receptor combination has fingerlike projections that interact with _(1)_ to regulate the synthesis of _(2)_, which moves into the cytoplasm and initiates _(3)_. The _(4)_ that result produce the response of the cell.

1. _____
2. _____
3. _____
4. _____

Chapter 17

Quick Recall

1. List categories of intercellular chemical signals.

2. Explain the three major patterns for regulation of hormone secretion, and give an example of each type.

3. List three important patterns of secretion of hormones.

4. List four means by which hormones are eliminated from the circulatory system.

5. List two means by which the half-life of hormones is prolonged.

6. List the two major classes of hormone receptors, and list the types of hormone molecules that bind to each.

7. Name the two major mechanisms by which membrane-bound receptors produce an intracellular response.

8. Name the two major effects of G proteins.

ANSWERS TO CHAPTER 17

CONTENT LEARNING ACTIVITY

General Characteristics of the Endocrine System
   A. 1. Endocrine system; 2. Hormones;
       3. Amplitude-modulated; 4. Frequency-modulated
   B. 1. Autocrine; 2. Paracrine; 3. Hormone;
       4. Neurohormone; 5. Neurotransmitter;
       6. Pheromone
   C. 1. Hypothalamus; 2. Pituitary; 3. Thymus;
       4. Adrenals; 5. Ovaries; 6. Testes; 7. Pancreas;
       8. Parathyroid glands; 9. Thyroid; 10. Pineal body

Control of Secretion Rate
   1. Nonhormone; 2. Neural; 3. Hormone;
   4. Negative-feedback

Transport and Distribution in the Body
   1. Plasma proteins; 2. Equilibrium; 3. Free hormone; 4. Target tissue

Metabolism and Excretion
   A. 1. Short half-life; 2. Short half-life; 3. Long half-life
   B. 1. Excretion; 2. Metabolism; 3. Conjugation;
       4. Reversible binding; 5. Structural protection

Interaction of Hormones with Their Target Tissues
   1. Receptor; 2. Receptor site; 3. Specificity;
   4. Receptor; 5. Agonist; 6. Down-regulation;
   7. Up-regulation

Classes of Hormone Receptors
   A. 1. Membrane-bound receptors;
       2. Intracellular receptors; 3. Membrane-bound receptors
   B. 1. G proteins; 2. Binds to; 3. Binds to; 4. Ion channels; 5. Enzymes; 6. Splits apart
   C. 1. Adenylyl cyclase; 2. Intracellular mediator;
       3. Phosphorylation; 4. Phosphodiesterase;
       5. Cascade effect
   D. 1. DNA; 2. mRNA; 3. Protein synthesis;
       4. Enzymes

QUICK RECALL

1. Autocrine, paracrine, hormone, neurohormone, neurotransmitter/neuromodulator; pheromone
2. Regulation by a nonhormone (glucose); by neural control (epinephrine from the adrenal gland); or by a hormone (thyroid hormones)
3. Relatively constant; sudden change in response to stimuli; and change occurring in cycles
4. Excretion, metabolism (enzymatic degradation), conjugation, and active transport
5. Reversible binding and structural protection
6. Membrane-bound receptors: water-soluble or large hormones; intracellular receptors: lipid-soluble hormones
7. Activate G proteins and alter the activity of enzymes
8. Regulate ion channels and regulate the synthesis of intracellular mediators

18 Endocrine Glands

CONTENT LEARNING ACTIVITY

Pituitary Gland and Hypothalamus

A. Match these terms with the correct statement or definition:

Action potentials
Anterior pituitary
Hypothalamohypophysial portal system
Hypothalamohypophysial tract
Hypothalamus
Infundibulum
Neurohormones
Posterior pituitary

_____ 1. Connection between the pituitary gland and hypothalamus.

_____ 2. Also called the neurohypophysis.

_____ 3. Develops from the roof of the embryonic oral cavity.

_____ 4. Vessels from the hypothalamus to the anterior pituitary.

_____ 5. Produced in the hypothalamus; travel to the anterior pituitary and act as releasing or inhibiting hormones.

_____ 6. Source of neurohormones secreted from the posterior pituitary.

_____ 7. Source of hormones secreted from the anterior pituitary.

_____ 8. Regulate the release of hormones from the posterior pituitary.

B. Match these releasing or inhibiting hormones with their action on the anterior pituitary:

CRH
GHIH (somatostatin)
GHRH
GnRH
PIH
PRH
TRH

_____ 1. Small peptide that inhibits the secretion of growth hormone.

_____ 2. Small peptide that stimulates the secretion of thyroid-stimulating hormone.

_____ 3. Peptide that stimulates the secretion of adrenocorticotropic hormone.

_____ 4. Small peptide that stimulates the secretion of luteinizing hormone and follicle-stimulating hormone.

_____ 5. Stimulates the secretion of prolactin.

C. Match these terms with the correct parts of the diagram labeled in figure 18.1:

Anterior pituitary
Hypothalamohypophysial portal system
Hypothalamohypophysial tract
Hypothalamus
Neurosecretory cell
Posterior pituitary

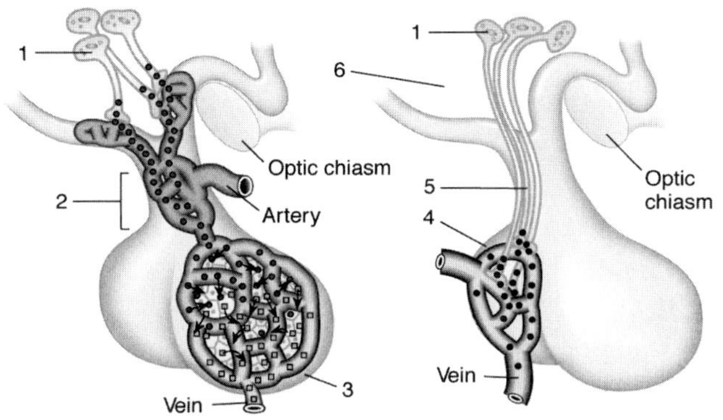

1. _____
2. _____
3. _____
4. _____
5. _____
6. _____

Figure 18.1

Posterior Pituitary Hormones

A. Match these terms with the correct statement as it pertains to the effect of increased ADH:

Decreases
Increases

_____ 1. Effect of increased ADH on urine volume.

_____ 2. Effect of increased ADH on blood osmolality.

_____ 3. Effect of increased ADH on blood volume.

_____ 4. Effect of increased ADH on blood vessel constriction and blood pressure.

B. Match these terms with the correct statement:

Blood osmolality
Blood pressure

_____ 1. Osmoreceptors in the hypothalamus respond to this factor.

_____ 2. A decrease in this factor usually accompanies a drop in blood volume.

_____ 3. An increase in this factor results in an increase of ADH secretion.

_____ 4. An increase in this factor results in a decrease of ADH secretion.

C. Match these terms with the correct statement:

Decreases
Increases

_____ 1. Effect of oxytocin on smooth muscle contraction in the uterus.

_____ 2. Effect of oxytocin on milk ejection in lactating females.

D. Match these terms with the correct statement as it pertains to <u>oxytocin secretion</u>:

Decreases
Increases

_____ 1. Effect of stretching the uterus.

_____ 2. Effect of stimulating the nipples by nursing.

Anterior Pituitary Hormones

A. Match these terms with the correct statement as it pertains to <u>increased growth hormone</u>:

Decreases
Increases

_____ 1. Effect on growth and metabolic rate.

_____ 2. Effect on amino acid uptake and protein synthesis by cells.

_____ 3. Effect on the rate of lipid breakdown and blood glucose levels.

_____ 4. Effect on somatomedin production by liver and skeletal muscle.

_____ 5. Effect of insulin-like growth factor I and II on cartilage and bone growth.

B. Match these terms with the correct statement as it pertains to <u>GH secretion</u>:

Decreases
Increases

_____ 1. Effect of low blood glucose levels.

_____ 2. Effect of stress.

_____ 3. Effect of GH-RH from the hypothalamus.

_____ 4. Effect of GH-IH (somatostatin) from the hypothalamus.

C. Match these terms with the correct statement as it pertains to GH secretion:

Hypersecretion
Hyposecretion

_____ 1. In children, this produces dwarfism.

_____ 2. In children, this produces giantism.

_____ 3. In adults, this produces acromegaly.

D. Match these terms with the correct statement or definition:

ACTH
β endorphins
Lipotropins
MSH
Proopiomelanocortin

_____ 1. Large precursor molecule that produces ACTH and others.

_____ 2. Cause fat breakdown and release of fatty acids.

_____ 3. Have the same effect as opiate drugs.

_____ 4. Two hormones that stimulate melanocytes in the skin to produce melanin.

_____

E. Match these terms with the correct statement or definition:

FSH and LH
GnRH
PIH
PRH
Prolactin

_____ 1. Secreted from the anterior pituitary; regulate gamete and reproductive hormone production.

_____ 2. Neurohormone that stimulates the secretion of LH and FSH.

_____ 3. Responsible for milk production in the mammary glands.

_____ 4. Neurohormone that stimulates prolactin production.

Thyroid Gland

A. Match these terms with the correct statement or definition:

Calcitonin
Follicle
Parafollicular cells
Thyroid hormones

_____ 1. Small sphere of cuboidal epithelium in the thyroid that is filled with thyroglobulin.

_____ 2. Scattered cells between follicles in the thyroid.

_____ 3. Product of follicular cells.

_____ 4. Product of parafollicular cells.

B. Match these terms with the correct statement or definition:

I⁻
Tetraiodothyronine (T₄)
Thyroglobulins
Thyroxine-binding globulin
Triiodothyronine (T₃)
Tyrosine

_____ 1. Large proteins that are synthesized in, and fill the lumen of, thyroid follicles.

_____ 2. Amino acid used in the synthesis of thyroid hormones.

_____ 3. Actively absorbed into thyroid follicles; oxidized and bound to tyrosines.

_____ 4. Also called thyroxine; the major secretory product of the thyroid gland.

_____ 5. Major thyroid hormone that interacts with target cells.

_____ 6. Transports most thyroid hormones; increases the half-life of thyroid hormones.

C. Match these terms with the correct statement as it pertains to the effects of increased thyroid hormones:

Decreases
Increases

_____ 1. Effect on glucose, fat, and protein metabolism.

_____ 2. Effect on body temperature.

_____ 3. Effect on blood levels of cholesterol.

D. Match these terms with the correct statement as it applies to thyroid hormone secretion:

Hypersecretion
Hyposecretion

_____ 1. Increased metabolic rate, weight loss, sweating.

_____ 2. Hyperactivity, rapid heart rate, exophthalmos.

_____ 3. Weight gain; reduced appetite; rough, dry skin.

_____ 4. Myxedema, decreased iodide uptake, cold intolerance.

_____ 5. Decreased iodide uptake resulting from iodine deficiency in the diet; goiter.

_____ 6. Cretin; a mentally retarded person of short stature.

_____ 7. Graves disease.

_____ 8. Thyroid storm.

E. Match these terms with the correct statement:

Decreases
Increases

_____ 1. Effect of stress or exposure to cold on TRH secretion.

_____ 2. Effect of prolonged fasting on TRH secretion.

_____ 3. Effect of increase of TRH on TSH secretion.

_____ 4. Effect of TSH increase on synthesis and secretion of T_3 and T_4.

_____ 5. Effect of increases in T_3 and T_4 on TRH secretion.

_____ 6. Effect of increases of T_3 and T_4 on TSH secretion.

F. Match these terms with the correct statement:

Decreases
Increases

_____ 1. Effect of calcitonin on the breakdown of bone by osteoclasts.

_____ 2. Effect of calcitonin on blood calcium levels.

_____ 3. Effect of increased blood calcium levels on calcitonin secretion.

Parathyroid Glands

A. Match these terms with the correct statement as it pertains to the effects of <u>increased parathyroid hormone</u>:

Decreases
Increases

_____ 1. Effect on osteoclast activity in bone.

_____ 2. Effect on calcium reabsorption in the kidneys.

_____ 3. Effect on the formation of active vitamin D synthesis, which increases the rate of calcium and phosphate absorption in the intestine.

_____ 4. Effect on blood calcium levels.

_____ 5. Effect of PTH on blood phosphate levels.

B. Match the correct term with this statement:

Decreases PTH secretion
Increases PTH secretion

_____ 1. Effect of low blood calcium levels on PTH secretion.

Chapter 18

C. Match these terms with the correct symptom as it pertains to <u>PTH</u> <u>secretion</u>:

Hypersecretion
Hyposecretion

_____

_____

_____

1. Kidney stones, eroded bones, muscular weakness, constipation.

2. Increased muscular excitability, tachycardia, muscle tetany, diarrhea.

3. Increased cell permeability to sodium ions, depolarization of the cell membrane.

Adrenal Glands

Match these terms with the correct statement or definition:

Adrenal cortex
Adrenal medulla

_____

_____

_____

1. Inner portion of the adrenal glands.

2. Derived from neural crest cells; part of the sympathetic division of the ANS.

3. Contains the zona glomerulosa, zona fasciculata, and zona reticularis layers.

Adrenal Medulla

A. Match these terms with the correct statement:

Epinephrine (adrenaline)   Norepinephrine (noradrenaline)
Neurohormone

_____

_____

1. Adrenal medullary hormone secreted in larger quantity.

2. General category of hormones produced by the adrenal medulla.

B. Match these terms with the correct statement as it pertains to the effects of <u>increased</u> <u>adrenal</u> <u>medullary</u> <u>hormones</u>:

Decreases
Increases

_____

_____

_____

_____

1. Effect on heart rate and the force of contraction of the heart.

2. Effect on blood glucose level.

3. Effect on blood flow to the skin, kidneys, and digestive system.

4. Effect on blood flow to heart and skeletal muscle.

C. Match these terms with the correct statement as it pertains to <u>adrenal</u> <u>medullary</u> <u>hormone</u> <u>secretion</u>:

Decreases
Increases

_____ 1. Effect of emotional excitement, stress, exercise, or injury.

_____ 2. Effect of stimulation of sympathetic neurons.

_____ 3. Effect of low blood glucose levels.

D. Match these terms with the correct statement as it pertains to <u>adrenal</u> <u>medullary</u> <u>hormone</u> <u>secretion</u>:

Hypersecretion
Hyposecretion

_____ 1. Result of pheochromocytoma.

_____ 2. Hypertension and pallor (pale skin).

Adrenal Cortex

A. Match these terms with the correct statement as it pertains to the adrenal cortex:

Androstenedione
Glucocorticoids
Mineralocorticoids

_____ 1. Secreted by the zona glomerulosa; an example is aldosterone.

_____ 2. Secreted by the zona fasciculata; an example is cortisol.

_____ 3. Androgen secreted by the zona reticularis; converted into testosterone.

B. Match these terms with the correct statement as it pertains to the effects of <u>increased</u> <u>aldosterone</u>:

Decreases
Increases

_____ 1. Effect on sodium ion concentration in the blood.

_____ 2. Effect on potassium ion concentration in the blood.

_____ 3. Effect on hydrogen ion concentration in the blood.

_____ 4. Effect on blood pH.

C. Match these terms with the correct statement as it pertains to <u>aldosterone secretion</u>:

Hypersecretion
Hyposecretion

_____ 1. High blood sodium levels and high blood pressure.

_____ 2. Alkalosis.

_____ 3. High blood potassium levels and skeletal muscle tetany.

D. Match these terms with the correct statement as it pertains to the effects of <u>increased cortisol</u>:

Decreases
Increases

_____ 1. Effect on fat and protein breakdown.

_____ 2. Effect on glucose and amino acid uptake by skeletal muscle.

_____ 3. Effect on glucose synthesis from amino acids (gluconeogenesis).

_____ 4. Effect on blood glucose levels and glycogen deposits in cells.

_____ 5. Effect on the intensity of the inflammatory response.

E. Match these terms with the correct statement or definition:

ACTH   Hypoglycemia or stress
Cortisol   Hypothalamus
CRH

_____ 1. Location of CRH production.

_____ 2. The production of ACTH is stimulated by this neurohormone.

_____ 3. Hormone that stimulates cortisol production.

_____ 4. Two hormones that inhibit CRH secretion.

_____

_____ 5. Hormone that inhibits ACTH production.

_____ 6. External factors that stimulate CRH production.

F. Match these terms with the correct statement as it pertains to <u>cortisol secretion</u>:

Hypersecretion
Hyposecretion

_____ 1. Hyperglycemia leading to diabetes mellitus.

_____ 2. Osteoporosis, muscle atrophy, and weakness.

_____ 3. Fat redistributed to face and neck.

_____ 4. Increased skin pigmentation.

G. Match these diseases with the correct conditions:

Addison disease Cushing syndrome
Aldosteronism

_____ 1. Hypersecretion of aldosterone.

_____ 2. Hypersecretion of cortisol and androgens.

_____ 3. Hyposecretion of aldosterone and cortisol.

H. Match these terms with the correct statement as it pertains to the effects of <u>increased adrenal androgens</u>:

Decreases
Increases

_____ 1. Effect on amount of pubic and axillary hair in women.

_____ 2. Effect on sex drive in women.

Pancreas and Pancreatic Hormones

A. Match these terms with the correct statement as it pertains to the pancreas:

Alpha cells   Ducts and acini
Beta cells   Pancreatic islets
Delta cells

_____ 1. Constitute the exocrine portion of the pancreas.

_____ 2. Islet cells that secrete glucagon.

_____ 3. Islet cells that secrete insulin.

_____ 4. Islet cells that secrete somatostatin.

B. Match these terms with the correct statement as it pertains to insulin or glucagon:

Decreases
Increases

_____ 1. Effect of insulin on the uptake and use of glucose and amino acids in muscle cells, as well as on the uptake of glucose into the satiety center.

_____ 2. Effect of insulin on glycogen and fat synthesis.

_____ 3. Effect of insulin on blood sugar levels.

_____ 4. Effect of glucagon on the breakdown of liver glycogen to glucose.

_____ 5. Effect of glucagon on glucose synthesis from amino acids and fats.

_____ 6. Effect of glucagon on blood sugar levels.

_____ 7. Effect of glucagon on fat breakdown and production of ketones.

Chapter 18

C. Match these terms with the correct statement as it pertains to insulin secretion:

Decreases
Increases

_____ 1. Hyperglycemia, certain amino acids.

_____ 2. Parasympathetic stimulation or gastrointestinal hormones.

D. Match these terms with the correct statement as it pertains to glucagon secretion:

Decreases
Increases

_____ 1. Hypoglycemia, certain amino acids.

_____ 2. Sympathetic stimulation.

Hormonal Regulation of Nutrients

A. Match these terms with the correct statement for conditions immediately after a meal:

Decreased
Increased

_____ 1. Levels of glucagon, cortisol, growth hormone, and epinephrine.

_____ 2. Insulin secretion.

_____ 3. Uptake of glucose, amino acids, and fats.

_____ 4. Glucose converted to glycogen.

B. Match these terms with the correct statement for conditions 2 hours after a meal:

Decreased
Increased

_____ 1. Levels of glucagon, cortisol, growth hormone, and epinephrine.

_____ 2. Insulin secretion.

_____ 3. Uptake of glucose.

_____ 4. Glycogen converted to glucose.

_____ 5. Fats and proteins used as an energy source for most tissues.

C. Match these terms with the correct statement for conditions during exercise:

Decreased
Increased

_____ 1. Sympathetic division stimulation.

_____ 2. Release of epinephrine and glucagon.

_____ 3. Release of insulin.

_____ 4. Fatty acids, triglycerides, and ketones in blood.

_____ 5. Fat and glycogen used in skeletal muscle as an energy source.

Hormones of the Reproductive System

Match these terms with the correct statement or definition:

Estrogen and progesterone Relaxin
FSH and LH Testosterone
Inhibin

_____ 1. Hormone that stimulates sperm production and maintains male reproductive organs and secondary sexual characteristics.

_____ 2. Hormone secreted by the testes or ovaries that inhibits the secretion of FSH from the anterior pituitary.

_____ 3. Main hormones secreted by the ovaries in females.

_____ 4. Hormones secreted by the anterior pituitary that regulate the functions of the ovaries and testes.

_____ 5. Hormone secreted by the ovaries and placenta during pregnancy that increases the flexibility of the symphysis pubis and helps dilate the cervix of the uterus.

Hormones of the Pineal Body, Thymus, and Others

Match these terms with the correct statement or definition:

Melatonin Thymosin
Pineal body

_____ 1. Endocrine gland in the epithalamus that secretes hormones that inhibit reproductive function.

_____ 2. One secretion of the pineal body; production of this hormone decreases as photoperiod increases.

_____ 3. Hormone produced by the thymus; affects the immune system.

Hormonelike Substances

A. Match these terms with the correct statement or definition:

Autocrine chemical signals
Paracrine chemical signals

_____ 1. Released from cells and influence the same cell type.

_____ 2. Released from cells of one cell type, diffuse a short distance, and influence the activity of another cell type.

Chapter 18

B. Using the terms provided, complete these statements:

Arachidonic acid
Aspirin
Endorphins
Inflammation
Pain receptors
Peptide growth factors
Platelet activating factor
Prostaglandins

Examples of autocrine chemical signals include chemical mediators of inflammation derived from the fatty acid (1) , such as eicosanoids and modified phospholipids. The eicosanoids include (2) , thromboxanes, prostacyclins, and leukotrienes. Prostaglandins and other eicosanoids are released from injured cells and are responsible for initiating some of the symptoms of (3) . Prostaglandins directly stimulate (4) , or cause dilation of blood vessels, which is associated with headache. Anti-inflammatory drugs, such as (5) , inhibit prostaglandin synthesis and reduce inflammation and pain. Modified phospholipids include (6) . Paracrine chemical signals include substances that play a role in moderating the sensation of pain, including (7) , enkephalins, and dynorphins, and several (8) , such as epidermal growth factor, fibroblast growth factor, and interleukin-2.

1. _____
2. _____
3. _____
4. _____
5. _____
6. _____
7. _____
8. _____

Effects of Aging on the Endocrine System

Match these terms with the correct statement as it pertains to the <u>effects of aging</u> on the endocrine system:

Decrease
Increase
No change

_____ 1. Effect of aging on the secretion of GH, melatonin, and thyroid hormones.

_____ 2. Effect of aging on parathyroid hormone secretion.

_____ 3. Effect of aging on secretion of renin by the kidneys.

_____ 4. Effect of aging on secretion of reproductive hormones.

_____ 5. Effect of aging on ability to regulate blood glucose levels.

_____ 6. Effect of aging on the secretion of thymosin and the ability of the immune system to protect the body.

Quick Recall

A. Match these endocrine glands with the correct hormone each secretes:

Adrenal cortex
Adrenal medulla
Anterior pituitary
Hypothalamus
Ovaries
Pancreas
Parathyroid glands
Pineal body
Posterior pituitary
Testes
Thymus
Thyroid gland (follicular cells)
Thyroid gland (parafollicular cells)

_____ 1. ACTH.

_____ 2. ADH.

_____ 3. Adrenal androgens.

_____ 4. Aldosterone.

_____ 5. Calcitonin.

_____ 6. Cortisol.

_____ 7. CRH.

_____ 8. Epinephrine.

_____ 9. Estrogen.

_____ 10. FSH.

_____ 11. GH.

_____ 12. GH-RH and GH-IH.

_____ 13. Glucagon.

_____ 14. GnRH.

_____ 15. Insulin.

_____ 16. LH.

_____ 17. Melatonin.

_____ 18. MSH.

_____ 19. Norepinephrine.

_____ 20. Oxytocin.

_____ 21. Parathyroid hormone.

_____ 22. Progesterone.

_____ 23. Prolactin.

_____ 24. PRH and PIH.

_____ 25. Testosterone.

_____ 26. Tetraiodothyronine and triiodothyronine.

_____ 27. Thymosin.

_____ 28. TRH.

_____ 29. TSH.

B. Match these target tissues with the correct hormone that affects them:

Adipose tissue, liver, and skeletal muscle
Adrenal cortex
Blood vessels and heart
Bone
Immune tissues
Kidneys
Liver
Mammary glands
Most tissues of the body
Ovaries or testes
Skin
Thyroid gland
Uterus

_____ 1. ADH.

_____ 2. Oxytocin (two).

_____

_____ 3. GH.

_____ 4. TSH.

_____ 5. ACTH.

_____ 6. MSH.

_____ 7. LH and FSH.

_____ 8. Prolactin.

_____ 9. Thyroid hormones.

_____ 10. Calcitonin.

_____
_____ 11. PTH (two).

_____
_____ 12. Epinephrine (two).

_____ 13. Aldosterone.

_____ 14. Glucocorticoids.

_____ 15. Insulin.

_____ 16. Glucagon.

_____ 17. Testosterone, progesterone, and estrogen.

_____ 18. Thymosin.

C. Simmond disease results from a tumor in the pituitary gland or from a lack of blood supply to the pituitary. Symptoms are similar to those that occur when the pituitary gland is removed. What effects (increase, decrease, or no effect) does Simmond disease have on these variables?

_____ 1. Metabolic rate.

_____ 2. Height, if condition developed in a child.

_____ 3. Ability to deal with stress.

_____ 4. Blood sugar levels.

_____ 5. Blood calcium levels.

_____ 6. Milk production.

_____ 7. Ability to produce gametes (sperm cells or oocytes).

ANSWERS TO CHAPTER 18

CONTENT LEARNING ACTIVITY

Pituitary Gland and Hypothalamus
A. 1. Infundibulum; 2. Posterior pituitary;
   3. Anterior pituitary;
   4. Hypothalamohypophysial portal system;
   5. Neurohormones; 6. Hypothalamus;
   7. Anterior pituitary; 8. Action potentials
B. 1. GHIH; 2. TRH; 3. CRH; 4. GnRH; 5. PRH
C. 1. Neurosecretory cell;
   2. Hypothalamohypophysial portal system;
   3. Anterior pituitary; 4. Posterior pituitary;
   5. Hypothalamohypophysial tract;
   6. Hypothalamus

Posterior Pituitary Hormones
A. 1. Decreases; 2. Decreases; 3. Increases;
   4. Increases
B. 1. Blood osmolality; 2. Blood pressure;
   3. Blood osmolality; 4. Blood pressure
C. 1. Increases; 2. Increases
D. 1. Increases; 2. Increases

Anterior Pituitary Hormones
A. 1. Increases; 2. Increases; 3. Increases;
   4. Increases; 5. Increases
B. 1. Increases; 2. Increases; 3. Increases;
   4. Decreases
C. 1. Hyposecretion; 2. Hypersecretion;
   3. Hypersecretion
D. 1. Proopiomelanocortin; 2. Lipotropins;
   3. β endorphins; 4. MSH and ACTH
E. 1. FSH and LH; 2. GnRH; 3. Prolactin; 4. PRH

Thyroid Gland
A. 1. Follicle; 2. Parafollicular cells; 3. Thyroid hormones; 4. Calcitonin
B. 1. Thyroglobulins; 2. Tyrosine; 3. I⁻;
   4. Tetraiodothyronine (T_4);
   5. Triiodothyronine (T_3); 6. Thyroxine-binding globulin
C. 1. Increases; 2. Increases; 3. Decreases
D. 1. Hypersecretion; 2. Hypersecretion;
   3. Hyposecretion; 4. Hyposecretion;
   5. Hyposecretion; 6. Hyposecretion;
   7. Hypersecretion; 8. Hypersecretion
E. 1. Increases; 2. Decreases; 3. Increases;
   4. Increases; 5. Decreases; 6. Decreases
F. 1. Decreases; 2. Decreases; 3. Increases

Parathyroid Glands
A. 1. Increases; 2. Increases; 3. Increases;
   4. Increases; 5. Decreases
B. 1. Increases
C. 1. Hypersecretion; 2. Hyposecretion;
   3. Hyposecretion

Adrenal Glands
1. Adrenal medulla; 2. Adrenal medulla;
3. Adrenal cortex

Adrenal Medulla
A. 1. Epinephrine; 2. Neurohormone
B. 1. Increases; 2. Increases; 3. Decreases;
   4. Increases
C. 1. Increases; 2. Increases; 3. Increases
D. 1. Hypersecretion; 2. Hypersecretion

Adrenal Cortex
A. 1. Mineralocorticoids; 2. Glucocorticoids;
   3. Androstenedione
B. 1. Increases; 2. Decreases; 3. Decreases;
   4. Increases
C. 1. Hypersecretion; 2. Hypersecretion;
   3. Hyposecretion
D. 1. Increases; 2. Decreases; 3. Increases;
   4. Increases; 5. Decreases
E. 1. Hypothalamus; 2. CRH; 3. ACTH;
   4. ACTH and cortisol; 5. Cortisol;
   6. Hypoglycemia or stress
F. 1. Hypersecretion of cortisol;
   2. Hypersecretion of cortisol;
   3. Hypersecretion of cortisol;
   4. Hyposecretion of cortisol;
G. 1. Aldosteronism; 2. Cushing syndrome;
   3. Addison disease
H. 1. Increases; 2. Increases

Pancreas and Pancreatic Hormones
A. 1. Ducts and acini; 2. Alpha cells; 3. Beta cells;
   4. Delta cells
B. 1. Increases; 2. Increases; 3. Decreases;
   4. Increases; 5. Increases; 6. Increases;
   7. Increases
C. 1. Increases; 2. Increases
D. 1. Increases; 2. Increases

Hormonal Regulation of Nutrients
A. 1. Decreased; 2. Increased; 3. Increased;
   4. Increased
B. 1. Increased; 2. Decreased; 3. Decreased;
   4. Increased; 5. Increased
C. 1. Increased; 2. Increased; 3. Decreased;
   4. Increased; 5. Increased

Hormones of the Reproductive System
1. Testosterone; 2. Inhibin; 3. Estrogen and progesterone; 4. FSH and LH; 5. Relaxin

Hormones of the Pineal Body, Thymus, and Others
1. Pineal body; 2. Melatonin; 3. Thymosin

Hormonelike Substances
A. 1. Autocrine chemical signals; 2. Paracrine chemical signals
B. 1. Arachidonic acid; 2. Prostaglandins;
   3. Inflammation; 4. Pain receptors; 5. Aspirin;
   6. Platelet activating factor; 7. Endorphins;
   8. Peptide growth factors

Effects of Aging on the Endocrine System
A. 1. Decrease; 2. No change; 3. Decrease;
   4. Decrease; 5. No change; 6. Decrease

Quick Recall

A.
1. Anterior pituitary
2. Posterior pituitary
3. Adrenal cortex
4. Adrenal cortex
5. Thyroid gland (parafollicular cells)
6. Adrenal cortex
7. Hypothalamus
8. Adrenal medulla
9. Ovaries
10. Anterior pituitary
11. Anterior pituitary
12. Hypothalamus
13. Pancreas (alpha cells)
14. Hypothalamus
15. Pancreas (beta cells)
16. Anterior pituitary
17. Pineal body
18. Anterior pituitary
19. Adrenal medulla
20. Posterior pituitary
21. Parathyroid glands
22. Ovaries
23. Anterior pituitary
24. Hypothalamus
25. Testes
26. Thyroid gland (follicular cells)
27. Thymus
28. Hypothalamus
29. Anterior pituitary

B.
1. Kidneys
2. Uterus; mammary glands
3. Most tissues of the body
4. Thyroid gland
5. Adrenal cortex
6. Skin
7. Ovaries or testes
8. Mammary glands
9. Most tissues of the body
10. Bone
11. Bone; kidneys
12. Blood vessels and heart; adipose tissue, liver, and skeletal muscle
13. Kidneys
14. Most tissues
15. Adipose tissue, liver, and skeletal muscle
16. Liver
17. Most tissues of the body
18. Immune cells

C.
1. Decrease (TSH, T_4, T_3)
2. Decrease (GH)
3. Decrease (ACTH, cortisol)
4. Possible decrease (ACTH, cortisol), but glucagon and epinephrine might stabilize
5. No effect (PTH and calcitonin)
6. Decrease (prolactin)
7. Decrease (FSH, LH)

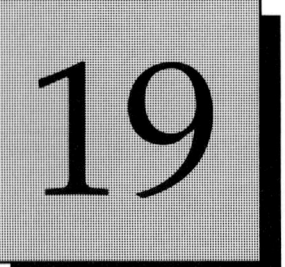

Cardiovascular System: Blood

CONTENT LEARNING ACTIVITY

Plasma

Match these terms with the correct statement or definition:

Albumin
Colloid
Fibrinogen
Globulins
Plasma
Serum
Water

1. Pale yellow fluid; makes up slightly more than half of blood volume.

2. Major (approximately 91%) component of plasma.

3. Fine particles suspended in a liquid and resistant to sedimentation or filtration.

4. Plasma proteins important in the regulation of water movement between tissues and blood.

5. Plasma proteins that include antibodies, complement, and transport molecules.

6. Plasma protein involved in blood clotting.

7. Plasma without the clotting factors.

Formed Elements

Match these terms with the correct statement or definition:

Agranulocytes
Granulocytes
Platelets
Red blood cells
White blood cells

1. Cells that constitute most of the formed elements; erythrocytes.

2. Formed elements with nuclei; leukocytes.

3. Cell fragments that are part of the formed elements; thrombocytes.

_____ 4. White blood cells containing large cytoplasmic granules; includes neutrophils, eosinophils, and basophils.

_____ 5. White blood cells with very small cytoplasmic granules; includes lymphocytes and monocytes.

Production of Formed Elements

Match these terms with the correct statement or definition:

Hematopoiesis
Lymphoblasts
Megakaryoblasts
Monoblasts
Myeloblasts
Proerythroblasts
Stem cells (hemocytoblasts)

_____ 1. Blood cell production process that takes place in red bone marrow and lymphoid tissue after birth; hemopoiesis.

_____ 2. Cells from which all the formed elements develop.

_____ 3. Cells from which granulocytes develop.

_____ 4. Cells from which lymphocytes develop.

_____ 5. Cells from which platelets develop.

Red Blood Cells

A. Using the terms provided, complete these statements:

Biconcave
Hemoglobin
Nucleus

A red blood cell is a __(1)__ disk that loses its __(2)__ and nearly all of its organelles during maturation. The main component of the red blood cell is the pigmented protein __(3)__, which occupies about one-third of the total cell volume.

1. _____
2. _____
3. _____

B. Match these terms with the correct statement or definition:

Bicarbonate ion
Carbonic acid
Carbonic anhydrase
Hemolysis

_____ 1. Process in which red blood cells rupture and hemoglobin is released.

_____ 2. Enzyme that catalyzes the reaction between carbon dioxide and water.

_____ 3. Product of the reaction between carbon dioxide and water.

_____ 4. Major form of carbon dioxide transported in the blood.

C. Match these terms with the correct statement or definition:

Carbaminohemoglobin
Deoxyhemoglobin
Globin
Heme
Iron
Nitric oxide
Oxyhemoglobin

_____ 1. One of four polypeptide chains in the hemoglobin molecule.

_____ 2. Red pigment molecule.

_____ 3. Element in the center of each heme molecule; binds with oxygen.

_____ 4. Hemoglobin that has oxygen associated with each heme group.

_____ 5. Hemoglobin with a darker red color; has no oxygen.

_____ 6. Hemoglobin with carbon dioxide attached to amino groups of the globin molecule.

_____ 7. Transported by hemoglobin; causes blood vessels to relax.

D. Using the terms provided, complete these statements:

Anemia
Iron
Less
More

Embryonic and fetal hemoglobins are _(1)_ effective at binding oxygen than is adult hemoglobin. Abnormal hemoglobins are _(2)_ effective at attracting oxygen than is normal hemoglobin and may result in _(3)_. _(4)_ is necessary for the normal function of hemoglobin. Dietary iron is absorbed into the circulation in the upper part of the digestive tract.

1. _____
2. _____
3. _____
4. _____

E. Using the terms provided, complete these statements:

Erythropoiesis
Hemoglobin
Late erythroblasts
Proerythroblasts
Red blood cells
Reticulocytes

The process by which new red blood cells are produced is called _(1)_. The cells from which red blood cells develop are _(2)_. They become early erythroblasts, which begin to produce _(3)_. The early erythroblasts become intermediate erythroblasts, which become _(4)_. These cells have almost a complete complement of hemoglobin. The late erythroblasts lose their nuclei by a process of extrusion, after which the cells are called _(5)_. These cells are released from red bone marrow into the blood, and within 1 or 2 days they lose their remaining ribosomes and become _(6)_.

1. _____
2. _____
3. _____
4. _____
5. _____
6. _____

F. Match these terms with the correct statement or definition:

Erythropoietin	Iron
Folate and vitamin B$_{12}$	Oxygen

_____ 1. Necessary for cell division in erythropoiesis.

_____ 2. Low levels of this substance stimulate the formation of erythropoietin.

_____ 3. Stimulates red bone marrow to produce more red blood cells.

_____ 4. Necessary for heme formation.

G. Match these terms with the correct parts of the diagram labeled in figure 19.1:

Bilirubin	Hemoglobin
Globin	Iron
Heme

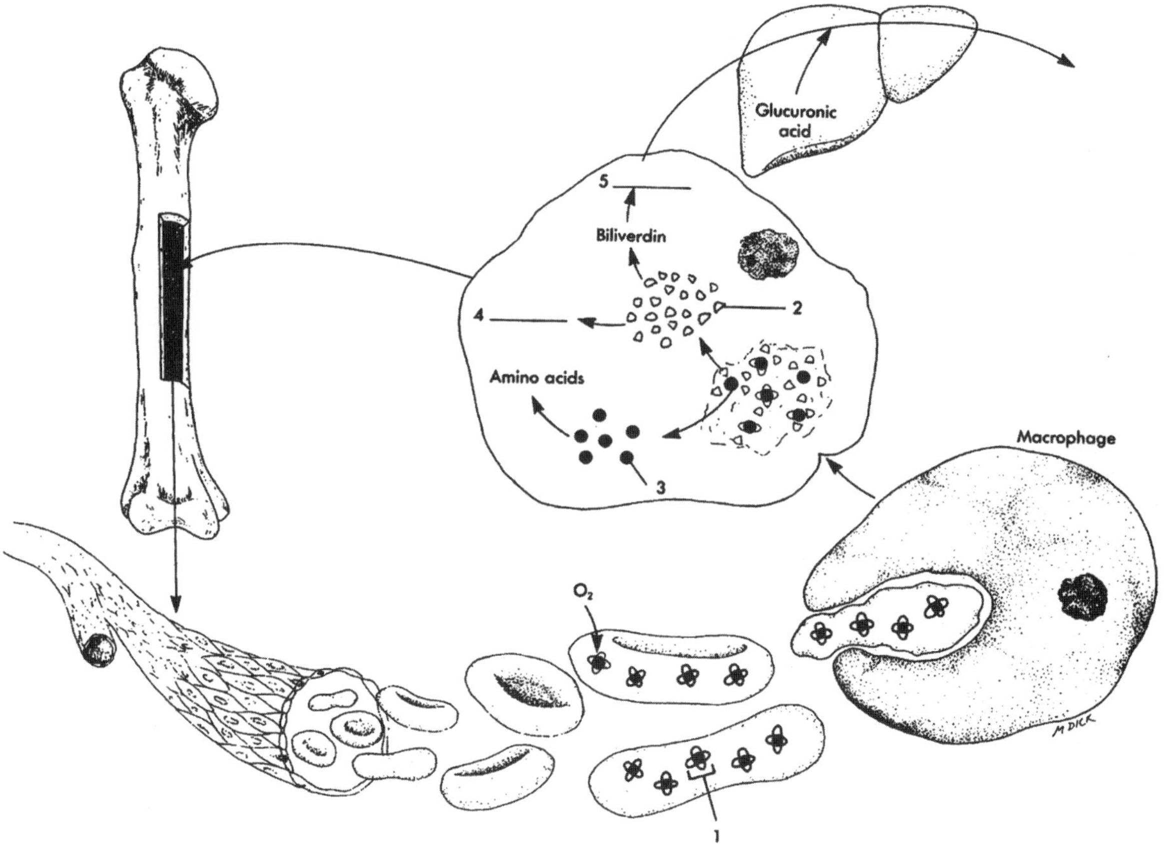

Figure 19.1

1. _____   3. _____   5. _____
2. _____   4. _____

Chapter 19

H. Match these terms with the correct statement or definition:

Bilirubin
Biliverdin
Conjugated
Free
Macrophages

_____ 1. Remove old or damaged red blood cells from the blood.

_____ 2. First breakdown product of heme groups.

_____ 3. Formed from biliverdin; excreted by the liver in bile; excess in blood causes jaundice.

_____ 4. Bilirubin that binds to albumin; transported to the liver.

_____ 5. Bilirubin that is water-soluble; excreted by the kidneys.

White Blood Cells

A. Match these terms with the correct statement or definition:

Chemotaxis
Diapedesis
Lysozyme
Phagocytosis
Polymorphonuclear neutrophils
Pus

_____ 1. Process by which white blood cells leave the circulation.

_____ 2. Process by which white blood cells find foreign material or dead cells.

_____ 3. Process by which white blood cells ingest bacteria, dirt, and dead cells.

_____ 4. Accumulation of dead white blood cells, fluid, and debris.

_____ 5. Another name for neutrophils, referring to their lobed nuclei.

_____ 6. Class of enzyme, secreted by neutrophils, capable of destroying certain bacteria.

B. Match these terms with the correct statement or definition:

Basophils
Eosinophils
Heparin
Histamine

_____ 1. White blood cells that accept an acid stain.

_____ 2. White blood cells that produce enzymes that destroy histamine.

_____ 3. White blood cells that contain histamine and heparin.

_____ 4. Substance that increases inflammation.

_____ 5. Substance that inhibits blood clotting.

C. Match these terms with the correct statement or definition:

B cell Monocyte
Lymphocyte T cell
Macrophage

_____ 1. Smallest of all white blood cells.

_____ 2. Type of lymphocyte that protects against bacteria and toxins by producing antibodies.

_____ 3. Type of lymphocyte that protects against viruses by destroying the cells in which they are found.

_____ 4. This cell enters tissues and is transformed into a macrophage.

_____ 5. Large "eating" cell that phagocytizes bacteria, dead cells, cell fragments, and debris; often associated with chronic infections.

Platelets

Match these terms with the correct statement or definition:

Megakaryocytes
Platelets

_____ 1. Minute fragments of cells consisting of a small amount of cytoplasm surrounded by a cell membrane; thrombocytes.

_____ 2. Cells from which platelets are formed.

Vascular Spasm and Platelet Plug Formation

A. Match these terms with the correct statement or definition:

Endothelin Thromboxanes
Hemostasis Vascular spasm

_____ 1. Can be divided into three stages: vascular spasm, platelet plug formation, and coagulation.

_____ 2. Immediate but temporary closure of a blood vessel resulting from the contraction of smooth muscles within the blood vessel's wall.

_____ 3. Released by platelets, these substances derived from prostaglandins stimulate vascular spasms.

_____ 4. Released by the lining of blood vessels, this peptide stimulates vascular spasms.

B. Match these terms with the correct statement or definition:

Platelet adhesion Platelet plug
Platelet aggregation Platelet release reaction

_____ 1. Platelet surface receptors bind to collagen through Von Willebrand's factor or bind directly to collagen.

_____ 2. Activated platelets release ADP and thromboxane, which activate other platelets.

_____ 3. Platelet surface receptors on one platelet bind to the surface receptors of another platelet using fibrinogen.

_____ 4. Cluster of platelets; seals small holes in blood vessels.

Coagulation

A. Match these terms with the correct statement or definition:

Coagulation Prothrombinase
Coagulation factors Thrombin
Fibrin Vitamin K

_____ 1. Formation of a blood clot.

_____ 2. Threadlike proteins that combine to form a network called a blood clot.

_____ 3. Plasma proteins involved in clot formation; normally in an inactive state.

_____ 4. Converts prothrombin to thrombin.

_____ 5. Converts fibrinogen to fibrin.

_____ 6. Compound required for the production of many blood clotting factors.

B. Match these stages of clotting with the processes:

Common pathway Intrinsic pathway
Extrinsic pathway

_____ 1. Thromboplastin (issue factor) is produced by damaged cells.

_____ 2. Platelet factor XII is activated in the plasma.

_____ 3. Prothrombinase is formed.

_____ 4. Prothrombin is converted to thrombin.

_____ 5. Fibrinogen is converted to fibrin.

C. Match these terms with the correct statement or definition:

Anticoagulant Heparin
Antithrombin Prostacyclin

_____ 1. Substance that prevents the initiation of clot formation.

_____ 2. Plasma protein produced by the liver; inactivates thrombin.

_____ 3. Increases the effectiveness of antithrombin.

_____ 4. Prostaglandin derivative that counteracts thrombin.

D. Match these terms with the correct statement or definition:

Clot retraction   Plasmin
Embolus   Serum
Fibrinolysis   Thrombus

_____ 1. Clot within a blood vessel.

_____ 2. Clot that has broken loose and is floating through the circulation.

_____ 3. Condensation of a clot into a denser, more compact structure.

_____ 4. Plasma with the clot-producing proteins removed.

_____ 5. Dissolution of a clot.

_____ 6. Enzyme that hydrolyzes fibrin.

E. Match these terms with the correct statement or definition:

Antithrombin   Sodium citrate
EDTA   Streptokinase
Heparin   Tissue plasminogen activator
Prostacyclin

_____ 1. Three chemicals commonly used to prevent blood from clotting outside the body.
_____
_____

_____ 2. Two chemicals used to dissolve clots.
_____

Blood Grouping

A. Match these terms with the correct statement or definition:

Agglutination   Hemolysis
Antibodies   Infusion
Antigens   Transfusion

_____ 1. Transfer of blood or blood components into the blood.

_____ 2. Introduction of a fluid other than blood into the blood.

_____ 3. Substances recognized by the immune system; agglutinogens.

_____ 4. Proteins that react with antigens on blood cells; agglutinins.

_____ 5. Clumping of cells.

_____ 6. Rupture of red blood cells.

B. Match these blood types with the correct statement:

Type A blood   Type B blood
Type AB blood Type O blood

_____ 1. In this blood type, red blood cells contain both A and B antigens.

_____ 2. These blood types have anti-A antibodies.

_____

_____ 3. This blood type is the most common throughout the world.

_____ 4. A person (universal donor) with this type of blood can usually give blood to a person with any blood type.

_____ 5. A person (universal recipient) with this type of blood can usually receive blood from anyone.

C. Match these blood types with the correct statement:

Rh-negative
Rh-positive

_____ 1. Blood type with no Rh antigens.

_____ 2. A fetus with this type of blood could develop hemolytic disease of the newborn.

_____ 3. A woman with this type of blood will never have a baby who develops hemolytic disease of the newborn.

Diagnostic Blood Tests

Match these terms with the correct statement or definition:

Blood chemistry   Platelet count
Complete blood count   Prothrombin time
Differential white   Red blood count
  blood count   Type and crossmatch
Hematocrit   White blood count
Hemoglobin

_____ 1. Test used to prevent transfusion reactions.

_____ 2. Includes a red blood cell count, hemoglobin and hematocrit measurements, and white blood cell count.

_____ 3. Test that can detect erythrocytosis.

_____ 4. Percentage of total blood volume composed of red blood cells.

_____ 5. Test that can detect leukemia.

_____ 6. Test that determines the percentages of each of the five kinds of white blood cells.

_____ 7. Test that can detect thrombocytopenia.

_____ 8. Measures how long it takes for blood to start clotting.

_____ 9. Determines the composition of materials dissolved or suspended in plasma, such as glucose and bilirubin.

Blood Disorders

A. Match these types of anemia with the correct description:

Aplastic anemia   Pernicious anemia
Hemolytic anemia   Sickle-cell disease
Hemorrhagic anemia   Thalassemia
Iron-deficiency anemia

_____ 1. Insufficient red blood cell production caused by abnormal red bone marrow or destruction of the red bone marrow.

_____ 2. Caused by insufficient vitamin B_{12} or intrinsic factor.

_____ 3. Deficiency of red blood cells caused by the loss of large quantities of blood.

_____ 4. Anemia in which red blood cells rupture or are destroyed at an excessive rate.

_____ 5. Hereditary disorder that causes insufficient production of the globin portion of hemoglobin.

_____ 6. Hereditary disorder that causes the formation of an abnormal hemoglobin.

B. Match these blood disorders with the correct description:

AIDS   Infectious mononucleosis
Erythrocytosis   Leukemia
Hemophilia   Malaria
Hepatitis   Septicemia

_____ 1. Condition characterized by an overabundance of red blood cells.

_____ 2. Genetic disorder in which coagulation factors are abnormal or absent.

_____ 3. Type of cancer in which abnormal production of one or more white blood cells occurs.

_____ 4. Spread of microorganisms and their toxins in the blood.

_____ 5. Protozoan infection of red blood cells; symptoms include fever and chills.

_____ 6. Viral infection of lymphocytes (B cells) that alters the B cells, resulting in their destruction by the immune system.

_____ 7. Two viral diseases that can be acquired through transfusions.

_____

Quick Recall

1. List seven major functions of blood.

2. Name the parts of a hemoglobin molecule, give the function of each part, and state the fate of each part when hemoglobin is broken down.

3. List the events that lead to increased red blood cell production when blood oxygen levels decrease.

4. List the functions of the five types of white blood cells.

5. Give two ways that platelets prevent blood loss.

6. List the three steps in platelet plug formation.

7. For the extrinsic and intrinsic pathways of clotting, list the starting and ending chemicals.

8. For the common pathway of clotting, list the chemical reactions that occur.

9. List the chemicals in the body that prevent clot formation or dissolve clots.

ANSWERS TO CHAPTER 19

CONTENT LEARNING ACTIVITY

Plasma
1. Plasma; 2. Water; 3. Colloid; 4. Albumin; 5. Globulins; 6. Fibrinogen; 7. Serum

Formed Elements
1. Red blood cells; 2. White blood cells; 3. Platelets; 4. Granulocytes; 5. Agranulocytes

Production of Formed Elements
1. Hematopoiesis; 2. Stem cells (hemocytoblasts); 3. Myeloblasts; 4. Lymphoblasts; 5. Megakaryoblasts

Red Blood Cells
A. 1. Biconcave; 2. Nucleus; 3. Hemoglobin
B. 1. Hemolysis; 2. Carbonic anhydrase; 3. Carbonic acid; 4. Bicarbonate ion
C. 1. Globin; 2. Heme; 3. Iron; 4. Oxyhemoglobin; 5. Deoxyhemoglobin; 6. Carbaminohemoglobin; 7. Carbaminohemoglobin
D. 1. More; 2. Less; 3. Anemia; 4. Iron
E. 1. Erythropoiesis; 2. Proerythroblasts; 3. Hemoglobin; 4. Late erythroblasts; 5. Reticulocytes; 6. Red blood cells
F. 1. Folate and vitamin B_{12}; 2. Oxygen; 3. Erythropoietin; 4. Iron
G. 1. Hemoglobin; 2. Heme; 3. Globin; 4. Iron; 5. Bilirubin
H. 1. Macrophages; 2. Biliverdin; 3. Bilirubin; 4. Free; 5. Conjugated

White Blood Cells
A. 1. Diapedesis; 2. Chemotaxis; 3. Phagocytosis; 4. Pus; 5. Polymorphonuclear neutrophils; 6. Lysozyme
B. 1. Eosinophils; 2. Eosinophils; 3. Basophils; 4. Histamine; 5. Heparin
C. 1. Lymphocyte; 2. B cells; 3. T cells; 4. Monocyte; 5. Macrophage

Platelets
1. Platelets; 2. Megakaryocytes

Vascular Spasm and Platelet Plug Formation
A. 1. Hemostasis; 2. Vascular spasm; 3. Thromboxanes; 4. Endothelin
B. 1. Platelet adhesion; 2. Platelet release reaction; 3. Platelet aggregation; 4. Platelet plug

Coagulation
A. 1. Coagulation; 2. Fibrin; 3. Coagulation factors; 4. Prothrombinase; 5. Thrombin; 6. Vitamin K
B. 1. Extrinsic pathway; 2. Intrinsic pathway; 3. Common pathway; 4. Common pathway; 5. Common pathway
C. 1. Anticoagulant; 2. Antithrombin; 3. Heparin; 4. Prostacyclin
D. 1. Thrombus; 2. Embolus; 3. Clot retraction; 4. Serum; 5. Fibrinolysis; 6. Plasmin
E. 1. Heparin, EDTA, and sodium citrate; 2. Streptokinase and tissue plasminogen activator

Blood Grouping
A. 1. Transfusion; 2. Infusion; 3. Antigens; 4. Antibodies; 5. Agglutination; 6. Hemolysis
B. 1. Type AB blood; 2. Type B blood, type O blood; 3. Type O blood; 4. Type O blood; 5. Type AB blood
C. 1. Rh-negative; 2. Rh-positive; 3. Rh-positive

Diagnostic Blood Tests
1. Type and crossmatch; 2. Complete blood count; 3. Red blood count; 4. Hematocrit; 5. White blood count; 6. Differential white blood count; 7. Platelet count; 8. Prothrombin time; 9. Blood chemistry

Blood Disorders
A. 1. Aplastic anemia; 2. Pernicious anemia; 3. Hemorrhagic anemia; 4. Hemolytic anemia; 5. Thalassemia; 6. Sickle-cell disease
B. 1. Erythrocytosis; 2. Hemophilia; 3. Leukemia; 4. Septicemia; 5. Malaria; 6. Infectious mononucleosis; 7. AIDS, hepatitis

QUICK RECALL

1. Transport of gases nutrients, and waste products; transport of processed molecules; transport of regulatory molecules; regulation of pH and osmosis; maintenance of body temperature; protection against foreign substances; clot formation
2. Globin: transports carbon dioxide, broken down to amino acids that are recycled; heme: transports oxygen, broken down to biliverdin, then to bilirubin, which is carried in the plasma to the liver, where it is incorporated into bile; iron is recycled
3. Decreased oxygen, increased erythropoietin production by kidney, erythropoiesis stimulated
4. Neutrophils: phagocytize foreign matter, secrete lysozyme; eosinophils: reduce inflammatory response; basophils: release histamine for inflammatory or allergic response; lymphocytes: immunity, including antibody production; monocytes: become macrophages
5. Formation of platelet plugs and formation of clots
6. Platelet adhesion; platelet release reaction; platelet aggregation
7. Extrinsic pathway: thromboplastin (tissue factor) to factor X, intrinsic pathway: factor XII (Hageman factor) to factor X
8. Factor X to prothrombinase, prothrombin to thrombin, fibrinogen to fibrin
9. Heparin, antithrombin, and prostacyclin prevent clots; plasmin dissolves clots

20 Cardiovascular System: The Heart

CONTENT LEARNING ACTIVITY

Size, Shape, and Location of the Heart

Match these terms with the correct statement or definition:
    Apex   Mediastinum
    Base

_____ 1. Blunt, rounded point of the heart.

_____ 2. End of the heart where veins enter and arteries exit; superior part of the heart.

_____ 3. Midline partition in the thorax that contains the heart.

Anatomy of the Heart

A. Match these terms with the correct statement or definition:
    Fibrous pericardium   Pericardium
    Parietal pericardium   Serous pericardium
    Pericardial cavity   Visceral pericardium
    Pericardial fluid   (epicardium)

_____ 1. Double-layered, closed sac that surrounds the heart; prevents overdistension of the heart and anchors it within the mediastinum.

_____ 2. Tough, outer layer of the pericardium.

_____ 3. Portion of the serous pericardium that lines the fibrous pericardium.

_____ 4. Portion of the serous pericardium that covers the heart surface.

_____ 5. Space between the visceral and parietal pericardium.

_____ 6. Fluid that fills the pericardial cavity.

B. Match these terms with the correct statement or definition:

Crista terminalis Myocardium
Endocardium Pectinate muscles
Epicardium Trabeculae carneae

_____ 1. Thin serous membrane that constitutes the outer surface of the heart.

_____ 2. Thick middle layer of the heart; composed of cardiac muscle cells.

_____ 3. Smooth inner surface of the heart.

_____ 4. Muscular ridges in both auricles.

_____ 5. Ridge that separates the pectinate muscles from the smooth portion of the right atrium.

_____ 6. Large ridges and columns in the ventricles.

C. Match these vessels or structures with the correct description:

Anterior interventricular Posterior interventricular
  sulcus   sulcus
Aorta Pulmonary trunk
Auricles Pulmonary veins
Coronary sulcus Venae cavae

_____ 1. Flaplike extensions of the atria.

_____ 2. Veins that carry blood from the body to the right atrium.

_____ 3. Veins that carry blood from the lungs to the left atrium.

_____ 4. Large artery that exits the right ventricle.

_____ 5. Large artery that exits the left ventricle.

_____ 6. Large groove that separates the atria and ventricles and runs obliquely around the heart.

_____ 7. Anterior groove between the right and left ventricles.

D. Match these vessels with the correct description:

Anterior interventricular Left marginal artery
  artery Posterior interventricular
Circumflex artery   artery
Coronary arteries Right marginal artery
Coronary sinus Small cardiac vein
Great cardiac vein

_____ 1. Two vessels that exit the aorta just above the point where the aorta leaves the heart.

_____ 2. Branch of the left coronary artery; supplies blood to most of the anterior part of the heart.

Chapter 20

_____ 3. Branch of the left coronary artery; supplies blood to the lateral wall of the left ventricle.

_____ 4. Branch of the left coronary artery; supplies blood to the posterior wall of the heart.

_____ 5. Branch of the right coronary artery; supplies blood to the lateral wall of the right ventricle.

_____ 6. Branch of the right coronary artery; supplies blood to the posterior and inferior part of the heart.

_____ 7. Major vein that drains the left side of the heart.

_____ 8. Vein that drains the right margin of the heart.

_____ 9. Large venous cavity that empties into the right atrium; carries blood from the walls of the heart.

E. Match these terms with the correct statement or definition:

Atrioventricular canal   Interatrial septum
Foramen ovale   Interventricular septum
Fossa ovalis

_____ 1. Wall separating the right and left atria.

_____ 2. Oval depression on the right side of the interatrial septum.

_____ 3. Opening between the right and left atria in embryonic and fetal stages of development.

_____ 4. Opening between an atrium and a ventricle.

_____ 5. Wall separating the two ventricles.

F. Match these terms with the correct statement or definition:

Atrioventricular valve   Papillary muscles
Bicuspid (mitral) valve Semilunar valve
Chordae tendineae   Tricuspid valve

_____ 1. General term for a one-way valve between the atrium and ventricle.

_____ 2. Valve between the right atrium and right ventricle.

_____ 3. Valve between the left atrium and left ventricle.

_____ 4. Cone-shaped muscular pillars in the ventricles.

_____ 5. Connective tissue strings between papillary muscles and atrioventricular valves.

_____ 6. One-way valve with three pocketlike cusps; located in the aorta or pulmonary trunk.

G. Match these terms with the correct parts of the diagram labeled in figure 20.1:

Aorta
Aortic semilunar valve
Bicuspid (mitral) valve
Chordae tendineae
Interventricular septum
Left atrium
Left ventricle
Papillary muscles
Pulmonary semilunar valve
Pulmonary trunk
Pulmonary veins
Right atrium
Right ventricle
Superior vena cava
Tricuspid valve

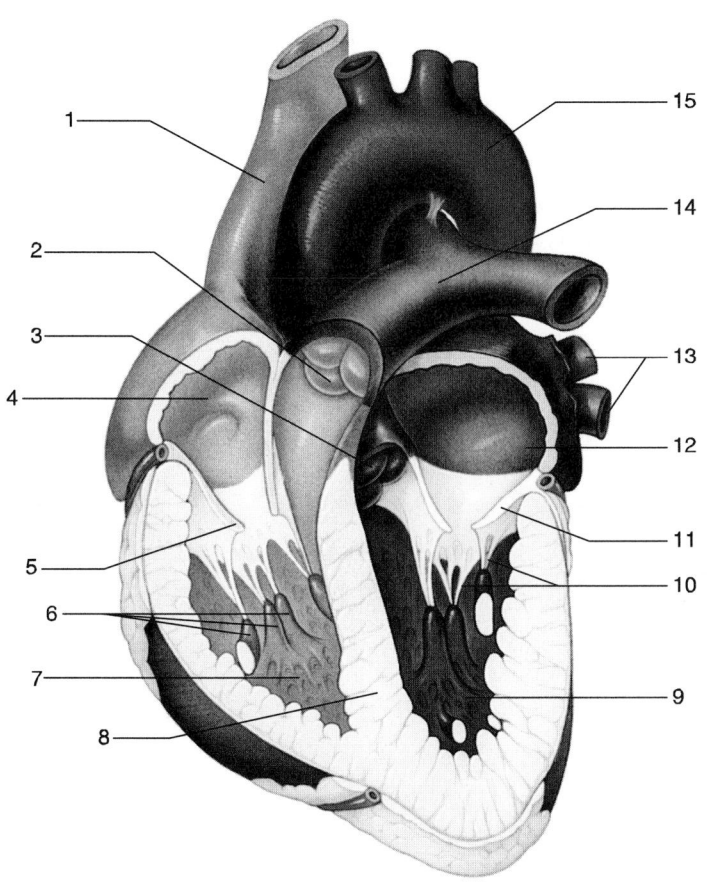

Figure 20.1

1. _____
2. _____
3. _____
4. _____
5. _____
6. _____
7. _____
8. _____
9. _____
10. _____
11. _____
12. _____
13. _____
14. _____
15. _____

Route of Blood Flow Through the Heart

Arrange these terms in sequential order as blood returns to the heart from the body and is pumped through the heart:

Aorta
Aortic semilunar valve
Bicuspid (mitral) valve
Left atrium
Left ventricle

Lungs
Pulmonary arteries
Pulmonary semilunar valve
Pulmonary trunk
Pulmonary veins

Right atrium
Right ventricle
Tricuspid valve
Venae cavae

1. _____
2. _____
3. _____
4. _____
5. _____
6. _____
7. _____
8. _____
9. _____
10. _____
11. _____
12. _____
13. _____
14. _____

Histology

A. Using the terms provided, complete these statements:

Contraction
Desmosomes
Gap junctions
Intercalated disks
Mitochondria
Oxygen deficit
Sarcomeres
Sarcoplasmic reticulum
Transverse tubules (T tubules)

Like skeletal muscle, cardiac muscle contains actin and myosin organized to form __(1)__ and myofibrils. Cardiac muscle also has a smooth __(2)__, which is loosely associated with membranes of the __(3)__. This loose association is partly responsible for the slow onset of __(4)__ in cardiac muscle. The production of energy in cardiac muscles depends on oxygen, and cardiac muscle cannot develop a significant __(5)__. An extensive capillary network and large numbers of __(6)__ help sustain the normal myocardial energy requirement. Cardiac muscle cells are bound end-to-end and laterally by specialized cell-to-cell contacts called __(7)__, which greatly increase contact between adjacent cells. Intercalated disks consist of __(8)__, which hold the cells together and __(9)__, which function as areas of low electric resistance between the cells, allowing cardiac muscle cells to function as a single unit.

1. _____
2. _____
3. _____
4. _____
5. _____
6. _____
7. _____
8. _____
9. _____

B. Match these terms with the correct statement or definition:

Apex
Atrioventricular (AV) bundle
Atrioventricular (AV) node
Base
Bundle branches
Purkinje fibers
Sinoatrial (SA) node

_____ 1. Cardiac muscle cells that generate spontaneous action potentials with the greatest frequency; the pacemaker.

_____ 2. Modified cardiac muscle cells that delay action potentials between the atria and the AV bundle; allow time for the atria to contract before the ventricles contract.

_____ 3. Conducting cells that arise from the AV node and extend through the interventricular septum.

_____ 4. Right and left subdivisions of the AV bundle.

_____ 5. Inferior, terminal branches of the bundle branches, composed of large-diameter cardiac muscle fibers.

_____ 6. Part of the heart where ventricular contraction begins.

C. Match these terms with the correct parts of the diagram labeled in figure 20.2:

AV bundle
AV node
Bundle branches
Purkinje fibers
SA node

1. _____
2. _____
3. _____
4. _____
5. _____

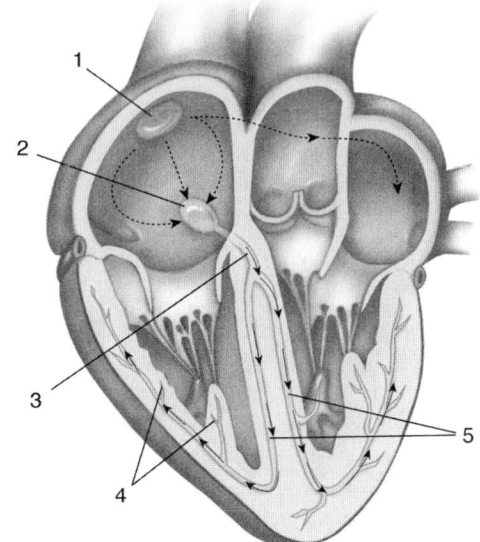

Figure 20.2

Electrical Properties

A. Match these terms with the correct statement or definition:

Calcium-induced calcium release
Depolarization phase
Early repolarization and plateau phase
Final repolarization phase
Resting membrane potential
Voltage-gated Ca^{2+} channels
Voltage-gated K$^+$ channels
Voltage-gated Na$^+$ channels

_____ 1. Condition necessary for electrically excitable cells to produce an action potential.

_____ 2. Phase in which voltage-gated Na$^+$ channels open, voltage-gated K$^+$ channels close, and voltage-gated Ca^{2+} channels begin to open.

_____ 3. Phase in which voltage-gated Na$^+$ channels close, some K$^+$ channels close, and voltage-gated Ca^{2+} channels remain open.

_____ 4. Phase in which voltage-gated Ca^{2+} channels close and many voltage-gated K$^+$ channels open.

_____ 5. At rest, the movement of ions through these channels establishes the resting membrane potential; depolarization causes them to close.

_____ 6. During the early repolarization and plateau phases, these channels open and cause the membrane to remain partly depolarized.

_____ 7. At the end of the plateau phase, these channels open and cause the membrane to return to its resting level.

_____ 8. The movement of Ca^{2+} through the plasma membrane into cardiac muscle cells causes the release of Ca^{2+} from the sarcoplasmic reticulum.

B. Match these terms with the correct statement or definition:

Cardiac muscle
Skeletal muscle

_____ 1. Depolarization phase of the action potential occurs because of Na$^+$ and Ca^{2+}.

_____ 2. Action potentials are faster.

_____ 3. Action potentials are conducted from cell to cell.

C. Match these terms with the correct statement or definition:

Depolarization   Repolarization
Prepotential

_____ 1. Spontaneously developing local potential in the conducting system of the heart.

_____ 2. The movement of Ca^{2+} into pacemaker cells is primarily responsible for this phase.

_____ 3. This occurs in pacemaker cells when voltage-gated Ca^{2+} channels close and voltage-gated K^+ channels open.

D. Match these terms with the correct statement or definition:

Absolute refractory period   Ectopic focus
Autorhythmic   Relative refractory period
AV node   SA node

_____ 1. Heart's ability to stimulate itself to contract at regular intervals.

_____ 2. Area of the heart that normally generates action potentials at the fastest rate; acts as the pacemaker.

_____ 3. Normally, the area of the heart with the second greatest frequency of action potentials.

_____ 4. Any area outside of the SA node that generates a heartbeat.

_____ 5. Time during which a cardiac muscle cell is completely insensitive to further stimulation; prevents tetany.

E. Match these terms with the correct statement or definition:

Electrocardiogram (ECG)   QRS complex
P wave   QT interval
PQ (PR) interval   T wave

_____ 1. Summated record of all cardiac action potentials that are transmitted through the heart during a given time period.

_____ 2. Record of action potentials that cause depolarization of the atrial myocardium.

_____ 3. Record of action potentials from ventricular depolarization.

_____ 4. Record of repolarization of the ventricles.

_____ 5. Approximate time of ventricular contraction.

Chapter 20

F. Match these terms with the correct location on the diagram in figure 20.3:

P wave
PQ (PR) interval
QRS complex
QT interval
T wave

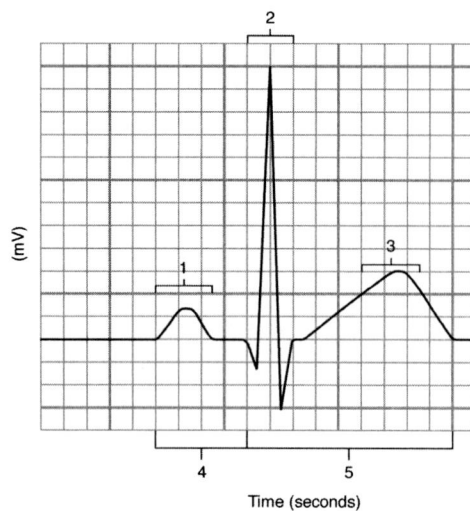

Figure 20.3

1. _____
2. _____
3. _____
4. _____
5. _____

Cardiac Cycle

A. Match these time periods with the correct statement or definition:

Ejection   Isovolumic relaxation
Isovolumic contraction

_____ 1. Period between atrioventricular valve closure and semilunar valve opening; no movement of blood out of the ventricles.

_____ 2. Period of time that blood flows from the ventricles.

_____ 3. Period between semilunar valve closure and atrioventricular valve opening; no blood flows from atria into the ventricles.

B. Match these terms with the correct statement or definition:

Active ventricular filling   Passive ventricular filling
End-diastolic volume   Ventricular diastole
End-systolic volume   Ventricular systole

_____ 1. Contraction of the ventricular myocardium.

_____ 2. Relaxation of the ventricular myocardium.

_____ 3. First third of ventricular diastole; responsible for 70% of ventricular filling.

_____ 4. Final third of ventricular diastole; responsible for 30% of ventricular filling.

_____ 5. Volume of blood in the ventricle when it is filled.

_____ 6. Volume of blood in the ventricle after ejection.

C. Match these terms with the correct statement or definition:

First heart sound
Second heart sound
Third heart sound

_____ 1. Vibrations associated with the atrioventricular valves closing; the low-pitched "lubb" sound.

_____ 2. Vibrations associated with the semilunar valves closing.

_____ 3. Sound of blood flowing in a turbulent fashion into the ventricles.

D. Match these terms with the correct statement or definition:

Blood pressure
Cardiac output
Cardiac reserve
Dicrotic notch (incisura)
Heart rate
Mean arterial pressure
Peripheral resistance
Stroke volume
Venous return

_____ 1. Increase in aortic pressure when the semilunar valve closes and blood flows back toward the ventricle from the aorta.

_____ 2. Force responsible for blood movement in vessels.

_____ 3. Average blood pressure between systolic and diastolic pressure in the aorta.

_____ 4. Volume of blood pumped during each cardiac cycle.

_____ 5. Total resistance against which blood must be pumped.

_____ 6. Total amount of blood pumped per minute.

_____ 7. Amount of blood returning to the heart.

_____ 8. Difference between cardiac output when a person is at rest and maximum cardiac output.

E. Match these terms with the correct location on graph A in figure 20.4:

Atrioventricular valves close.   Semilunar valves close.
Atrioventricular valves open.   Semilunar valves open.

Match these terms with the correct location on graph B in figure 20.4:

Ejection   Isovolumic contraction
End-diastolic volume   Isovolumic relaxation
End-systolic volume   Stroke volume

1. _____
2. _____
3. _____
4. _____
5. _____
6. _____
7. _____
   (The difference between 5 and 6)
8. _____
9. _____
10. _____

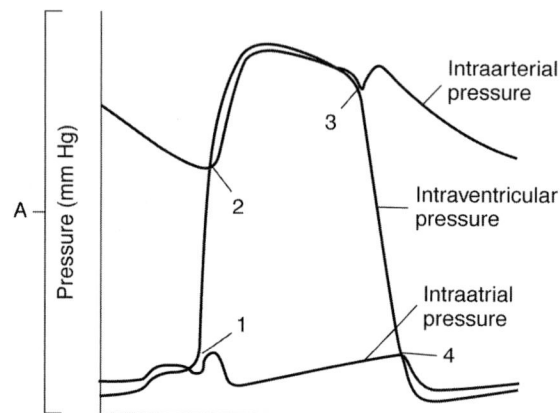

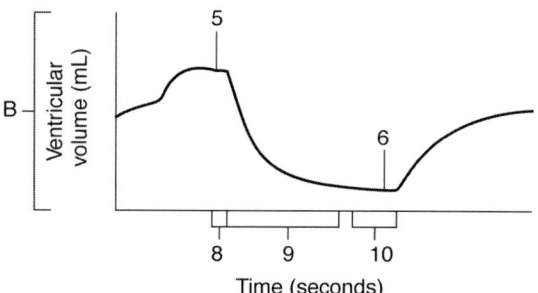

Figure 20.4

F. Match these terms with the correct statement or definition:

Incompetent valve   Stenosed valve
Murmur

_____ 1. Abnormal heart sound.

_____ 2. Valve that leaks significantly.

_____ 3. Valve with an abnormally narrow opening.

300 Chapter 20

Regulation of the Heart

A. Match these terms with the correct statement or definition:

Extrinsic regulation
Intrinsic regulation

_____ 1. Regulation by the heart's normal functional characteristics.

_____ 2. Regulation of the heart by neural or hormonal control.

_____ 3. Regulation that involves venous return and Starling's law of the heart.

B. Match these terms with the correct statement or definition:

Decrease   Little effect
Increase

_____ 1. Effect of decreased venous return on cardiac output.

_____ 2. Effect of decreased preload on cardiac output.

_____ 3. Effect of changes in afterload on heart pumping efficiency.

_____ 4. Effect of stimulating the vagus nerve on heart rate.

_____ 5. Effect of acetylcholine on the permeability of the cardiac muscle cell membrane to K^+.

_____ 6. Effect of sympathetic stimulation on heart rate and the force of heart muscle contraction.

_____ 7. Effect on stroke volume if heart rate becomes too great.

_____ 8. Effect of norepinephrine and epinephrine on heart rate and force of contraction.

Heart and Homeostasis

A. Match these terms with the correct statement or definition:

Baroreceptors   Cardioinhibitory center
Cardioacceleratory center   Cardioregulatory center

_____ 1. Sensory receptors that measure blood pressure; located in the walls of the aorta and internal carotid arteries.

_____ 2. General area of the medulla oblongata where action potentials from the baroreceptors are integrated.

_____ 3. Part of the cardioregulatory center that increases heart rate.

_____ 4. Part of the cardioregulatory center that decreases heart rate.

B. Match these terms with the correct statement or definition:

Decreases
Increases

_____ 1. Effect on heart rate when arterial blood pressure increases.

_____ 2. Effect on heart rate of an increase in blood pH and a decrease of blood carbon dioxide level.

_____ 3. Effect on heart rate of a sudden large decrease in blood oxygen level; mediated through carotid body chemoreceptor reflex.

_____ 4. Effect on heart rate of a large, prolonged decrease in blood oxygen level; results from increased respiratory movements.

_____ 5. Effect on heart rate and stroke volume of excess potassium ions.

_____ 6. Effect on force of contraction of increased extracellular calcium ions.

_____ 7. Effect on heart rate of increased extracellular calcium ions.

_____ 8. Effect on heart rate and force of contraction of increased temperature.

Effects of Aging on the Heart

Match these terms as they pertain to the effects of aging on the heart:

Decreases   No effect
Increases

_____ 1. Effects of aging on arterial elasticity.

_____ 2. Effect of aging on the accumulation of lipid in myocardial cells and the number of collagen fibers in cardiac tissue.

_____ 3. Effects of aging on maximum heart rate.

_____ 4. Effect of aging on the tendency for heart valves to become incompetent or stenosed.

_____ 5. Effects of aging on the likelihood of cardiac arrhythmias.

_____ 6. Effect of aging on the oxygen required by the left ventricle to pump a given amount of blood.

_____ 7. Effect of regular aerobic exercise on the functional capacity of the heart.

Quick Recall

1. List the parts of the serous pericardium, and describe their function.

2. Name the major veins that enter the right and left atria.

3. Name the four valves that regulate the direction of blood flow in the heart, and give their location.

4. Give the two nodes of the conducting system of the heart and their functions.

5. List four differences between skeletal and cardiac muscle in regard to action potentials.

6. State the cause of the P wave, the QRS complex, and the T wave of the ECG. Name the contraction events associated with each wave.

7. List the two major normal heart sounds, and give the reason for each.

8. List the effects of sympathetic and parasympathetic stimulation of the heart.

9. List the locations where the nervous system detects changes in blood pressure, carbon dioxide, pH, and oxygen that affect the heart.

ANSWERS TO CHAPTER 20

CONTENT LEARNING ACTIVITY

Size, Shape, and Location of the Heart
 1. Apex; 2. Base; 3. Mediastinum

Anatomy of the Heart
 A. 1. Pericardium; 2. Fibrous pericardium; 3. Parietal pericardium; 4. Visceral pericardium; 5. Pericardial cavity; 6. Pericardial fluid
 B. 1. Epicardium; 2. Myocardium; 3. Endocardium; 4. Pectinate muscles; 5. Crista terminalis; 6. Trabeculae carneae
 C. 1. Auricles; 2. Venae cavae; 3. Pulmonary veins; 4. Pulmonary trunk; 5. Aorta; 6. Coronary sulcus; 7. Anterior interventricular sulcus
 D. 1. Coronary arteries; 2. Anterior interventricular artery; 3. Left marginal artery; 4. Circumflex artery; 5. Right marginal artery; 6. Posterior interventricular artery; 7. Great cardiac vein; 8. Small cardiac vein; 9. Coronary sinus
 E. 1. Interatrial septum; 2. Fossa ovalis; 3. Foramen ovale; 4. Atrioventricular canal; 5. Interventricular septum
 F. 1. Atrioventricular valve; 2. Tricuspid valve; 3. Bicuspid (mitral) valve; 4. Papillary muscles; 5. Chordae tendineae; 6. Semilunar valve
 G. 1. Superior vena cava; 2. Pulmonary semilunar valve; 3. Aortic semilunar valve; 4. Right atrium; 5. Tricuspid valve; 6. Papillary muscles; 7. Right ventricle; 8. Interventricular septum; 9. Left ventricle; 10. Chordae tendineae; 11. Bicuspid (mitral) valve; 12. Left atrium; 13. Pulmonary veins; 14. Pulmonary trunk; 15. Aorta

Route of Blood Flow Through the Heart
 1. Venae cavae; 2. Right atrium; 3. Tricuspid valve; 4. Right ventricle; 5. Pulmonary semilunar valve; 6. Pulmonary trunk; 7. Pulmonary arteries; 8. Lungs; 9. Pulmonary veins; 10. Left atrium; 11. Bicuspid (mitral) valve; 12. Left ventricle; 13. Aortic semilunar valve; 14. Aorta

Histology
 A. 1. Sarcomeres; 2. Sarcoplasmic reticulum; 3. Transverse tubules (T tubules); 4. Contraction; 5. Oxygen deficit; 6. Mitochondria; 7. Intercalated disks; 8. Desmosomes; 9. Gap junctions
 B. 1. Sinoatrial (SA) node; 2. Atrioventricular (AV) node; 3. Atrioventricular (AV) bundle; 4. Bundle branches; 5. Purkinje fibers; 6. Apex
 C. 1. SA node; 2. AV node; 3. AV bundle; 4. Purkinje fibers; 5. Bundle branches

Electrical Properties
 A. 1. Resting membrane potential; 2. Depolarization phase; 3. Early repolarization and plateau phase; 4. Final repolarization phase; 5. Voltage-gated K^+ channels; 6. Voltage-gated Ca^{2+} channels; 7. Voltage-gated K^+ channels; 8. Calcium-induced calcium release
 B. 1. Cardiac muscle; 2. Skeletal muscle; 3. Cardiac muscle
 C. 1. Prepotential; 2. Depolarization; 3. Repolarization
 D. 1. Autorhythmic; 2. SA node; 3. AV node; 4. Ectopic focus; 5. Absolute refractory period
 E. 1. Electrocardiogram (ECG); 2. P wave; 3. QRS complex; 4. T wave; 5. QT interval
 F. 1. P wave; 2. QRS complex; 3. T wave; 4. PQ (PR) interval; 5. QT interval

Cardiac Cycle
 A. 1. Isovolumic contraction; 2. Ejection; 3. Isovolumic relaxation
 B. 1. Ventricular systole; 2. Ventricular diastole; 3. Passive ventricular filling; 4. Active ventricular filling; 5. End-diastolic volume; 6. End-systolic volume
 C. 1. First heart sound; 2. Second heart sound; 3. Third heart sound
 D. 1. Dicrotic notch (incisura); 2. Blood pressure; 3. Mean arterial pressure; 4. Stroke volume; 5. Peripheral resistance; 6. Cardiac output; 7. Venous return; 8. Cardiac reserve
 E. 1. Atrioventricular valves close; 2. Semilunar valves open; 3. Semilunar valves close; 4. Atrioventricular valves open; 5. End-diastolic volume; 6. End-systolic volume; 7. Stroke volume; 8. Isovolumic contraction; 9. Ejection; 10. Isovolumic relaxation
 F. 1. Murmur; 2. Incompetent valve; 3. Stenosed valve

Regulation of the Heart
 A. 1. Intrinsic regulation; 2. Extrinsic regulation; 3. Intrinsic regulation
 B. 1. Decrease; 2. Decrease; 3. Little effect; 4. Decrease; 5. Increase; 6. Increase; 7. Decrease; 8. Increase

Heart and Homeostasis
 A. 1. Baroreceptors; 2. Cardioregulatory center; 3. Cardioaccelleratory center; 4. Cardioinhibitory center
 B. 1. Decreases; 2. Decreases; 3. Decreases; 4. Increases; 5. Decreases; 6. Increases; 7. Decreases; 8. Increases

Effects of Aging on the Heart
 1. Decreases; 2. Increases; 3. Decreases; 4. Increases; 5. Increases; 6. Increases; 7. Increases

Quick Recall

1. Parietal pericardium and visceral pericardium: reduce friction
2. Right atrium: inferior and superior venae cavae and coronary sinus; left atrium: four pulmonary veins
3. Tricuspid valve: between right atrium and right ventricle; bicuspid (mitral) valve: between left atrium and left ventricle; aortic semilunar valve: in the aorta; pulmonary semilunar valve: in the pulmonary trunk
4. SA node: pacemaker of the heart; AV node: slows action potentials, allowing the atria to contract and move blood into the ventricles
5. Depolarization caused by sodium *and* calcium in cardiac muscle; rate of action potential propagation is slower in cardiac muscle; action potentials are propagated cell-to-cell in cardiac muscle but not in skeletal muscle; cardiac muscle has spontaneous generation of action potentials, but skeletal muscle action potentials result from nervous system stimulation; cardiac muscle has a prolonged repolarization (plateau) phase
6. P wave: atrial depolarization, atrial systole; QRS complex: ventricular depolarization, ventricular systole; T wave: ventricular repolarization, ventricular diastole
7. First heart sound: closing of atrioventricular valves; second heart sound: closing of semilunar valves
8. Parasympathetic stimulation: decreased heart rate; sympathetic stimulation: increased heart rate, force of contraction, and stroke volume
9. Baroreceptors in aorta and carotid arteries monitor blood pressure; chemoreceptors in medulla oblongata monitor carbon dioxide and pH changes; chemoreceptors in the carotid and aortic bodies monitor oxygen changes

21 Cardiovascular System: Peripheral Circulation and Regulation

CONTENT LEARNING ACTIVITY

General Features of Blood Vessel Structure

A. Match these terms with the correct statement or definition:

Continuous capillaries Sinusoidal capillaries
Endothelium Sinusoids
Fenestrated capillaries Venous sinuses
Pericapillary cells

_____ 1. Layer of simple squamous epithelium lining all blood vessels.

_____ 2. Scattered cells that lie between the basement membrane and endothelial cells; fibroblasts, macrophages, or undifferentiated smooth muscle cells.

_____ 3. Capillaries with no gaps between endothelial cells; present in nervous and muscle tissue.

_____ 4. Capillaries with endothelial cells possessing numerous fenestrae; present in intestinal villi and glomeruli of the kidney.

_____ 5. Capillaries with large diameters, large fenestrae, and a less prominent basement membrane; present in endocrine glands.

_____ 6. Large-diameter sinusoidal capillaries; present in the liver and bone marrow.

_____ 7. Even larger than sinusoids; found in the spleen.

B. Match these terms with the correct statement or definition:

Arterial capillaries Thoroughfare channels
Metarterioles Venous capillaries
Precapillary sphincters

_____ 1. End of capillaries closest to the arterioles.

_____ 2. Arterioles with isolated smooth muscle cells along their walls.

_____ 3. Channels through which blood flow is relatively continuous and that extend from a metarteriole to a venule.

_____ 4. Smooth muscle cells that regulate blood flow from the thoroughfare channel into capillaries.

C. Match these terms with the correct statement or definition:

   Tunica adventitia   Tunica media
   Tunica intima

_____ 1. Tunic closest to the lumen of blood vessels, consisting of endothelium, a basement membrane, lamina propria, and a layer of elastic fibers (internal elastic membrane).

_____ 2. Middle layer of blood vessel walls, consisting of smooth muscle cells and elastic and collagen fibers.

_____ 3. Regulates blood flow by vasoconstriction and vasodilation.

_____ 4. Outer layer of blood vessel walls, composed of connective tissue that varies from dense to loose, depending on the blood vessel.

D. Match these types of arteries with the correct statement or definition:

   Arterioles   Muscular arteries
   Elastic arteries

_____ 1. Largest-diameter arteries; have more elastic and less smooth muscle than other arteries.

_____ 2. Have relatively thick walls because of smooth muscle layers in the tunica media.

_____ 3. Often called conducting arteries.

_____ 4. Often called distributing arteries.

_____ 5. Transport blood from the small arteries to the capillaries.

_____ 6. Best adapted for vasoconstriction and vasodilation.

E. Match these terms with the correct statement or definition:

   Medium and large veins   Venules
   Small veins

_____ 1. The structure of these veins is similar to capillaries; the smallest are capable of nutrient exchange.

_____ 2. Smallest veins to have a continuous layer of smooth muscle.

_____ 3. Veins in this category can have valves (folds in the tunica intima) that prevent the backflow of blood.

F. Match these terms with the correct statement or definition: Arteriovenous anastomoses   Vasa vasorum
Glomus

_____ 1. Small vessels that supply blood to the walls of veins and arteries.

_____ 2. Arterioles that allow blood to flow directly into small veins without an intermediate capillary.

_____ 3. Arteriovenous anastomosis consisting of branched and coiled arterioles; function in temperature regulation.

G. Match these terms with the correct statement or definition: Arteriosclerosis   Phlebitis
Atherosclerosis   Varicose veins

_____ 1. Dilated veins with incompetent valves.

_____ 2. Inflammation of the veins.

_____ 3. Degenerative changes in arteries that make them less elastic.

_____ 4. Deposition of a fatlike substance containing cholesterol in the walls of arteries to form plaques.

Pulmonary Circulation

Match these terms with the correct statement or definition: Pulmonary arteries   Pulmonary veins
Pulmonary trunk

_____ 1. Blood passes from the right ventricle directly into this vessel.

_____ 2. These vessels transport blood to each lung.

_____ 3. There are four of these vessels from the lungs; they enter the left atrium.

Systemic Circulation: Arteries

A. Match these parts of the aorta with the correct statement: Aortic arch   Descending aorta
Ascending aorta

_____ 1. Portion of the aorta that gives rise to the right and left coronary arteries.

_____ 2. Three major branches of this part of the aorta are the brachiocephalic, the left common carotid, and the left subclavian arteries.

_____ 3. Longest part of the aorta, running from the aortic arch to the common iliac arteries; divided into the thoracic and abdominal aorta.

B. Match these arteries with the correct parts of the diagram labeled in figure 21.1:

Brachiocephalic artery
Left common carotid artery
Left subclavian artery
Left vertebral artery
Right common carotid artery
Right subclavian artery
Right vertebral artery

1. _____
2. _____
3. _____
4. _____
5. _____
6. _____
7. _____

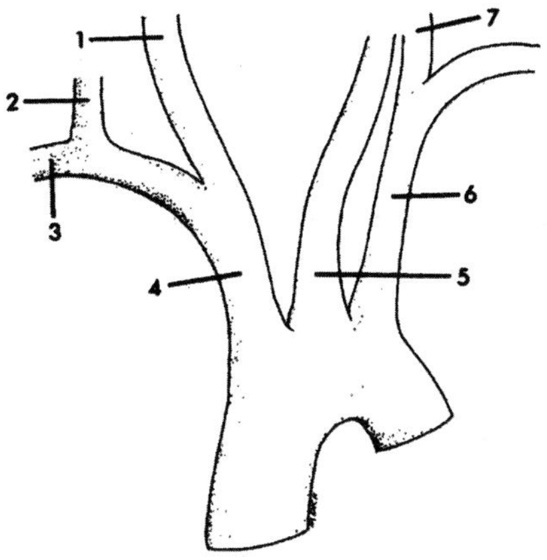

Figure 21.1

C. Match these arteries with the correct parts of the diagram labeled in figure 21.2:

Basilar artery
Cerebral arterial circle (circle of Willis)
External carotid artery
Internal carotid artery
Vertebral artery

1. _____
2. _____
3. _____
4. _____
5. _____

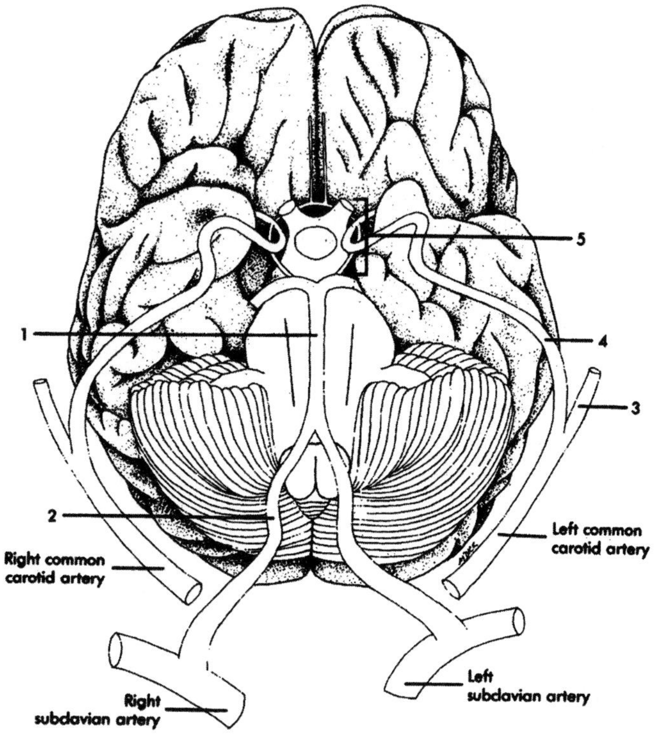

Figure 21.2

Chapter 21

D. Match these arteries with the correct parts of the diagram labeled in figure 21.3:

Axillary artery
Brachial artery
Digital artery
Palmar arches
Radial artery
Ulnar artery

1. _____
2. _____
3. _____
4. _____
5. _____
6. _____

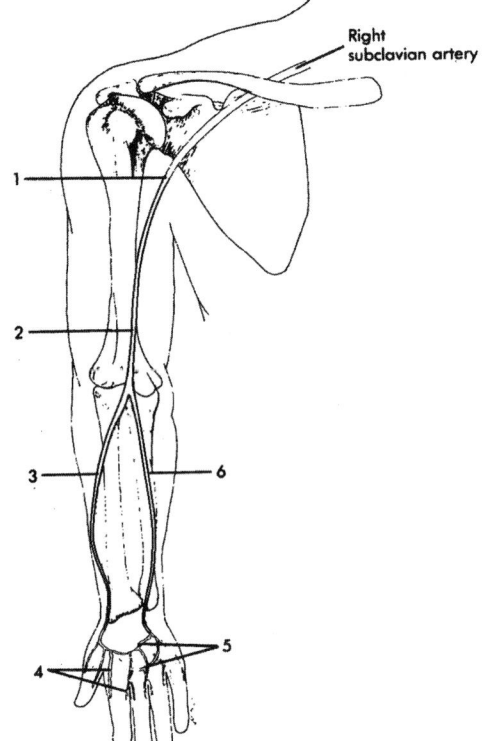

Figure 21.3

E. Match these arteries with the correct parts of the diagram labeled in figure 21.4:

Anterior intercostal artery
Internal thoracic artery
Posterior intercostal artery
Visceral arteries

1. _____
2. _____
3. _____
4. _____

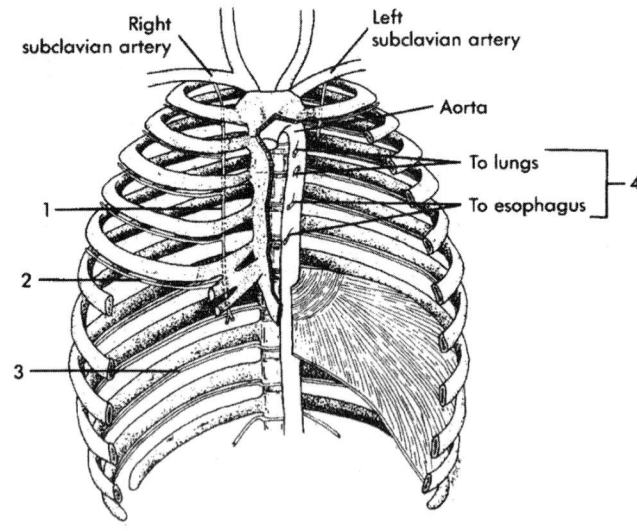

Figure 21.4

F. Match these arteries with the correct parts of the diagram labeled in figure 21.5:

Celiac trunk
Common hepatic artery
Common iliac artery
Gonadal artery
Inferior mesenteric artery
Left gastric artery
Renal artery
Splenic artery
Superior mesenteric artery
Suprarenal artery

1. _____
2. _____
3. _____
4. _____
5. _____
6. _____
7. _____
8. _____
9. _____
10. _____

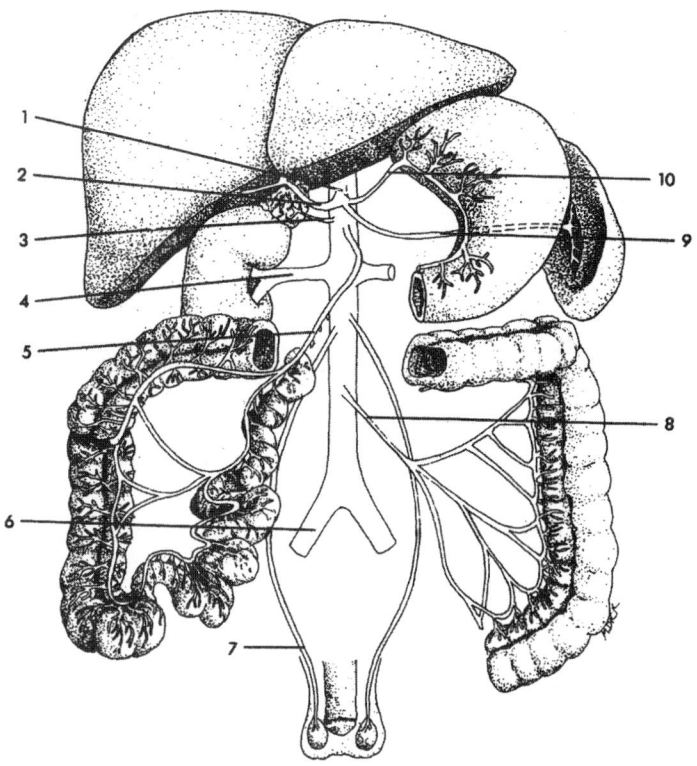

Figure 21.5

Chapter 21

G. Match these arteries with the correct parts of the diagram labeled in figure 21.6:

Anterior tibial artery
Digital arteries
Dorsalis pedis artery
Femoral artery
Fibular (peroneal) artery
Lateral plantar artery
Medial plantar artery
Popliteal artery
Posterior tibial artery

1. _____
2. _____
3. _____
4. _____
5. _____
6. _____
7. _____
8. _____
9. _____

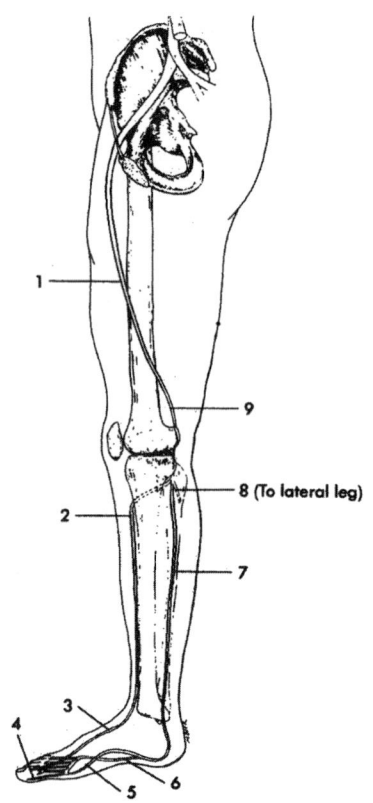

Figure 21.6

Systemic Circulation: Veins

A. Match these major veins with the correct description:

Coronary sinus
Inferior vena cava
Internal jugular vein
Superior vena cava
Venous sinuses

_____ 1. Vein that returns blood from the walls of the heart.

_____ 2. Vein that returns blood from the head, neck, thorax, and upper limbs to the heart.

_____ 3. Vein that returns blood from the abdomen, pelvis, and lower limbs to the heart.

_____ 4. Spaces within the dura mater surrounding the brain; the superior sagittal sinus is an example.

_____ 5. Vein that drains blood from the venous sinuses of the brain.

B. Match these veins with the correct parts of the diagram labeled in figure 21.7:

Inferior vena cava
Pulmonary veins
Right brachiocephalic vein
Right external jugular vein
Right internal jugular vein
Right subclavian vein
Superior vena cava

1. _____
2. _____
3. _____
4. _____
5. _____
6. _____
7. _____

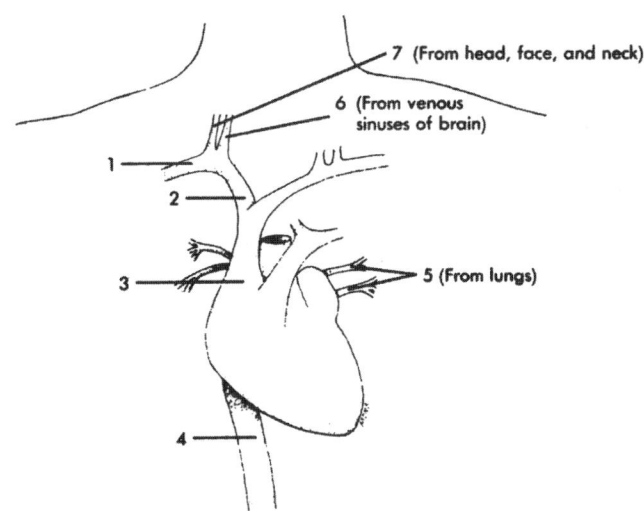

Figure 21.7

C. Match these veins with the correct parts of the diagram labeled in figure 21.8:

Axillary vein
Basilic vein
Brachial veins
Cephalic vein
Digital vein
Median cubital vein
Venous arch

1. _____
2. _____
3. _____
4. _____
5. _____
6. _____
7. _____

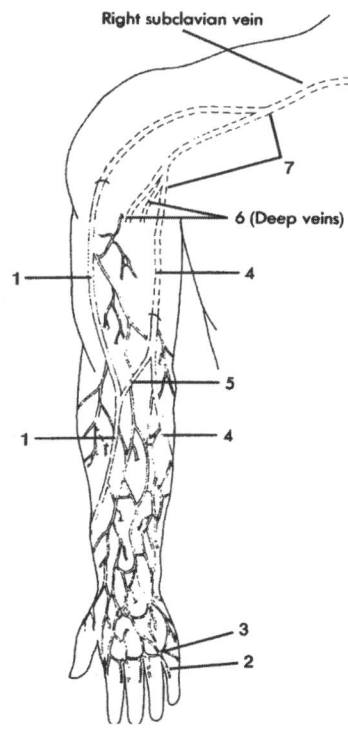

Figure 21.8

Chapter 21

D. Match these veins with the correct parts of the diagram labeled in figure 21.9:

Accessory hemiazygos vein
Azygos vein
Hemiazygos vein

1. _____

2. _____

3. _____

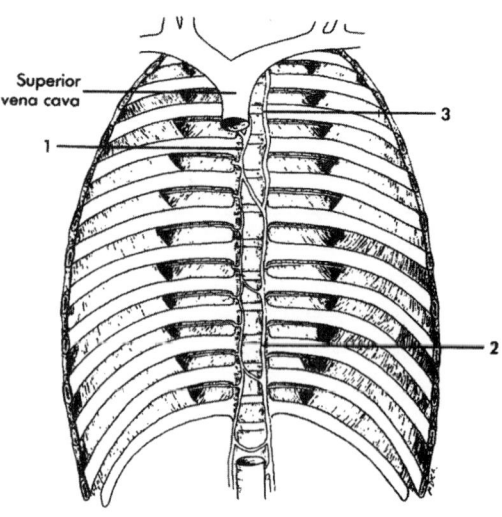

Figure 21.9

E. Match these veins with the correct parts of the diagram labeled in figure 21.10:

Common iliac vein
External iliac vein
Gonadal vein
Hepatic veins
Internal iliac vein
Renal vein
Suprarenal vein

1. _____

2. _____

3. _____

4. _____

5. _____

6. _____

7. _____

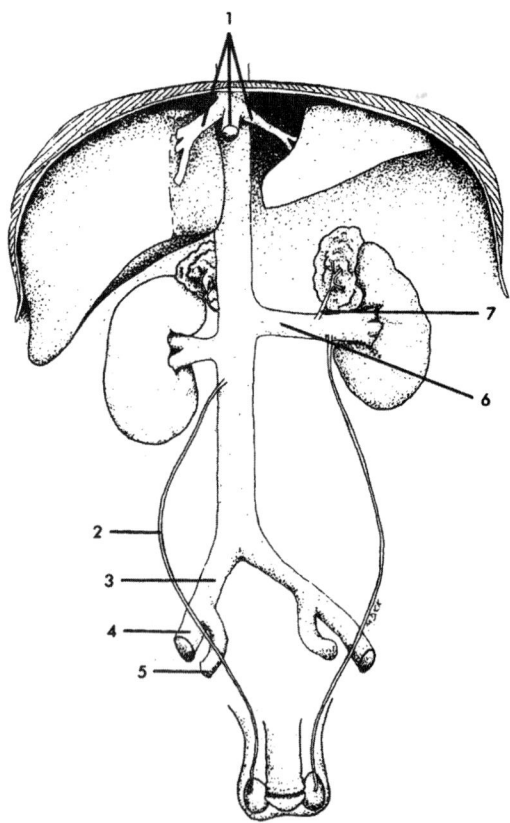

Figure 21.10

314 Chapter 21

F. Match these veins with the correct parts of the diagram labeled in figure 21.11:

Gastric veins
Hepatic portal vein
Hepatic veins
Inferior mesenteric vein
Inferior vena cava
Splenic vein
Superior mesenteric vein

1. _____
2. _____
3. _____
4. _____
5. _____
6. _____
7. _____

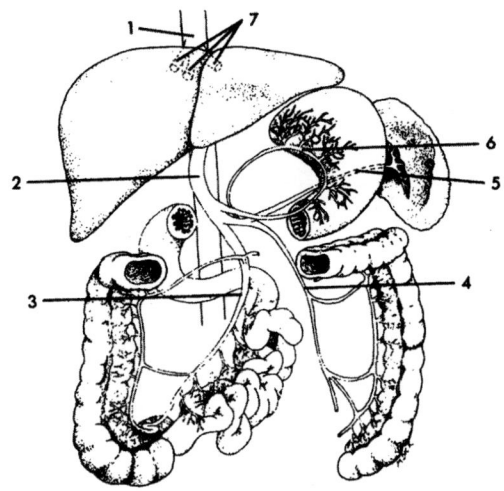

Figure 21.11

G. Match these veins with the correct parts of the diagram labeled in figure 21.12:

Femoral vein
Great saphenous vein
Popliteal vein
Small saphenous vein

1. _____
2. _____
3. _____
4. _____

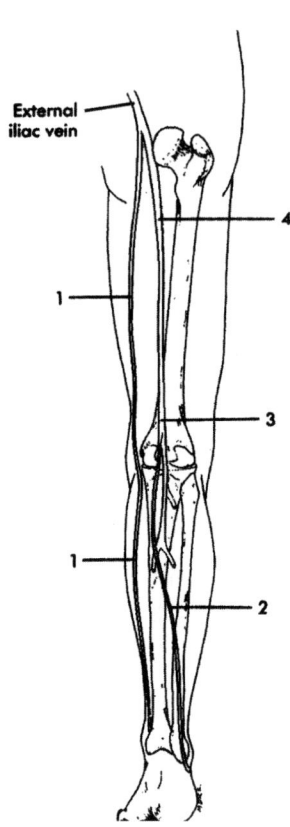

Figure 21.12

Dynamics of Blood Circulation

A. Match these terms with the correct statement or definition:

Auscultatory
Blood pressure
Diastolic pressure
Korotkoff sounds
Laminar flow
Sphygmomanometer
Systolic pressure
Turbulent flow

_____ 1. Tendency for a fluid to flow through tubes as if the fluid were composed of concentric layers.

_____ 2. Numerous small currents flowing crosswise or obliquely to the long axis of a vessel.

_____ 3. Measure of the force blood exerts against the blood vessel walls.

_____ 4. Device that uses an inflatable cuff to measure blood pressure.

_____ 5. Method of determining blood pressure by listening to the sound of blood flowing through the arteries.

_____ 6. Produced by turbulent blood flow.

_____ 7. Pressure at which Korotkoff sounds are first heard when taking a blood pressure measurement.

_____ 8. Pressure at which Korotkoff sounds completely disappear when taking a blood pressure measurement.

B. Using the terms provided, complete these statements:

Decreases   Increases

According to Poiseuille's law, blood flow through a blood vessel decreases when resistance to blood flow _(1)_. Resistance to blood flow increases dramatically when the radius of the blood vessel _(2)_. Resistance to blood flow also increases when blood viscosity _(3)_. An increase in hematocrit _(4)_ blood viscosity. Therefore, an increase in hematocrit _(5)_ flow, unless the workload of the heart increases to compensate. If the pressure gradient between the ends of a blood increases, then blood flow _(6)_.

1. _____
2. _____
3. _____
4. _____
5. _____
6. _____

C. Match these terms with the correct statement or definition:

Aneurysm
Compliance
Critical closing pressure
Decreases
Increases
Laplace's law

_____ 1. Blood pressure below which a blood vessel will collapse.

_____ 2. The force that stretches the vascular wall is proportional to the diameter of the vessel times blood pressure.

_____ 3. What happens to the force acting on the wall of a blood vessel when the diameter of the vessel decreases (e.g., with sympathetic stimulation)?

_____ 4. Bulge in a weakened blood vessel wall.

_____ 5. Tendency for blood vessel volume to change as blood pressure increases.

_____ 6. Change in blood vessel volume when blood pressure increases.

Physiology of Systemic Circulation

A. Match these terms with the correct statement or definition:

Aorta
Arteries
Arterioles
Capillaries
Veins

_____ 1. The largest percentage of blood volume is in these vessels.

_____ 2. Vessels with the slowest velocity of blood flow but the greatest cross-sectional area.

_____ 3. Blood flowing through this vessel has the greatest velocity and the greatest pressure.

_____ 4. Vessels with the highest resistance to flow.

_____ 5. Vessels with the lowest resistance to flow.

B. Using the terms provided, complete the following statements:

Decreased
Increased
Pulse pressure

The difference between the systolic and diastolic pressure is called the _(1)_. When stroke volume is increased, pulse pressure is _(2)_. For a given stroke volume, pulse pressure is _(3)_ when vascular compliance is decreased (e.g., with age). Pulse pressure produces a pressure wave that can be monitored as the pulse. Weak pulses usually indicate a(n) _(4)_ stroke volume or a(n) _(5)_ constriction of the arteries as a result of intense sympathetic stimulation.

1. _____
2. _____
3. _____
4. _____
5. _____

C. Match these terms with the correct statement or definition:

Arterial
Diffusion
Interstitial fluid pressure
Lymphatic system
Net filtration pressure
Net hydrostatic pressure
Net osmotic pressure
Venous

_____ 1. Major means by which nutrients and waste products are exchanged across capillary surfaces.

_____ 2. Force responsible for moving fluid across capillary walls; equal to net hydrostatic pressure minus net osmotic pressure.

_____ 3. Small, negative pressure within tissue spaces.

_____ 4. Equal to blood pressure minus interstitial fluid pressure.

_____   5. Difference in osmotic pressure between the blood and the interstitial fluid; equal to blood colloid osmotic pressure minus interstitial colloid osmotic pressure.

_____   6. At this end of a capillary, there is net movement of fluid out of the capillary.

_____   7. At this end of a capillary, net filtration pressure decreases.

_____   8. System that picks up excess tissue fluid and returns it to the general circulation.

D. Match these terms with the correct statement or definition:

Decreases
Increases

_____   1. Effect on venous return to the heart when blood volume increases.

_____   2. Effect on venous return to the heart when sympathetic stimulation increases venous tone.

_____   3. Effect on venous return to the heart when exercising.

_____   4. Effect on cardiac output when venous return increases.

_____   5. Effect on venous pressure in the legs when standing still.

Local Control of Blood Flow by the Tissues

Using the terms provided, complete these statements:

Autoregulation
Decrease
Increase
Metabolism
Metarterioles and precapillary sphincters
Nutrients
Vasodilator substances
Vasomotion

Control of local blood flow occurs through _(1)_. As the rate of _(2)_ of a tissue increases, blood flow through its capillaries increases. _(3)_, including carbon dioxide and lactic acid, increase in tissues as the rate of metabolism increases. Lack of _(4)_ can also be important in regulating local blood flow; for instance, increased metabolism reduces the amount of oxygen and other nutrients in the tissues. In response to a(n) _(5)_ in vasodilator substances or a(n) _(6)_ in nutrients, metarterioles and precapillary sphincters dilate, resulting in an increased blood flow. Periodic contraction and relaxation of the precapillary sphincters, which is called _(7)_, produces cyclic blood flow through tissues. _(8)_ is the local control mechanisms determining blood flow to tissues, despite large changes in systemic blood pressure. In long-term regulation, if the metabolic activity of a tissue remains elevated, the number of capillaries in the tissue will _(9)_.

1. _____
2. _____
3. _____
4. _____
5. _____
6. _____
7. _____
8. _____
9. _____

Nervous and Hormonal Regulation of Local Circulation

Match these terms with the correct statement or definition:

Acetylcholine
Epinephrine
Hypothalamus and cerebral cortex
Norepinephrine
Parasympathetic
Sympathetic
Vasomotor center
Vasomotor tone

_____ 1. Most important ANS division for nervous control of blood flow.

_____ 2. Brain area controlling sympathetic action potentials to blood vessels.

_____ 3. Brain areas that can inhibit or stimulate the vasomotor center.

_____ 4. Condition of partial constriction of peripheral blood vessels.

_____ 5. Neurotransmitter causing vasoconstriction in most blood vessels.

_____ 6. Substance causing vasodilation in skeletal muscle blood vessels.

Regulation of Mean Arterial Pressure

Match these terms with the correct statement or definition:

Cardiac output
Decreases
Heart rate
Increases
Mean arterial pressure
Peripheral resistance
Stroke volume

_____ 1. Equal to heart rate times stroke volume.

_____ 2. Volume of blood pumped by the heart during each contraction.

_____ 3. Equal to heart rate × stroke volume × peripheral resistance.

_____ 4. Resistance to the flow of blood in the blood vessels.

_____ 5. Effect on mean arterial pressure when heart rate decreases.

_____ 6. Effect on mean arterial pressure when peripheral resistance increases.

_____ 7. Effect on mean arterial pressure when stroke volume increases.

Short-Term Regulation of Blood Pressure

A. Match these types of short-term blood pressure regulation with the correct statement or definition:

Adrenal medullary mechanism
Baroreceptor reflexes
Central nervous system ischemic response
Chemoreceptor reflexes

_____ 1. Mechanism that involves sensory receptors sensitive to stretch; receptors are in the carotid sinus and aortic arch.

_____ 2. Mechanism that is important in regulating blood pressure on a moment-to-moment basis—for example, when body position changes.

_____ 3. Involves the secretion of epinephrine and norepinephrine.

_____ 4. Mechanism that involves sensory receptors sensitive to decreased oxygen or increased carbon dioxide and hydrogen ion levels (i.e., decreased pH); receptors are in the carotid bodies and aortic bodies.

_____ 5. Mechanism that responds to increased carbon dioxide and H^+ (i.e., decreased pH) in the medulla oblongata.

B. Match these terms with the correct statements:

Decreased   Vasoconstriction
Increased   Vasodilation

_____ 1. Effect on heart rate and stroke volume by the baroreceptor reflexes when arterial pressure suddenly decreases.

_____ 2. Effect on blood vessels produced by the baroreceptor reflexes when blood pressure increases.

_____ 3. Effect on blood vessels produced by the chemoreceptor reflexes when arterial oxygen levels decrease markedly.

_____ 4. Effect on blood vessels produced by the CNS ischemic response when medullary carbon dioxide levels increase markedly.

Long-Term Regulation of Blood Pressure

A. Match these terms with the correct statement:

Decrease
Increase

_____ 1. Effect on urine production of an increase in blood pressure.

_____ 2. Effect on blood volume of an increase in urine production.

_____ 3. Effect on blood pressure of a decrease in blood volume.

B. Match these types of hormonal control with the correct statement or definition:

Atrial natriuretic mechanism
Renin-angiotensin-aldosterone mechanism
Vasopressin (ADH) mechanism

_____ 1. Involves the release of an enzyme from the juxtaglomerular apparatuses in the kidneys.

_____ 2. Involves a hormone from the posterior pituitary.

_____ 3. Involves the release of a polypeptide from the heart.

C. Using the terms provided, complete the following statements:

Aldosterone
Angiotensin II (active angiotensin)
Angiotensinogen
Decreased
Increased
Renin

The kidneys release an enzyme called __(1)__ into the circulatory system from the juxtaglomerular apparatuses. Renin acts on a plasma protein called __(2)__ to split a fragment off one end. The fragment, called angiotensin I, has two more amino acids cleaved from it by angiotensin converting enzyme to become __(3)__. Angiotensin II causes __(4)__ vasoconstriction in arterioles and veins, which raises blood pressure. Angiotensin II also stimulates __(5)__ secretion from the adrenal cortex. Aldosterone acts on the kidneys, resulting in __(6)__ urine production. Angiotensin II also stimulates thirst, increases salt appetite, and stimulates ADH secretion. Stimuli that increase renin secretion include __(7)__ blood pressure, __(8)__ blood K^+ levels, and __(9)__ blood Na^+ levels.

1. _____
2. _____
3. _____
4. _____
5. _____
6. _____
7. _____
8. _____
9. _____

D. Match these terms with the correct parts of the diagram labeled in figure 21.13:

Aldosterone
Angiotensin I
Angiotensin II

Angiotensinogen
Renin

1. _____
2. _____
3. _____
4. _____
5. _____

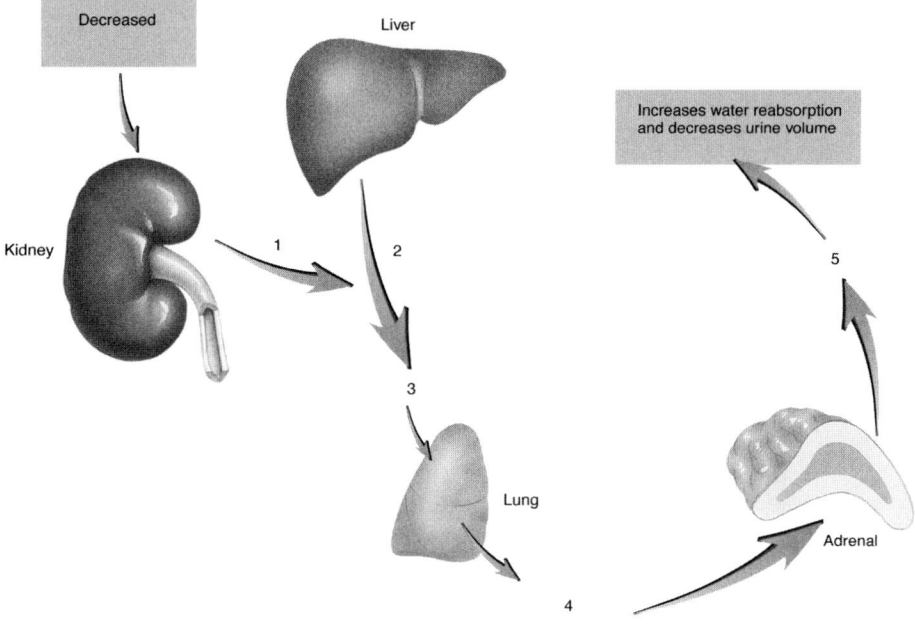

Figure 21.13

Chapter 21

E. Match these terms with the correct statements as they apply to hormonal mechanisms that control arterial pressure:

Decrease
Increase

_____ 1. Effect of epinephrine and norepinephrine on heart rate and stroke volume.

_____ 2. Effect of epinephrine on vasoconstriction (of skin and visceral blood vessels).

_____ 3. Effect of ADH on vasoconstriction.

_____ 4. Effect of ADH on urine production.

_____ 5. Effect of a decrease in blood pressure on ADH secretion.

_____ 6. Effect of atrial natriuretic hormone on urine production.

_____ 7. Effect of increased atrial blood pressure on atrial natriuretic hormone secretion.

Fluid Shift Mechanism and Stress–Relaxation Response

Match these terms with the correct statement or definition:

Decrease   Increase
Fluid shift mechanism   Stress–relaxation response

_____ 1. When blood pressure increases, fluid moves from the blood vessels into the interstitial spaces; when blood pressure decreases, the opposite effect occurs.

_____ 2. When blood pressure declines, smooth muscle cells in blood vessel walls contract; when blood pressure increases, the smooth muscle cells relax.

_____ 3. Change in the amount of interstitial fluid moved into capillaries when blood pressure decreases.

_____ 4. Effect on blood pressure when fluid moves out of tissues into capillaries.

_____ 5. Change in the contraction of smooth muscle in blood vessel walls when blood volume decreases.

_____ 6. Effect on blood pressure when smooth muscle in blood vessel walls contracts.

QUICK RECALL

1. Name the types of blood vessels, starting and ending at the heart.

2. List the three types of capillaries.

3. Name the subdivisions of the aorta.

4. Name the three major arteries that branch from the aorta to supply the head and upper limbs.

5. List the three unpaired arteries and the four major paired arteries that branch from the abdominal aorta.

6. List three veins that return blood to the superior vena cava.

7. Name the veins that drain blood from the head and upper limbs.

8. Name the veins that join the inferior vena cava to return blood from the kidneys and the lower limbs.

9. List the two major superficial veins of the upper limbs and the two major superficial veins of the lower limbs.

10. Name four factors in Poiseuille's law that influence blood flow.

11. List two factors that influence the force acting on the wall of a blood vessel.

12. List the major force responsible for the movement of fluid out of capillaries and the major force responsible for the movement of fluid into capillaries.

13. List four mechanisms responsible for the short-term regulation of blood pressure.

14. List three hormonal mechanisms responsible for the long-term control of blood pressure.

15. List two mechanisms, in addition to nervous and hormonal mechanisms, that help regulate systemic blood pressure.

ANSWERS TO CHAPTER 21

CONTENT LEARNING ACTIVITY

General Features of Blood Vessel Structure
- A. 1. Endothelium; 2. Pericapillary cells; 3. Continuous capillaries; 4. Fenestrated capillaries; 5. Sinusoidal capillaries; 6. Sinusoids; 7. Venous sinuses
- B. 1. Arterial capillaries; 2. Metarterioles; 3. Thoroughfare channels; 4. Precapillary sphincters
- C. 1. Tunica intima; 2. Tunica media; 3. Tunica media; 4. Tunica adventitia
- D. 1. Elastic arteries; 2. Muscular arteries; 3. Elastic arteries; 4. Muscular arteries; 5. Arterioles; 6. Muscular arteries and arterioles
- E. 1. Venules; 2. Small veins; 3. Medium and large veins
- F. 1. Vasa vasorum; 2. Arteriovenous anastomoses; 3. Glomus
- G. 1. Varicose veins; 2. Phlebitis; 3. Arteriosclerosis; 4. Atherosclerosis

Pulmonary Circulation
1. Pulmonary trunk; 2. Pulmonary arteries; 3. Pulmonary veins

Systemic Circulation: Arteries
- A. 1. Ascending aorta; 2. Aortic arch; 3. Descending aorta
- B. 1. Right common carotid artery; 2. Right vertebral artery; 3. Right subclavian artery; 4. Brachiocephalic artery; 5. Left common carotid artery; 6. Left subclavian artery; 7. Left vertebral artery
- C. 1. Basilar artery; 2. Vertebral artery; 3. External carotid artery; 4. Internal carotid artery; 5. Cerebral arterial circle (circle of Willis)
- D. 1. Axillary artery; 2. Brachial artery; 3. Radial artery; 4. Digital artery; 5. Palmar arches; 6. Ulnar artery
- E. 1. Internal thoracic artery; 2. Anterior intercostal artery; 3. Posterior intercostal artery; 4. Visceral arteries
- F. 1. Celiac trunk; 2. Common hepatic artery; 3. Suprarenal artery; 4. Renal artery; 5. Superior mesenteric artery; 6. Common iliac artery; 7. Gonadal artery; 8. Inferior mesenteric artery; 9. Splenic artery; 10. Left gastric artery
- G. 1. Femoral artery; 2. Anterior tibial artery; 3. Dorsalis pedis artery; 4. Digital arteries; 5. Medial plantar artery; 6. Lateral plantar artery; 7. Posterior tibial artery; 8. Fibular (peroneal) artery; 9. Popliteal artery

Systemic Circulation: Veins
- A. 1. Coronary sinus; 2. Superior vena cava; 3. Inferior vena cava; 4. Venous sinuses; 5. Internal jugular vein
- B. 1. Right subclavian vein; 2. Right brachiocephalic vein; 3. Superior vena cava; 4. Inferior vena cava; 5. Pulmonary veins; 6. Right internal jugular vein; 7. Right external jugular vein
- C. 1. Cephalic vein; 2. Digital vein; 3. Venous arch; 4. Basilic vein; 5. Median cubital vein; 6. Brachial veins; 7. Axillary vein
- D. 1. Azygos vein; 2. Hemiazygos vein; 3. Accessory hemiazygos vein
- E. 1. Hepatic veins; 2. Gonadal vein; 3. Common iliac vein; 4. External iliac vein; 5. Internal iliac vein; 6. Renal vein; 7. Suprarenal vein
- F. 1. Inferior vena cava; 2. Hepatic portal vein; 3. Superior mesenteric vein; 4. Inferior mesenteric vein; 5. Splenic vein; 6. Gastric veins; 7. Hepatic veins
- G. 1. Great saphenous vein; 2. Small saphenous vein; 3. Popliteal vein; 4. Femoral vein

Dynamics of Blood Circulation
- A. 1. Laminar flow; 2. Turbulent flow; 3. Blood pressure; 4. Sphygmomanometer; 5. Auscultatory; 6. Korotkoff sounds; 7. Systolic; 8. Diastolic
- B. 1. Increases; 2. Decreases; 3. Increases; 4. Increases; 5. Decreases; 6. Increases
- C. 1. Critical closing pressure; 2. Laplace's law; 3. Decreases; 4. Aneurysm; 5. Compliance; 6. Increases

Physiology of Systemic Circulation
- A. 1. Veins; 2. Capillaries; 3. Aorta; 4. Arterioles; 5. Veins
- B. 1. Pulse pressure; 2. Increased; 3. Increased; 4. Decreased; 5. Increased
- C. 1. Diffusion; 2. Net filtration pressure; 3. Interstitial fluid pressure; 4. Net hydrostatic pressure; 5. Net osmotic pressure; 6. Arterial; 7. Venous; 8. Lymphatic system
- D. 1. Increases; 2. Increases; 3. Increases; 4. Increases; 5. Increases

Local Control of Blood Flow by the Tissues
1. Metarterioles and precapillary sphincters; 2. Metabolism; 3. Vasodilator substances; 4. Nutrients; 5. Increase; 6. Decrease; 7. Vasomotion; 8. Autoregulation; 9. Increase

Nervous and Hormonal Regulation of Local Circulation
1. Sympathetic; 2. Vasomotor center; 3. Hypothalamus and cerebral cortex; 4. Vasomotor tone; 5. Norepinephrine; 6. Epinephrine

Regulation of Mean Arterial Pressure
1. Cardiac output; 2. Stroke volume; 3. Mean arterial pressure; 4. Peripheral resistance; 5. Decreases; 6. Increases; 7. Increases

Short-Term Regulation of Blood Pressure
- A. 1. Baroreceptor reflexes; 2. Baroreceptor reflexes; 3. Adrenal medullary mechanism; 4. Chemoreceptor reflex; 5. Central nervous system ischemic response
- B. 1. Increased; 2. Vasodilation; 3. Vasoconstriction; 4. Vasoconstriction

Long-Term Regulation of Blood Pressure
A. 1. Increase; 2. Decrease; 3. Decrease
B. 1. Renin-angiotensin-aldosterone mechanism; 2. Vasopressin mechanism (ADH); 3. Atrial natriuretic mechanism
C. 1. Renin; 2. Angiotensinogen; 3. Angiotensin II (active angiotensin); 4. Increased; 5. Aldosterone; 6. Decreased; 7. Decreased; 8. Increased; 9. Decreased
D. 1. Renin; 2. Angiotensinogen; 3. Angiotensin I; 4. Angiotensin II; 5. Aldosterone
E. 1. Increase; 2. Increase; 3. Increase; 4. Decrease; 5. Increase; 6. Increase; 7. Increase

Fluid Shift Mechanism and Stress–Relaxation Response
1. Fluid shift mechanism; 2. Stress–relaxation response; 3. Increase; 4. Increase; 5. Increase; 6. Increase

QUICK RECALL

1. Elastic arteries, muscular arteries, arterioles, capillaries, venules, small veins, medium and large veins
2. Continuous capillaries, fenestrated capillaries, and sinusoidal capillaries
3. Ascending aorta, aortic arch, descending aorta (thoracic aorta and abdominal aorta)
4. Brachiocephalic artery, left common carotid artery, and left subclavian artery
5. Unpaired: celiac trunk, superior mesenteric artery, inferior mesenteric artery; paired: suprarenal arteries, renal arteries, gonadal arteries, common iliac arteries
6. Left and right brachiocephalic veins and azygos vein
7. From head and neck: internal and external jugular veins; from upper limbs: subclavian veins
8. From kidneys: renal veins; from lower limbs: common iliac veins
9. Basilic and cephalic veins in the upper limbs, small and great saphenous veins in the lower limbs
10. Viscosity, diameter of blood vessel, length of blood vessel, and pressure gradient
11. Force = diameter × blood pressure (Laplace's law)
12. Out of capillaries: net hydrostatic pressure; into capillaries: net osmotic pressure
13. Baroreceptor reflexes, adrenal medullary mechanism, chemoreceptor reflexes, and the central nervous system ischemic response
14. Renin-angiotensin-aldosterone mechanism, vasopressin mechanism, and atrial natriuretic hormone mechanism
15. Fluid shift mechanism and stress–relaxation response

22 Lymphatic System and Immunity

CONTENT LEARNING ACTIVITY

Lymphatic System

Match these terms with the correct statement or definition:

Chyle Lymph
Lacteal Lymphocyte

_____ 1. Interstitial fluid that is returned to the circulatory system through lymphatic vessels.

_____ 2. Special lymphatic vessel in the small intestine that transports fats.

_____ 3. Lymph with a milky appearance because of its high fat content.

_____ 4. Leukocyte in lymphatic tissue that is capable of destroying microorganisms and foreign substances.

Lymphatic Vessels

A. Match these terms with the correct statement or definition:

Lymph nodes Lymphatic trunks
Lymphatic capillaries Lymphatic vessels
Lymphatic ducts

_____ 1. Small, dead-end tubes that excess fluid enters to become lymph.

_____ 2. Lymphatic capillaries join to produce these structures; resemble small veins and contain one-way valves.

_____ 3. Round, oval, or bean-shaped bodies distributed along lymphatic vessels; filter lymph.

_____ 4. Larger vessels formed from lymphatic vessels that converge after passing through lymph nodes.

_____ 5. Vessels formed from lymphatic trunks that join together and enter large veins.

B. Match these terms with the correct statement or definition:

Bronchomediastinal trunks Lumbar trunks
Intestinal trunks Subclavian trunks
Jugular trunks

_____ 1. Lymphatic trunks that drain the head and neck.

_____ 2. Lymphatic trunks that drain the upper limbs, superficial thoracic wall, and mammary glands.

_____ 3. Lymphatic trunks that drain thoracic organs and the deep thoracic wall.

_____ 4. Lymphatic trunks that drain the intestines, stomach, pancreas, spleen, and liver.

_____ 5. Lymphatic trunks that drain the lower limbs, pelvic and abdominal walls, pelvic organs, kidneys, adrenal glands, and ovaries or testes.

C. Match these terms with the correct statement or definition:

Cisterna chyli Thoracic duct
Right lymphatic duct

_____ 1. Lymphatic duct formed from the jugular, subclavian, and bronchomediastinal trunks on the right side of the body.

_____ 2. Lymphatic duct that drains the left jugular, left subclavian, and both intestinal and lumbar trunks.

_____ 3. Sac formed from lymphatic trunks; present in a small proportion of individuals.

Lymphatic Tissue and Organs

A. Match these terms with the correct statement or definition:

Diffuse lymphatic tissue Mucosa-associated lymphoid
Lingual tonsil   tissue (MALT)
Lymph nodules Palatine tonsils
Lymphatic follicles Peyer's patches
Lymphocytes Pharyngeal tonsil
  Reticular fibers

_____ 1. White blood cells produced in red bone marrow and carried to lymphatic organs; an important part of the immune response.

_____ 2. Fine collagen fibers that form a network that traps microorganisms and serves as a point of attachment for lymphocytes.

_____ 3. Aggregates of nonencapsulated lymphatic tissue found in and beneath mucous membranes lining the digestive, respiratory, urinary, and reproductive tracts.

_____ 4. Lymphatic tissue that has no clear boundary and blends with surrounding tissues.

_____ 5. Denser lymphoid tissue organized into compact, spherical structures; found in loose connective tissue.

_____ 6. Aggregations of lymph nodules in the distal half of the small intestine and the appendix.

_____ 7. Lymph nodules in the lymph nodes and spleen.

_____ 8. Large, oval, lymphoid masses on each side of the junction between the oral cavity and the pharynx; "the tonsils."

_____ 9. Aggregation of lymphatic tissue near the internal opening of the nasal cavity; called the adenoid when enlarged.

B. Match these terms with the correct statement or definition:

Deep lymph nodes
Superficial lymph nodes

_____ 1. Located in the hypodermis beneath the skin.

_____ 2. Located in parts of the body other than the hypodermis.

C. Match these terms with the correct statement or definition:

Afferent lymphatic vessel   Lymphatic sinus
Capsule   Lymphatic tissue
Cortex   Medulla
Efferent lymphatic vessel   Medullary cords
Germinal center   Trabeculae

_____ 1. Extensions of the capsule; form an internal skeleton in the lymph nodes.

_____ 2. Areas of the lymph node in which lymphocytes and macrophages are packed around the reticular fibers.

_____ 3. Lymph node space through which reticular fibers extend.

_____ 4. Lymph node outer layer; contains lymph nodules and sinuses.

_____ 5. Structures consisting of branching, irregular strands of diffuse lymphatic tissue; separated by sinuses.

_____ 6. Vessel that carries lymph to the lymph node.

_____ 7. Vessel that carries lymph away from the lymph node.

_____ 8. Areas of rapid lymphocyte division in the lymph node.

D. Match these terms with the correct parts of the diagram labeled in figure 22.1:

Afferent lymphatic vessel
Capsule
Cortex
Diffuse lymphatic tissue
Efferent lymphatic vessel
Germinal center
Lymph nodule
Lymphatic sinus
Medulla
Medullary cord
Trabecula

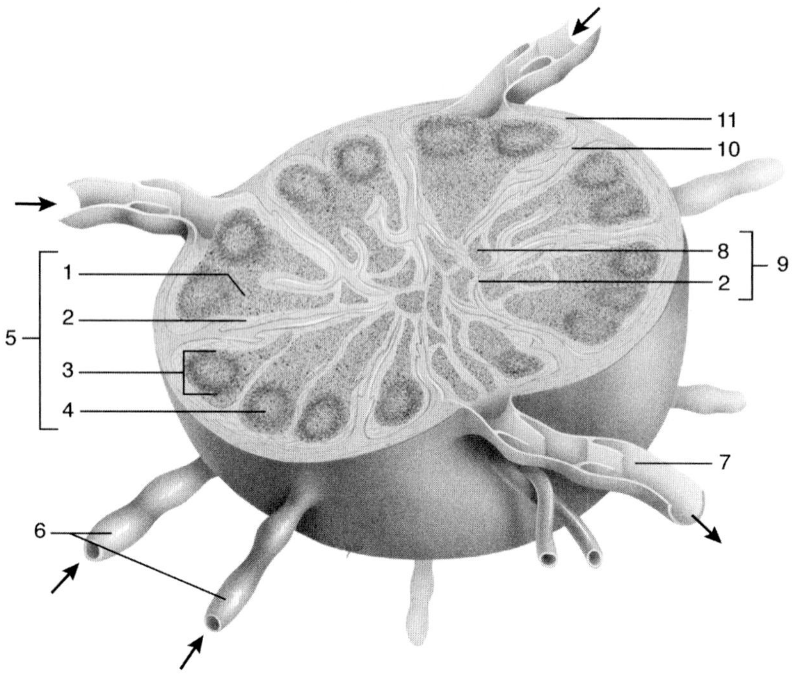

1. _____
2. _____
3. _____
4. _____
5. _____
6. _____
7. _____
8. _____
9. _____
10. _____
11. _____

Figure 22.1

E. Match these terms with the correct statement as it applies to the spleen:

Capsule
Hilum
Periarterial lymphatic sheath
Red pulp
Splenic cords
Trabeculae
Venous sinuses
White pulp

_____ 1. Fibrous outer covering of the spleen.

_____ 2. Bundles of connective tissue fibers from the capsule that extend into the spleen and subdivide it into small, interconnected compartments.

_____ 3. Location where splenic arteries enter the spleen.

_____ 4. Lymphatic tissue associated with the arterial supply of the spleen; consists of the periarterial lymphatic sheath and lymph nodules.

_____ 5. Diffuse lymphatic tissue surrounding arteries and arterioles extending to lymph nodules in the spleen.

_____ 6. Associated with the veins of the spleen; consists of splenic cords and venous sinuses.

_____ 7. Network of reticular cells that produce reticular fibers; spaces between fibers are occupied by macrophages and blood cells that have come from the capillaries.

_____ 8. Enlarged capillaries between the splenic cords.

F. Match these terms with the correct parts of the diagram labeled in figure 22.2:

Lymph nodule
Periarterial lymphatic sheath
Red pulp
Splenic cord
Venous sinuses
White pulp

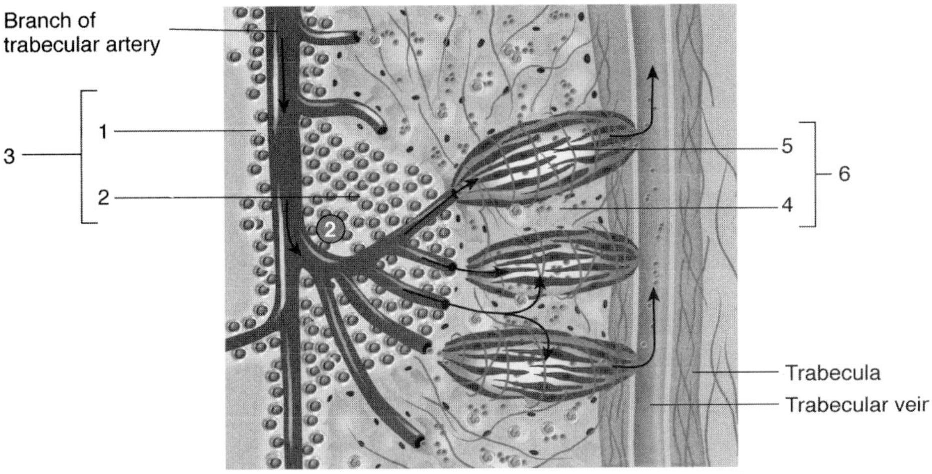

1. _____
2. _____
3. _____
4. _____
5. _____
6. _____

Figure 22.2

G. Using the terms provided, complete these statements:

Foreign substances
Rapid
Red blood cells
Reservoir
Slow
Splenic cords
Venous sinuses

Most blood flow through the spleen is __(1)__. Rapid flow results from the movement of blood from the ends of capillaries into the beginning of __(2)__. __(3)__ flow of blood occurs when blood leaves the ends of capillaries, enters the __(4)__, percolates through them, and passes through the walls of the venous sinuses. The spleen destroys defective __(5)__, detects and responds to __(6)__ in the blood, and acts as a blood __(7)__.

1. _____
2. _____
3. _____
4. _____
5. _____
6. _____
7. _____

H. Match these terms with the correct statement as it applies to the thymus:

Capsule
Cortex
Lobules
Medulla
Thymic corpuscles

_____ 1. Thin connective tissue covering of each lobe of the thymus.

_____ 2. Subdivisions of the thymus formed by the inward extension of trabeculae.

_____ 3. Outer portion of the thymus, that contains lymphocytes.

_____ 4. Relatively lymphocyte-free core of each lobule of the thymus.

_____ 5. Rounded epithelial structures of unknown function in the medulla.

Chapter 22

I. Match these lymphatic organs with the correct description or function:

Lymph node   Thymus
Spleen

_____ 1. Filters lymph and provides a source of lymphocytes; the only structure that has both afferent and efferent lymphatic vessels.

_____ 2. Detects and responds to foreign substances in the blood, destroys worn-out red blood cells, and acts as a blood reservoir.

_____ 3. Produces lymphocytes that move to other lymphatic tissue, where they can respond to foreign substances.

_____ 4. Located primarily in the superior mediastinum.

Immunity

Match these terms with the correct description or function:

Adaptive immunity
Innate immunity

_____ 1. For this type of immunity, the ability to destroy foreign organisms does not improve each time the body is exposed to them.

_____ 2. This type of immunity involves specificity and memory.

Innate Immunity

Using the terms provided, complete these statements:

Cells   Mechanical
Chemical mediators   mechanisms
Inflammatory
  response

Innate immunity consists of several important components. The skin and mucous membrane are _(1)_ that prevent the entry of microorganisms or other foreign substances, whereas _(2)_ are substances that directly kill microorganisms or activate other mechanisms that result in the destruction of microorganisms. _(3)_ utilize phagocytosis and the production of chemicals to provide innate immunity. The _(4)_ mobilizes the immune system and isolates micro-organisms until they are destroyed.

1. _____
2. _____
3. _____
4. _____

Mechanical Mechanisms

Using the terms provided, complete these statements:

Chemicals Skin
Ciliated Tears
Respiratory Urine
Saliva

The (1) and mucous membranes form barriers to the entry of microorganisms and (2) into the body. In addition, microorganisms and other substances are washed from the eyes by (3), from the mouth by (4), and from the urinary tract by (5). In the respiratory tract, mucous membranes are (6), and microbes trapped in mucus are swept to the back of the throat and swallowed. Coughing and sneezing also remove microorganisms from the (7) tract.

1. _____
2. _____
3. _____
4. _____
5. _____
6. _____
7. _____

Chemical Mediators

Using the terms provided, complete these statements:

Alternative pathway Cytokines
Classical pathway Inflammation
Complement Intensity
Complement   Microorganisms
 cascade Proliferation

Some chemical mediators on the surface of cells, such as lysozyme, sebum, and mucus, kill (1) or prevent their entry into cells. Other chemicals, such as histamine, complement, prostaglandins, and leukotrienes, promote (2) by causing vasodilation, increasing vascular permeability, attracting white blood cells, and stimulating phagocytosis. (3) are proteins or peptides secreted by cells; they bind to receptors on cell surfaces, stimulating a response. Cytokines regulate the (4) and duration of immune responses and stimulate the (5) and differentiation of cells. (6) is a group of about 20 proteins that make up approximately 10% of the globulin portion of serum. These proteins become activated in the (7), a series of reactions in which each component of the series activates the next component. These reactions begin through one of two pathways. The (8) pathway, part of innate immunity, is initiated when the complement protein C3 becomes spontaneously active.

1. _____
2. _____
3. _____
4. _____
5. _____
6. _____
7. _____
8. _____

Chapter 22

Cells

A. Match these terms with the correct statement or definition:

Chemotactic factors　　Phagocytosis
Chemotaxis　　White blood cells

_____ 1. Most important cellular component of the immune system.

_____ 2. Parts of microbes or chemicals released by tissue cells; act as chemical signals to attract white blood cells.

_____ 3. Ability to detect and move toward chemotactic factors.

_____ 4. Endocytosis and destruction of particles by cells.

B. Match these terms with the correct statement or definition:

Macrophages　　Neutrophils
Mononuclear phagocytic system　　Reticuloendothelial system

_____ 1. Usually the first cells to enter infected tissues; often die after a single phagocytic event.

_____ 2. Cells derived from monocytes that leave the blood, enlarge, and increase their numbers of lysosomes and mitochondria.

_____ 3. Responsible for most phagocytosis in the late stages of infection.

_____ 4. White blood cells beneath free surfaces or in sinuses, where they trap and destroy microbes.

_____ 5. Original name for macrophages found on reticular fibers and endothelium of sinuses.

_____ 6. Collective name for monocytes and macrophages.

_____ 7. Dust cells in the lungs, Kupffer cells in the liver, or microglia in the central nervous system.

C. Match these types of white blood cells with the correct function or description:

Basophils　　Mast cells
Eosinophils　　Natural killer cells

_____ 1. Motile white blood cells that become activated and secrete chemicals that promote inflammation or cause smooth muscle contraction.

_____ 2. Nonmotile cells located in connective tissue; when activated, produce chemicals that promote inflammation or cause smooth muscle contraction.

_____ 3. White blood cells that produce enzymes that reduce inflammation; secrete enzymes that kill some parasites.

_____ 4. White blood cells that kill certain tumor and virus-infected cells; exhibit no memory response.

Inflammatory Response

A. Using the terms provided, complete these statements:

Chemical mediators
Chemotactic
Complement
Fibrin
Vascular permeability
Vasodilation

Inflammation results when a microbe or damage to tissues causes the release or activation of _(1)_, such as histamine, prostaglandins, leukotrienes, complement, kinins, and others. Mediators cause _(2)_, which increases blood flow and brings phagocytes and other white blood cells to the area. Some of the mediators are _(3)_ factors that stimulate phagocytes to leave the blood. Mediators also increase _(4)_, allowing fibrinogen and complement to enter the tissue. Fibrinogen is converted to _(5)_, which prevents the spread of infection by walling off the infected area. _(6)_ enhances the inflammatory response and attracts additional phagocytes.

1. _____
2. _____
3. _____
4. _____
5. _____
6. _____

B. Using the terms provided, complete these statements:

Blood flow
Local inflammation
Loss of function
Neutrophils
Pain
Pyrogens
Systemic inflammation
Vascular permeability

An inflammatory response confined to a specific area of the body is a _(1)_. Symptoms of local inflammation include redness, heat, and swelling, which result from increased _(2)_ and increased vascular permeability. Other symptoms are _(3)_, resulting from swelling and chemicals acting on sensory or pain receptors, and _(4)_ caused by tissue destruction, swelling, and pain. An inflammatory response that occurs in many parts of the body is a _(5)_. In addition to local symptoms, additional features may be present in systemic inflammation. These include the production and release of large numbers of _(6)_ that promote phagocytosis, the release of _(7)_ by microorganisms or leukocytes to produce fever, and in severe cases a great increase in _(8)_, which causes a large amount of fluid loss from the blood and can lead to shock and death.

1. _____
2. _____
3. _____
4. _____
5. _____
6. _____
7. _____
8. _____

Chapter 22

Adaptive Immunity

A. Match these terms with the correct statement or definition:

Allergic reaction
Antigen
Foreign antigen
Hapten
Self-antigen

_____ 1. Substance that stimulates adaptive immunity.

_____ 2. Small molecule capable of combining with a larger molecule to stimulate adaptive immunity.

_____ 3. Molecule introduced from outside the body that stimulates adaptive immunity—e.g., components of bacteria and viruses.

_____ 4. Molecule, produced by the body, that stimulates adaptive immunity; stimulates autoimmune disease.

_____ 5. Overreaction of the immunie system triggered by foreign antigens.

B. Match these terms with the correct statement or definition:

Antibody-mediated immunity
B cells
Cell-mediated immunity
Effector T cells
Regulatory T cells

_____ 1. Lymphocytes that give rise to cells that produce antibodies.

_____ 2. Immunity produced by plasma antibodies.

_____ 3. Lymphocytes that produce cell-mediated immunity; include cytotoxic T cells and delayed hypersensitivity T cells.

_____ 4. Lymphocytes that control the activities of cell-mediated or antibody-mediated immunity; include helper T cells and suppressor T cells.

Origin and Development of Lymphocytes

Match these terms with the correct statement or definition:

Clones
Negative selection
Positive selection
Primary lymphatic organs
Red bone marrow
Secondary lymphatic organs
Thymus

_____ 1. Where hormones are produced for T-cell maturation.

_____ 2. Where pre-B cells are processed into B cells.

_____ 3. Survival of pre-B and pre-T cells that are capable of an immune response.

_____ 4. Small groups of identical lymphocytes produced during embryonic development.

_____ 5. Elimination or suppression of clones acting against self-antigens.

_____ 6. Sites where lymphocytes mature into functional cells; include red bone marrow and thymus.

_____ 7. Sites where lymphocytes interact with each other, antigen-presenting cells, and antigens to produce an immune response.

Activation of Lymphocytes

A. Match these terms with the correct statement or definition:

    Antigen receptors   B-cell receptor
    Antigenic determinants   T-cell receptor

_____ 1. Specific regions of a given antigen recognized by a lymphocyte; an epitope.

_____ 2. Proteins on the surface of lymphocytes that combine with specific antigenic determinants.

_____ 3. Consists of two polypeptide chains, subdivided into a variable and constant region; the variable region can bind to an antigen.

_____ 4. Consists of four polypeptide chains with two identical variable regions; it is a type of antibody.

B. Match these terms with the correct statement or definition:

    Antigen-presenting cells   MHC class II molecule
    Major histocompatibility   MHC restricted
      complex (MHC) molecules
    MHC class I molecule

_____ 1. General term for cell surface glycoproteins that are involved in lymphocyte activation.

_____ 2. Molecule that combines with antigens produced inside the cell; the combined antigen complex is then displayed on the cell surface.

_____ 3. Process in which the MHC class I/antigen complex is required to activate a T cell.

_____ 4. Cells specialized to take in foreign antigens, process the antigens, and display the antigens to other immune system cells.

_____ 5. B cells, macrophages, monocytes, and dendritic cells.

_____ 6. Molecule that combines with a foreign antigen taken in by antigen-presenting cells; located in antigen complex that is displayed on the cell surface.

_____ 7. An antigen displayed with this type of MHC molecule results in the destruction of the displaying cell.

_____ 8. An antigen displayed with this type of MHC molecule results in the activation of other immune system cells.

Chapter 22

C. Match these terms with the correct statement or definition:

B7, CD4, CD8, or CD28 Cytokines
Costimulation

_____ 1. Response from a B cell or T cell that requires both cytokines and a MHC class II/antigen complex.

_____ 2. Proteins or peptides secreted by one cell as a regulator of neighboring cells; also called lymphokines.

_____ 3. Surface proteins involved in costimulation; help connect T cells to other cells.

D. Using the terms provided, complete these statements:

Antibodies   Effector T cells
Antigen   Helper T cells
Decreases   Increases
Dividing   MHC class II

Exposure to a(n) (1) results in an increase in lymphocyte numbers. First, there is an increase in the number of (2). Antigen-presenting cells use (3) molecules to present processed antigens to helper T cells. These helper T cells respond to the MHC class II/antigen complex by (4). As a result, the number of helper T cells that recognize the antigen (5). These helper T cells then stimulate B cells to produce (6) or activate T cells to become (7).

1. _____
2. _____
3. _____
4. _____
5. _____
6. _____
7. _____

Inhibition of Lymphocytes

Using the terms provided, complete these statements:

Activation   Self-reactive
Anergy   Suppressor T cells
Costimulation   Tolerance
Self-antigens

A state of unresponsiveness of lymphocytes to a specific antigen is (1). The most important function of tolerance is to prevent the immune system from responding to (2). One way tolerance can be induced is the deletion of (3) lymphocytes during prenatal development and after birth. Also, blocking, altering, or deleting an antigen receptor prevents (4) of lymphocytes. A condition of inactivity in which a B cell or T cell does not respond to an antigen is called (5); this condition occurs when an MHC–antigen complex binds to an antigen receptor and there is no (6). In addition, (7) can prevent the activity of helper T cells, B cells, and effector T cells.

1. _____
2. _____
3. _____
4. _____
5. _____
6. _____
7. _____

Antibody-Mediated Immunity

A. Match these terms with the correct statement or definition:
Constant region
Gamma globulins
Immunoglobulins
Opsonins
Variable region

_____ 1. Also called antibodies.

_____

_____ 2. Part of the antibody that combines with antigenic determinant of the antigen; determines specificity of the antibody.

_____ 3. Part of the antibody that activates complement and attaches the antibody to cells, such as macrophages, basophils, mast cells, and eosinophils.

_____ 4. Substances that make an antigen more susceptible to phagocytosis.

B. Match these terms with the correct parts of the diagram labeled in figure 22.3:
Antigen-binding site
Complement-binding site
Constant region
Heavy chain
Light chain
Macrophage-binding site
Variable region

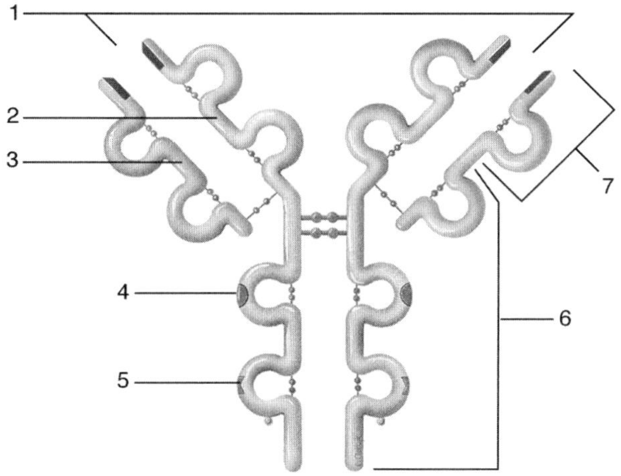

1. _____
2. _____
3. _____
4. _____
5. _____
6. _____
7. _____

Figure 22.3

C. Match these terms with the correct statement or definition:
Memory B cells
Plasma cells
Primary response
Secondary (memory) response

_____ 1. Cell division, cell differentiation, and antibody production from the first exposure of a B cell to an antigen.

_____ 2. Derived from activated B cells, these cells produce antibodies.

_____ 3. Cells that divide and produce plasma and memory cells when exposed to a previously encountered antigen.

_____ 4. Faster antibody-mediated response; produces more antibodies.

Chapter 22

Cell-Mediated Immunity

Using the terms provided, complete these statements:

Allergic
Cytokines
Cytotoxic T cells
Inflammation
Lyse
Memory T cells
Perforin
Secondary response

Once activated, T cells undergo a series of divisions and produce cytotoxic T cells and _(1)_. Memory T cells can provide a(n) _(2)_ and long-lasting immunity. Cytotoxic T cells have two main effects: They _(3)_ cells and they produce _(4)_ that activate additional components of the immune system. Cell lysis involves a protein called _(5)_, which forms a channel in the plasma membrane of the target cell. Delayed hypersensitivity T cells respond to antigens by releasing _(6)_; consequently, they promote phagocytosis and _(7)_, especially in _(8)_ reactions.

1. _____
2. _____
3. _____
4. _____
5. _____
6. _____
7. _____
8. _____

Immunotherapy

Using the terms provided, complete these statements:

Humanization
Inflammation
MHC molecules
Monoclonal antibodies
Tumor cells
Vaccination

Administering cytokines or other agents can promote _(1)_ and the activation of immune cells that can help in the destruction of _(2)_. In other cases, inhibiting the immune system may be helpful. For example, interferon beta blocks the expression of _(3)_ and is used to treat multiple sclerosis. _(4)_ is used to prevent many diseases and, in the future, _(5)_ could be used to deliver radioactive isotopes or chemicals to kill tumor cells. In a process called _(6)_, monoclonal antibodies are modified to prevent destruction by the immune system.

1. _____
2. _____
3. _____
4. _____
5. _____
6. _____

Acquired Immunity

Match these terms with the correct statement or definition:

Active artificial immunity
Active natural immunity
Immunization
Passive artificial immunity
Passive natural immunity

_____ 1. Deliberate introduction of an antigen or antibody into the body.

_____ 2. Natural exposure to an antigen, which causes the body to mount an immune system response against the antigen.

340 Chapter 22

_____ 3. Deliberate introduction of an antigen into the body; the type of immunity produced by vaccination.

_____ 4. Type of immunity produced by the transfer of antibodies from mother to child.

_____ 5. Type of immunity produced by antivenins, antisera, and antitoxins.

Effects of Aging on the Lymphatic System and Immunity

Match these terms with the correct statement or definition:

Decreases   Little effect
Increases

_____ 1. Effect of aging on the ability of the lymphatic system to remove fluid from tissues, absorb fats, or remove defective red blood cells from the blood.

_____ 2. Effect of aging on the number of B cells and T cells.

_____ 3. Effect of aging on the ability of helper T cells to proliferate in response to an antigen.

_____ 4. Effect of aging on primary and secondary antibody responses.

_____ 5. Effect of aging on the ability of cell-mediated immunity to resist intracellular pathogens.

QUICK RECALL

1. List three basic functions of the lymphatic system.

2. List the functions of the lymph nodes, spleen, and thymus.

3. Name the two ways that complement is activated.

4. List the steps that occur when a cell is protected against a viral infection by interferons.

Chapter 22

5. List the two cell types responsible for most of the phagocytosis in the body.

6. Name the function that basophils and mast cells have in common. Name the cell type that counteracts this function.

7. Name the cell type that produces antibodies and the cell type that produces most cytokines.

8. Explain the function of MHC class I and class II molecules.

9. List four types of antigen-presenting cells.

10. Contrast costimulation and anergy.

11. Name three ways to achieve immunologic tolerance.

12. Give the two basic ways that antibody-mediated immunity (antibodies) act against an antigen.

13. Name the two cell types responsible for the secondary response.

14. Give the two basic effects of cytotoxic cells.

ANSWERS TO CHAPTER 22

CONTENT LEARNING ACTIVITY

Lymphatic System
   1. Lymph; 2. Lacteal; 3. Chyle; 4. Lymphocyte
Lymphatic Vessels
  A. 1. Lymphatic capillaries; 2. Lymphatic vessels; 3. Lymph nodes; 4. Lymphatic trunks; 5. Lymphatic ducts
  B. 1. Jugular trunks; 2. Subclavian trunks; 3. Bronchomediastinal trunks; 4. Intestinal trunks; 5. Lumbar trunks
  C. 1. Right lymphatic duct; 2. Thoracic duct; 3. Cisterna chyli
Lymphatic Tissue and Organs
  A. 1. Lymphocytes; 2. Reticular fibers; 3. Mucous-associated lymphoid tissue (MALT); 4. Diffuse lymphatic tissue; 5. Lymph nodules; 6. Peyer's patches; 7. Lymphatic follicles; 8. Palatine tonsils; 9. Pharyngeal tonsils
  B. 1. Superficial lymph nodes; 2. Deep lymph nodes
  C. 1. Trabeculae; 2. Lymphatic tissue; 3. Lymphatic sinus; 4. Cortex; 5. Medullary cords; 6. Afferent lymphatic vessel; 7. Efferent lymphatic vessel; 8. Germinal center
  D. 1. Diffuse lymphatic tissue; 2. Lymphatic sinus; 3. Lymph nodule; 4. Germinal center; 5. Cortex; 6. Afferent lymphatic vessel; 7. Efferent lymphatic vessel; 8. Medullary cord; 9. Medulla; 10. Trabecula; 11. Capsule
  E. 1. Capsule; 2. Trabeculae; 3. Hilum; 4. White pulp; 5. Periarterial lymphatic sheath; 6. Red pulp; 7. Splenic cords; 8. Venous sinuses
  F. 1. Periarterial lymphatic sheath; 2. Lymph nodule; 3. White pulp; 4. Splenic cord; 5. Venous sinus; 6. Red pulp
  G. 1. Rapid; 2. Venous sinuses; 3. Slow; 4. Splenic cords; 5. Red blood cells; 6. Foreign substances; 7. Reservoir
  H. 1. Capsule; 2. Lobules; 3. Cortex; 4. Medulla; 5. Thymic corpuscles
  I. 1. Lymph node; 2. Spleen; 3. Thymus; 4. Thymus
Immunity
   1. Innate immunity; 2. Adaptive immunity
Innate Immunity
   1. Mechanical mechanisms; 2. Chemical mediators; 3. Cells; 4. Inflammatory response
Mechanical Mechanisms
   1. Skin; 2. Chemicals; 3. Tears; 4. Saliva; 5. Urine; 6. Ciliated; 7. Respiratory
Chemical Mediators
   1. Microorganisms; 2. Inflammation; 3. Cytokines; 4. Intensity; 5. Proliferation; 6. Complement; 7. Complement cascade; 8. Alternative pathway
Cells
  A. 1. White blood cells; 2. Chemotactic factors; 3. Chemotaxis; 4. Phagocytosis
  B. 1. Neutrophils; 2. Macrophages; 3. Macrophages; 4. Macrophages; 5. Reticuloendothelial system; 6. Mononuclear phagocytic system; 7. Macrophages
  C. 1. Basophils; 2. Mast cells; 3. Eosinophils; 4. Natural killer cells

Inflammatory Response
  A. 1. Chemical mediators; 2. Vasodilation; 3. Chemotactic; 4. Vascular permeability; 5. Fibrin; 6. Complement
  B. 1. Local inflammation; 2. Blood flow; 3. Pain; 4. Loss of function; 5. Systemic inflammation; 6. Neutrophils; 7. Pyrogens; 8. Vascular permeability
Adaptive Immunity
  A. 1. Antigen; 2. Hapten; 3. Foreign antigen; 4. Self-antigen; 5. Allergic reaction
  B. 1. B cells; 2. Antibody-mediated immunity; 3. Effector T cells; 4. Regulatory T cells
Origin and Development of Lymphocytes
   1. Thymus; 2. Red bone marrow; 3. Positive selection; 4. Clones; 5. Negative selection; 6. Primary lymphatic organs; 7. Secondary lymphatic organs
Activation of Lymphocytes
  A. 1. Antigenic determinants; 2. Antigen receptors; 3. T-cell receptor; 4. B-cell receptor
  B. 1. Major histocompatibility complex (MHC) molecules; 2. MHC class I molecule; 3. MHC restricted; 4. Antigen-presenting cells; 5. Antigen-presenting cells; 6. MHC class II molecule; 7. MHC class I molecule; 8. MHC class II molecule
  C. 1. Costimulation; 2. Cytokines; 3. B7, CD4, CD8, or CD28
  D. 1. Antigen; 2. Helper T cells; 3. MHC class II; 4. Dividing; 5. Increases; 6. Antibodies; 7. Effector T cells
Inhibition of Lymphocytes
   1. Tolerance; 2. Self-antigens; 3. Self-reactive; 4. Activation; 5. Anergy; 6. Costimulation; 7. Suppressor T cells
Antibody-Mediated Immunity
  A. 1. Gamma globulins or immunoglobulins; 2. Variable region; 3. Constant region; 4. Opsonins
  B. 1. Antigen-binding site; 2. Heavy chain; 3. Light chain; 4. Complement-binding site; 5. Macrophage-binding site; 6. Constant region; 7. Variable region
  C. 1. Primary response; 2. Plasma cells; 3. Memory B cells; 4. Secondary (memory) response
Cell-Mediated Immunity
   1. Memory T cells; 2. Secondary response; 3. Lyse; 4. Cytokines; 5. Perforin; 6. Cytokines; 7. Inflammation; 8. Allergic
Immunotherapy
   1. Inflammation; 2. Tumor cells; 3. MHC molecules; 4. Vaccination; 5. Monoclonal antibodies; 6. Humanization
Acquired Immunity
   1. Immunization; 2. Active natural immunity; 3. Active artificial immunity; 4. Passive natural immunity; 5. Passive artificial immunity
Effects of Aging on the Lymphatic System and Immunity
   1. Little effect; 2. Little effect; 3. Decreases; 4. Decreases; 5. Decreases

Quick Recall

1. Maintains fluid balance, absorbs fat and other substances from the intestines, and defends against microorganisms and other foreign substances
2. Lymph nodes: filter lymph and remove substances by phagocytosis, stimulate and release lymphocytes; spleen: foreign substances stimulate lymphocytes in white pulp, foreign substances and erythrocytes are phagocytized in the red pulp; thymus: processes T cells and produces and releases T cells
3. Alternative pathway: when activated, C3 protein combines with a foreign substance; classical pathway: by antibody activation
4. Interferons are produced by a virally infected cell, move to other cells, and protect them from viral infection.
5. Neutrophils and macrophages
6. Basophils and mast cells release inflammatory chemicals and cause smooth muscle contraction. Eosinophils release enzymes that reduce inflammation.
7. B cells (plasma cells) produce antibodies. T cells produce most cytokines.
8. MHC class I molecule: combines with antigens produced inside a cell and the combined antigen complex is displayed on the cell surface; results in destruction of the displaying cell; MHC class II molecule: combines with foreign antigen taken in by antigen-presenting cells and forms an antigen complex that is displayed on the cell surface; results in activation of other immune system cells.
9. B cells, macrophages, monocytes, and dendritic cells
10. Costimulation is activation of a B cell or T cell that requires both cytokines and a MHC class II/antigen complex; anergy is a condition of inactivity in which a B cell or T cell does not respond to an antigen. Anergy develops when an MHC–antigen complex binds to an antigen receptor but there is no costimulation.
11. Deletion of self-reactive lymphocytes, prevention of the activation of lymphocytes, and suppression of lymphocytes by suppressor T cells
12. Direct effects: inactivate antigen or bind antigens together; indirect effects (activates other mechanisms): act as opsonins, activate complement, and increase inflammation
13. Memory B cells and memory T cells
14. Effector T cells lyse cells and produce cytokines.

23 Respiratory System

CONTENT LEARNING ACTIVITY

Nose

A. Match these terms with the correct statement or definition:

Choanae Nares (nostrils)
Conchae Nasal septum
Hard palate Nasus (nose)
Meatus Vestibule

_____ 1. Consists of the external nose and nasal cavity.

_____ 2. External openings of the nasal cavity.

_____ 3. Openings from the nasal cavity into the pharynx.

_____ 4. Divides the nasal cavity into right and left parts.

_____ 5. Anterior part of the nasal cavity, just inside a naris.

_____ 6. Floor of the nasal cavity.

_____ 7. Three bony ridges on the lateral wall of each nasal cavity.

_____ 8. Passageway between the conchae.

B. Match these terms with the correct statement or definition:

Cilia Mucus
Mucous membrane Nasal hair

_____ 1. Structures located in the vestibule that trap large dust particles.

_____ 2. Substance secreted by goblet cells that traps debris in the incoming air.

_____ 3. Structures on the surface of the mucous membrane that move mucus posteriorly to the pharynx.

_____ 4. Structure that adds moisture and warmth to inspired air.

Pharynx

A. Match these terms with the correct statement or definition:

Fauces Nasopharynx
Laryngopharynx Oropharynx

_____ 1. Portion of the pharynx posterior to the choanae and superior to the soft palate.

_____ 2. Portion of the pharynx that extends from the soft palate to the epiglottis.

_____ 3. Portion of the pharynx that extends from the epiglottis to the esophagus.

_____ 4. Portion of the pharynx that contains the openings of the auditory tubes.

_____ 5. Location of the pharyngeal tonsil (adenoid).

_____ 6. Location of the palatine and lingual tonsils.

_____ 7. Opening of the oral cavity into the oropharynx.

B. Match these terms with the correct parts labeled on the diagram in figure 23.1:

Choana Nasopharynx
Concha Oropharynx
Hard palate Soft palate
Laryngopharynx Trachea
Larynx Uvula
Meatus Vestibule
Naris (nostril)

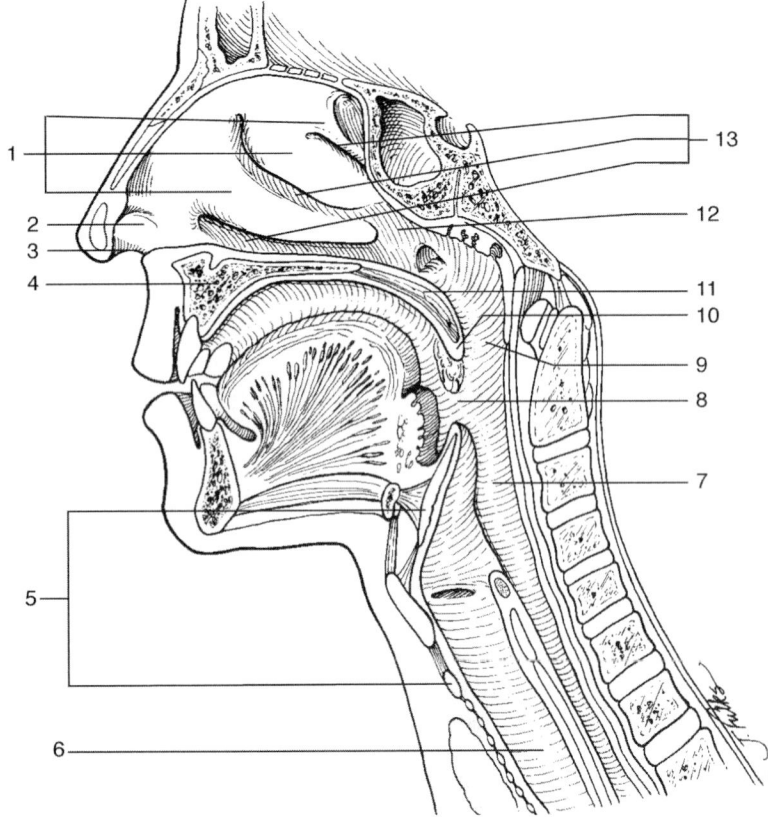

1. _____
2. _____
3. _____
4. _____
5. _____
6. _____
7. _____
8. _____
9. _____
10. _____
11. _____
12. _____
13. _____

Figure 23.1

Larynx

A. Match these terms with the correct statement or definition:

Cricoid cartilage Thyroid cartilage
Epiglottis Vestibular folds
Laryngitis Vocal folds

_____ 1. Largest, most superior laryngeal cartilage; the Adam's apple.

_____ 2. Forms the base of the larynx.

_____ 3. Covers the opening into the larynx during swallowing.

_____ 4. Superior pair of ligaments that close to prevent the movement of air, food, or liquids through the larynx.

_____ 5. Inferior pair of ligaments involved in sound production; these ligaments and the space between them are called the glottis.

_____ 6. Inflammation of the mucosal epithelium of the vocal cords.

B. Match these terms with the correct parts labeled on the diagram in figure 23.2:

Arytenoid cartilage
Corniculate cartilage
Cricoid cartilage
Cuneiform cartilage
Epiglottis
Hyoid bone
Thyroid cartilage
Tracheal cartilage

1. _____
2. _____
3. _____
4. _____
5. _____
6. _____
7. _____
8. _____

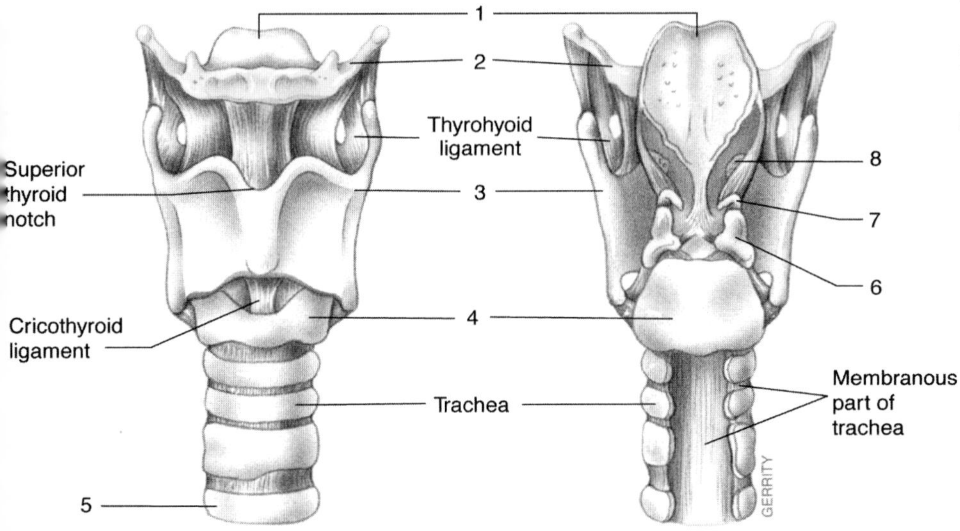

Figure 23.2

Chapter 23

Trachea

Match these terms with the correct statement or definition:

C-shaped cartilage
Carina
Main (primary) bronchi
Trachealis muscle

_____ 1. Holds the trachea open.

_____ 2. Changes the diameter of the trachea—for example, during coughing.

_____ 3. The trachea divides to form these tubes.

_____ 4. Cartilage ridge that separates the openings into the main bronchi.

Tracheobronchial Tree

A. Match these terms with the correct statement or definition:

Alveolar ducts
Alveolar sac
Alveoli
Bronchioles
Lobar (secondary) bronchi
Respiratory bronchiole
Segmental (tertiary) bronchi
Terminal bronchioles

_____ 1. Branch from main (primary) bronchi.

_____ 2. Branch from lobar (secondary) bronchi.

_____ 3. Branch from bronchi; less than 1 mm in diameter.

_____ 4. Branch from bronchioles.

_____ 5. Branch from terminal bronchioles.

_____ 6. Small, air-filled chambers that are the major site of gas exchange between the air and the blood.

_____ 7. Branch from respiratory bronchioles and are connected to many alveoli.

_____ 8. End of an alveolar duct; a chamber connected to two or more alveoli.

B. Using figure 23.3, arrange the structures in the order that a molecule of oxygen would pass through them to enter the blood:

Alveolar duct
Alveolar sac
Alveoli
Main (primary) bronchi
Respiratory bronchiole
Terminal bronchiole
Trachea

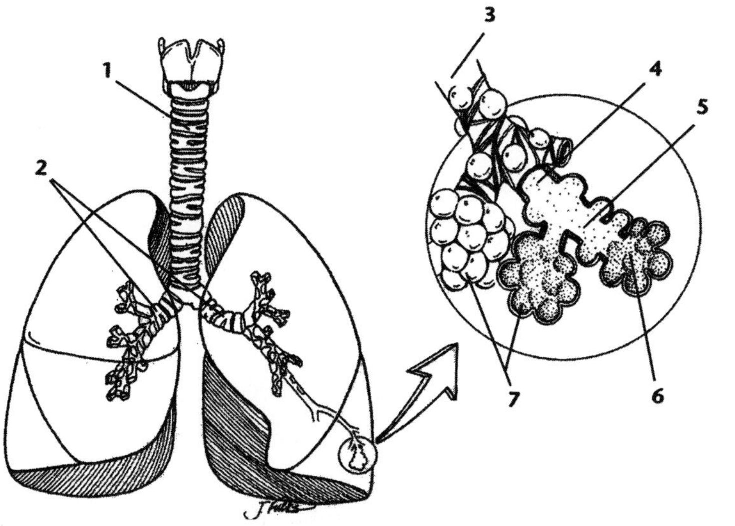

Figure 23.3

1. _____
2. _____
3. _____
4. _____
5. _____
6. _____
7. _____

C. Match the part of the tracheo-bronchial tree with the structure that forms the wall of that part:

Lobar (secondary) bronchi   Terminal bronchioles
Mail (primary) bronchi

_____ 1. C-shaped cartilages.

_____ 2. Cartilage plates.

_____ 3. Only smooth muscle.

D. Match the part of the tracheo-bronchial tree with the type of epithelium lining that part:

Alveolar ducts and alveoli   Terminal bronchioles
Bronchioles   Trachea and bronchi
Respiratory bronchioles

_____ 1. Pseudostratified ciliated columnar epithelium.

_____ 2. Ciliated simple columnar epithelium.

_____ 3. Ciliated simple cuboidal epithelium.

_____ 4. Simple cuboidal epithelium.

_____ 5. Simple squamous epithelium.

Chapter 23

E. Match these terms with the correct statement or definition:

Respiratory membrane   Type II pneumocytes
Type I pneumocytes

_____ 1. Where gas exchange between the air and blood takes place.

_____ 2. Most gas exchange between air and the blood takes place through these cells.

_____ 3. Produce surfactant.

F. Using the diagram in figure 23.4, arrange each of the structures of the respiratory membrane in the correct order from inside the alveolus to the blood:

Alveolar epithelium
Alveolar fluid
Basement membrane of alveolar epithelium
Basement membrane of capillary endothelium
Capillary endothelium
Interstitial space

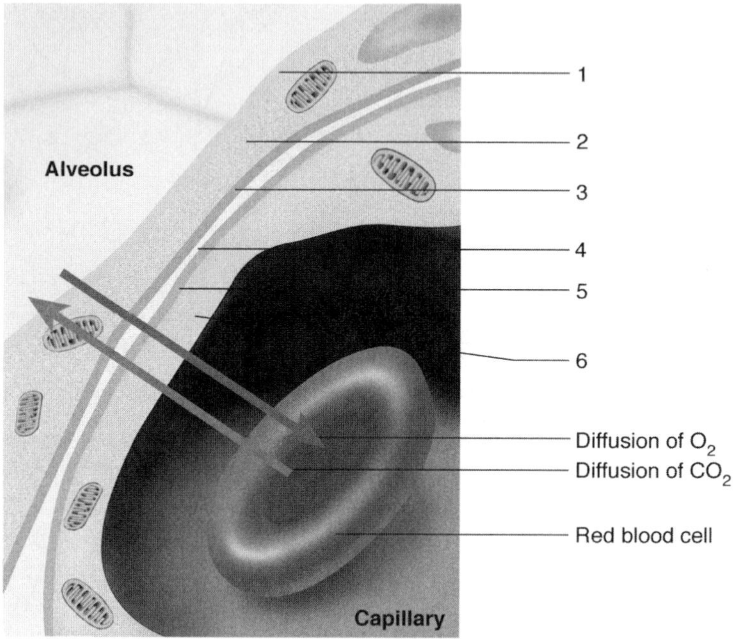

Figure 23.4

1. _____   3. _____   5. _____

2. _____   4. _____   6. _____

Lungs

Match these terms with the correct statement or definition:

Bronchopulmonary segments   Lobules
Hilum   Root of lung
Lobes

_____ 1. Region on the medial surface of the lung where structures, such as the bronchi and blood vessels, enter or exit the lung.

_____ 2. Collective name for all the structures that pass through the hilum.

_____ 3. The right lung has three and the left lung two; supplied by the lobar (secondary) bronchi.

_____ 4. Separated by the fissures of the lung.

_____ 5. Subdivisions of the lobes; the right lung has 10 and the left lung has 9; supplied by the segmental (tertiary) bronchi.

_____ 6. Separated by complete connective tissue partitions.

_____ 7. Supplied by the bronchioles; separated by incomplete connective tissue walls.

Thoracic Wall and Muscles of Respiration

A. Match these terms with the correct statement or definition:

Anterior
Diaphragm
Lateral
Muscles of expiration
Muscles of inspiration

_____ 1. Includes the diaphragm, external intercostal muscles, pectoralis minor, and scalenes.

_____ 2. Includes the abdominal muscles and the internal intercostal muscles.

_____ 3. Large dome of skeletal muscles that, when contracted, expands the superior-inferior dimension of the thoracic cavity.

_____ 4. The "bucket-handle" movement of the ribs increases this dimension of the thoracic cavity.

_____ 5. The "pump-handle" movement of the sternum increases this dimension of the thoracic cavity.

B. Using the terms provided, complete these statements:

Contract
Greater
Less
More
Relax
Smaller

During quiet inspiration, thoracic volume increases when the muscles of inspiration _(1)_ and the abdominal muscles _(2)_. During quiet expiration, passive recoil of the thorax and lungs occurs as the muscles of inspiration _(3)_. The abdominal muscles _(4)_, which pushes the abdominal organs and diaphragm superiorly. During labored inspiration, all the muscles of inspiration are active and they contract _(5)_ forcefully, causing a _(6)_ change in thoracic volume. During labored expiration, contraction of the muscles of expiration causes a _(7)_ decrease in thoracic volume than would be produced by the passive recoil of the thorax and lungs.

1. _____
2. _____
3. _____
4. _____
5. _____
6. _____
7. _____

Chapter 23

Pleura

Match these terms with the correct statement or definition:

Parietal pleura Pleural fluid
Pleural cavity Visceral pleura

_____ 1. Space surrounding each lung.

_____ 2. Covers the inner thoracic wall, superior surface of the diaphragm, and mediastinum.

_____ 3. Covers the surface of the lungs.

_____ 4. Lubricant that allows the pleural membranes to slide past each other and helps hold the pleural membranes together.

Blood Supply

Match these terms with the correct statement or definition:

Bronchial arteries Pulmonary arteries
Bronchial veins Pulmonary veins

_____ 1. Blood vessels that carry deoxygenated blood to the lungs.

_____ 2. Blood vessels that carry oxygenated blood from the lungs to the heart.

_____ 3. Blood vessels that branch from the thoracic aorta and carry oxygenated blood to the bronchi.

_____ 4. Blood vessels that carry deoxygenated blood from the bronchi to the azygos venous system.

Lymphatic Supply

Match these terms with the correct statement or definition:

Deep lymphatic vessels
Superficial lymphatic vessels

_____ 1. Located deep to the visceral pleura.

_____ 2. Follows the bronchi but does not supply alveoli.

Pressure, Airflow, and Volume Changes

Using the terms provided, complete these statements:

Boyle's law
Decrease
Higher
Increase
Length
Lower
Radius

Airflow through tubes, such as those in the tracheobronchial tree, flows from a region of __(1)__ pressure to a region of __(2)__ pressure. The resistance to airflow is proportional to the __(3)__ of the tube raised to the fourth power. Thus, a small decrease in the radius of a tube results in a large __(4)__ in the resistance to airflow and a large __(5)__ in airflow. According to __(6)__, an increase in thoracic volume results in a(n) __(7)__ in thoracic pressure.

1. _____
2. _____
3. _____
4. _____
5. _____
6. _____
7. _____

Airflow Into and Out of Alveoli

Match these terms with the correct statement or definition:

Alveolar pressure
Barometric air pressure
During expiration
During inspiration
End of expiration
End of inspiration

_____ 1. Atmospheric pressure outside the body; assigned a value of zero.

_____ 2. Pressure inside an alveolus.

_____ 3. Two times at which barometric air pressure and alveolar pressure are equal to each other; there is no air movement into or out of the alveoli.

_____ 4. Time during which barometric air pressure is greater than alveolar pressure.

_____ 5. Time during which barometric air pressure is lower than alveolar pressure.

Changing Alveolar Volume

A. Match these terms with the correct statement or definition:

Elastic recoil
Pleural pressure
Surface tension of alveolar fluid
Surfactant

_____
_____ 1. Two factors responsible for lung recoil.

_____ 2. Mixture of lipoprotein molecules produced by the type II pneumocytes of the alveolar epithelium; reduces the tendency for the lungs to collapse.

Chapter 23

_____ 3. Pressure within the pleural cavity.

_____ 4. Two factors that keep the alveoli from collapsing.

_____

B. Match these terms with the correct statement:

Decreases
Increases

_____ 1. Change in pleural pressure as thoracic volume increases.

_____ 2. Change in pleural pressure as the expanding lungs tend to recoil.

_____ 3. Change in alveolar volume as pleural pressure decreases.

_____ 4. Change in alveolar pressure as alveolar volume increases.

_____ 5. Change in air movement into the alveoli as alveolar pressure decreases below barometric air pressure.

Compliance of the Lungs and Thorax

Match these terms with the correct statement:

Decreases
Increases

_____ 1. Effect on the ease of lung expansion when compliance increases.

_____ 2. Effect on compliance when lung recoil increases, as in respiratory distress syndrome.

_____ 3. Effect on compliance of emphysema that destroys elastic lung tissue.

Pulmonary Volumes and Capacities

A. Match these terms with the correct statement or definition:

Pulmonary capacity   Spirometer
Pulmonary volume   Spirometry

_____ 1. Process of measuring volumes of air that move into and out of the respiratory system.

_____ 2. Device used to measure pulmonary volumes.

_____ 3. Examples are tidal volume, inspiratory reserve volume, expiratory reserve volume, and residual volume.

_____ 4. Sum of two or more pulmonary volumes.

_____ 5. Vital capacity is an example.

B. Match these terms with the correct statement or definition:

Decreases
Increases

_____ 1. Effect of asthma or collapse of the bronchi on FEV_1.

_____ 2. Effect of pulmonary fibrosis or kyphosis on FEV_1.

C. Match these terms with the correct definition and figure 23.5:

Expiratory reserve volume
Functional residual capacity
Inspiratory capacity
Inspiratory reserve volume
Residual volume
Tidal volume
Total lung capacity
Vital capacity

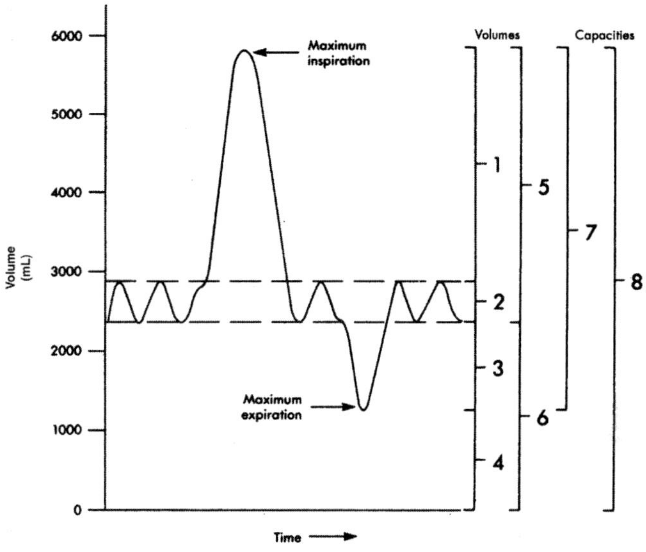

Figure 23.5

_____ 1. Volume of air that can be forcefully inspired after inspiration of the normal tidal volume.

_____ 2. Volume of air inspired or expired with each breath.

_____ 3. Volume of air that can be forcefully expired after expiration of the tidal volume.

_____ 4. Volume of air still in the respiratory passages after the most forceful expiration.

_____ 5. Tidal volume plus the inspiratory reserve volume.

_____ 6. Expiratory reserve volume plus the residual volume.

_____ 7. Sum of expiratory reserve, tidal, and inspiratory reserve volumes.

_____ 8. Sum of the inspiratory reserve, expiratory reserve, tidal, and residual volumes.

Minute Ventilation and Alveolar Ventilation

Match these terms with the correct statement or definition:

Alveolar ventilation
Anatomical dead space
Minute ventilation
Physiological dead space
Respiratory rate

_____ 1. Total amount of air moved in and out of the respiratory system each minute; tidal volume times respiratory rate.

_____ 2. Number of breaths taken per minute; the respiratory frequency.

_____ 3. Nasal cavity, pharynx, larynx, trachea, bronchi, bronchioles, and terminal bronchioles; gas exchange does not take place here.

_____ 4. Anatomical dead space plus the volume of any alveoli in which gas exchange is less than normal.

_____ 5. Volume of air that is available for gas exchange per minute.

Partial Pressure

A. Match these terms with the correct statement or definition:

Dalton's law
Partial pressure
Water vapor pressure

_____ 1. In a mixture of gases, the portion of the total pressure resulting from each type of gas is determined by the percentage of the total volume represented by each gas type.

_____ 2. Pressure exerted by each type of gas in a mixture.

_____ 3. Partial pressure of water molecules in the gaseous form.

B. Using the terms provided, complete these statements:

Greater
Smaller

Compared with atmospheric air, alveolar air has a (1) partial pressure of oxygen, a (2) partial pressure of carbon dioxide, and a (3) water vapor pressure.

1. _____
2. _____
3. _____

Diffusion of Gases Into and Out of Liquids

Match these terms with the correct statement or definition:

Henry's law
Solubility coefficient

_____ 1. The concentration of a dissolved gas is its partial pressure multiplied by its solubility coefficient.

_____ 2. Measure of how easily a gas dissolves in a liquid.

Diffusion of Gases Through the Respiratory Membrane

Match these terms with the correct statements:

Decreases
Increases

_____ 1. Effect on diffusion rate if the thickness of the respiratory membrane increases.

_____ 2. Effect on diffusion rate of a larger diffusion coefficient.

_____ 3. Effect on diffusion rate if the respiratory surface area decreases.

_____ 4. Effect on diffusion rate if the partial pressure difference across the respiratory membrane increases.

Relationship Between Alveolar Ventilation and Pulmonary Capillary Perfusion

A. Match these terms with the correct statement or definition:

Anatomical shunt   Inadequate cardiac output
Constriction of bronchioles   Physiologic shunt

_____ 1. In this situation, ventilation can exceed the ability of the blood to pick up oxygen.

_____ 2. In this situation, ventilation is not great enough to oxygenate the blood in the pulmonary capillaries.

_____ 3. Results from deoxygenated blood from the bronchi and bronchioles mixing with blood in the pulmonary veins.

_____ 4. Anatomical shunt plus deoxygenated blood from the pulmonary capillaries.

B. Using the terms provided, complete these statements:

Decreases   Less
Increases   More

In a standing person at rest, blood flow in the base of the lung is _(1)_ than the blood flow in the top of the lung because the effects of gravity cause blood vessels in the base of the lung to be _(2)_ expanded. During exercise, blood flow in the top of the lung increases _(3)_ than at the base of the lung because increased blood pressure during exercise expands the _(4)_ distended or collapsed vessels in the top of the lung. When the P_{O_2} of lung tissue decreases, blood flow to the tissue _(5)_.

1. _____
2. _____
3. _____
4. _____
5. _____

Chapter 23

Oxygen and Carbon Dioxide Partial Pressure Gradients

Match these partial pressures with with the correct location in figure 23.6:   104 mm Hg   45 mm Hg
95 mm Hg   40 mm Hg

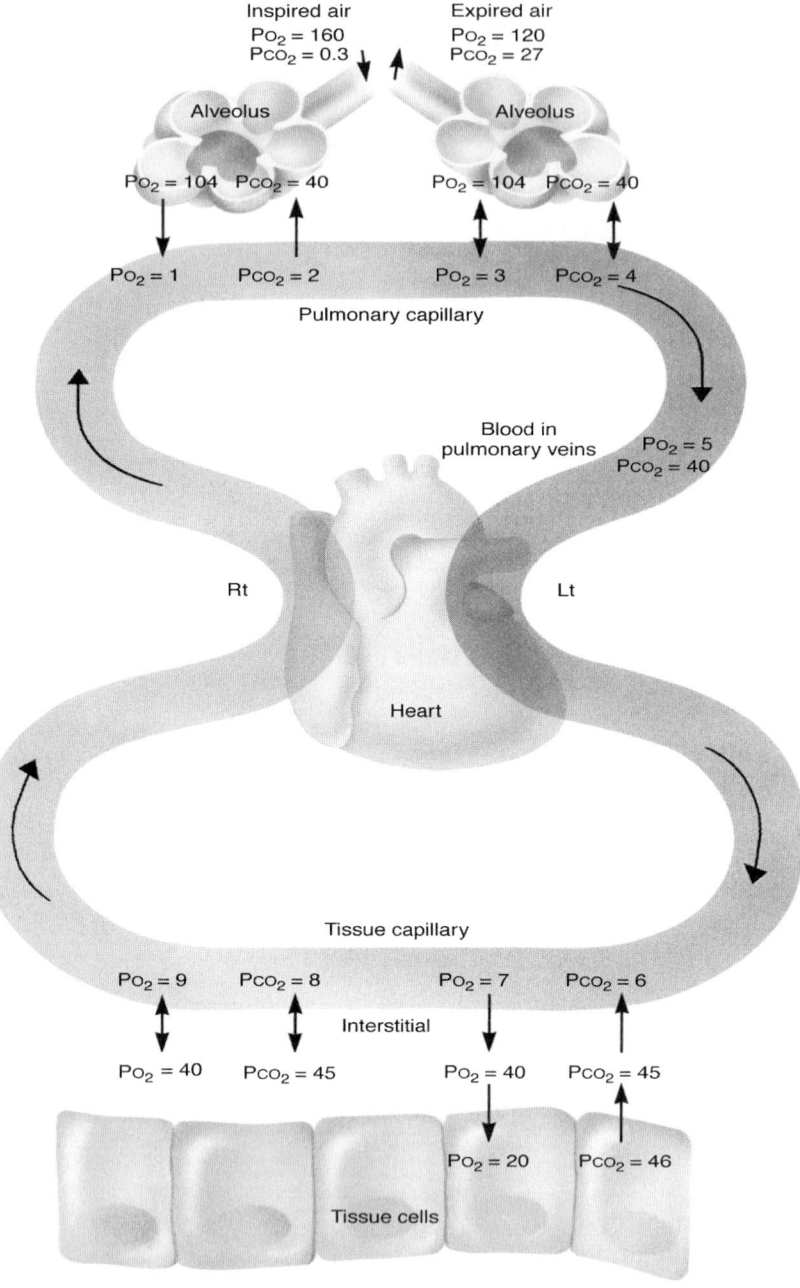

Figure 23.6

1. _____
2. _____
3. _____
4. _____
5. _____
6. _____
7. _____
8. _____
9. _____

Hemoglobin and Oxygen Transport

A. Match these terms with the correct statement or definition:

23% Decreases
73% Increases
98% Saturated hemoglobin

_____ 1. Hemoglobin molecule with an oxygen molecule bound to each of its four heme groups.

_____ 2. Effect of decreasing the partial pressure of oxygen on the ability of hemoglobin to bind to oxygen.

_____ 3. Percent saturation of hemoglobin in the lungs.

_____ 4. Percent decrease in hemoglobin saturation in resting tissues; a measure of how much oxygen is released into tissues.

_____ 5. Percent decrease in hemoglobin saturation in exercising tissues.

B. Match these terms with the correct statement or definition:

Bohr effect   Shift to the left
Decreases   Shift to the right
Increases

_____ 1. Change in the ability of hemoglobin to bind oxygen when H^+ bind to hemoglobin.

_____ 2. Change in the oxygen–hemoglobin dissociation curve resulting from changes in blood pH.

_____ 3. Change in the ability of hemoglobin to bind oxygen when pH decreases as a result of an increase in carbon dioxide.

_____ 4. Shift of the oxygen–hemoglobin dissociation curve resulting in hemoglobin releasing more oxygen; normally occurs in the tissues as pH decreases and carbon dioxide increases.

_____ 5. Shift of the oxygen–hemoglobin dissociation curve resulting in hemoglobin binding more oxygen; normally occurs in the lungs as pH increases and carbon dioxide decreases.

_____ 6. Effect on the amount of oxygen bound to hemoglobin when temperature increases.

C. Match these terms with the correct statement or definition:

Decreases   Increases
Greater   Less

_____ 1. Effect of increased levels of 2,3-bisphosphoglycerate (BPG) on the release of oxygen from hemoglobin.

_____ 2. Effect of high altitude on BPG levels.

_____ 3. Ability of fetal hemoglobin to pick up oxygen, compared with maternal hemoglobin.

Chapter 23

Transport of Carbon Dioxide

A. Match these terms with the correct statement or definition:

Dissolved in plasma
HCO_3^-
Haldane effect
Hemoglobin and blood proteins

_____ 1. Seventy percent of carbon dioxide is transported in this form.

_____ 2. Twenty-three percent of carbon dioxide is transported bound to these.

_____ 3. Seven percent of carbon dioxide is transported in this form.

_____ 4. Hemoglobin that has released oxygen binds more readily to carbon dioxide than hemoglobin that has oxygen bound to it.

B. Using the terms provided, complete these statements:

Bicarbonate ions
Bohr effect
Buffer
Carbonic acid
Carbonic anhydrase
Chloride ions
Chloride shift
Decreases
Hemoglobin
Increases

Most of the carbon dioxide inside red blood cells reacts with water to form __(1)__; the reaction is catalyzed by __(2)__ inside the red blood cells. The carbonic acid then dissociates to for __(3)__ and H^+. Because more HCO_3^- are inside the cells than outside, the HCO_3^- readily diffuse out of the red blood cells into the plasma. In response to this movement of negatively charged ions out of the red blood cells, __(4)__ move from the plasma into the red blood cells. The exchange of Cl^- for the HCO_3^- across the membranes of the red blood cells is called the __(5)__. The H^+ formed by the dissociation of carbonic acid bind to __(6)__ in the red blood cells. Thus, hemoglobin acts as a(n) __(7)__ to prevent a decrease in pH within the red blood cells. The removal of HCO_3^- from red blood cells and the binding of H^+ to hemoglobin __(8)__ carbon dioxide transport. The binding of H^+ to hemoglobin is also responsible for the __(9)__.

1. _____
2. _____
3. _____
4. _____
5. _____
6. _____
7. _____
8. _____
9. _____

C. Match these terms with the correct statement:

Decreases
Increases

_____ 1. Effect of increased plasma carbon dioxide on blood (plasma) H^+ levels.

_____ 2. Effect of increased H^+ levels on blood (plasma) pH.

_____ 3. Effect of hyperventilation on blood (plasma) pH.

Regulation of Rhythmic Ventilation

A. Match these terms with the correct statement or definition:

    Dorsal respiratory groups   Pontine respiratory group
    Medullary respiratory center   Ventral respiratory groups

_____ 1. Consists of two dorsal respiratory groups and two ventral respiratory groups.

_____ 2. Most active during inspiration; controls the diaphragm.

_____ 3. Active during inspiration and expiration; controls the external intercostal, internal intercostals, and abdominal muscles.

_____ 4. Appears to play a role in switching between inspiration and expiration.

B. Using the terms provided, complete these statements:

    Dorsal   Pontine
    Fewer   Stretch receptors
    More   Threshold

Inspiration begins when the input from many sources, such as from receptors that monitor blood gas levels or body movements, reach a (1). Once inspiration begins, (2) inspiratory neurons are gradually activated, resulting in the stimulation of the muscles of inspiration for approximately 2 seconds. Neurons responsible for stopping inspiration receive input from the neurons stimulating respiration, the (3) respiratory group, and (4) in the lungs. When the input to these neurons exceeds threshold, inspiration stops.

1. _____
2. _____
3. _____
4. _____

C. Match these terms with the correct statement or definition:

    Apnea   Limbic system
    Cerebral cortex

_____ 1. Part of the brain that is able to consciously or unconsciously change the rate or depth of respiration, such as talking or holding one's breath.

_____ 2. Part of the brain through which emotions alter respiratory activity, such as hyperventilation or crying.

_____ 3. Absence of breathing.

D. Match these terms with the correct statement or definition:

Carbon dioxide
Central
Decreases
Hering-Breuer reflex
Hypercapnia
Hypocapnia
Hypoxia
Increases
Peripheral
Touch, pain, and thermal

1. Chemoreceptors of the chemosensitive area of the medulla oblongata; detect changes in cerebrospinal fluid pH.
2. Chemoreceptors in the carotid and aortic bodies; detect changes in blood pH and P_{O_2}.
3. Major regulator of respiration because of its effect on pH.
4. Lower than normal blood carbon dioxide levels.
5. Effect of decreased blood pH, caused by increased blood carbon dioxide, on breathing rate and depth.
6. Chemoreceptors responsible for most of the response to changes in pH or P_{CO_2}.
7. Lower than normal blood oxygen levels.
8. Effect of greatly decreased blood oxygen levels, detected by peripheral chemoreceptors, on breathing rate and depth.
9. Limits the degree to which inspiration proceeds and prevents overinflation of the lungs.
10. When these receptors are activated, respiratory activity can be altered—e.g., sneezing and coughing.

E. Match these terms with the correct statement or definition:

Anaerobic threshold
Decreases
Increases
No significant change

1. Effect of action potentials, traveling from collateral fibers of motor pathways, on breathing rate during exercise.
2. Effect of the stimulation of proprioceptors on respiratory rate during exercise.
3. Changes in average arterial P_{O_2}, P_{CO_2}, and pH during exercise.
4. Highest level of exercise that can be performed without causing a significant change in blood gases and pH.

Respiratory Adaptations to Exercise

Match these terms with the
effect produced by training:

Decreases
Increases

No change
Slight decrease

_____ 1. Change in tidal volume at rest and during submaximal exercise.

_____ 2. Change in tidal volume during maximal exercise.

_____ 3. Change in respiratory rate at rest and during submaximal exercise.

_____ 4. Change in respiratory rate during maximal exercise.

_____ 5. Change in minute ventilation at rest or during submaximal exercise.

_____ 6. Change in minute ventilation during maximal exercise.

_____ 7. Change in gas exchange between alveoli and blood during maximal exercise.

Effects of Aging on the Respiratory System

Match these terms with the
effect produced by aging:

Decreases
Increases

No change
Slight decrease

_____ 1. Effect of aging on vital capacity.

_____ 2. Effect of aging on maximum minute ventilation

_____ 3. Effect of aging on dead space.

_____ 4. Effect of aging on resting tidal volume.

_____ 5. Effect of aging on the ability to remove mucus from the tracheobronchial tree.

QUICK RECALL

1. List five functions performed by the respiratory system.

2. List five functions performed by the nasal cavity.

3. List three function performed by the larynx.

4. Trace the path of inspired air from the trachea to the alveoli by naming the structures through which the air passes.

5. Describe the relationship between the tracheobronchial tree and the lungs and the parts of the lungs.

6. Name the two factors that determine the rate of airflow in a tube. Describe the relationship between the volume and the pressure of a gas.

7. List two factors that cause the lungs to recoil and two factors that prevent the alveoli from collapsing.

8. List the four pulmonary volumes and define pulmonary capacity.

9. Give the formula for calculating alveolar ventilation.

10. List the six layers of the respiratory membrane.

11. Name four factors that influence the rate of gas diffusion across the respiratory membrane.

12. List two ways oxygen is transported in the blood, and state their relative importance.

13. List three ways that carbon dioxide is transported in the blood, and indicate their relative importance.

14. Describe the chemical events that result in a decrease in blood pH when blood carbon dioxide levels increase.

15. Name the factors that have the greatest effect on the regulation of respiration at rest and during exercise.

ANSWERS TO CHAPTER 23

CONTENT LEARNING ACTIVITY

Nose
A. 1. Nasus (nose); 2. Nares (nostrils); 3. Choanae; 4. Nasal septum; 5. Vestibule; 6. Hard palate; 7. Conchae; 8. Meatus
B. 1. Nasal hair; 2. Mucus; 3. Cilia; 4. Mucous membrane

Pharynx
A. 1. Nasopharynx; 2. Oropharynx; 3. Laryngopharynx; 4. Nasopharynx; 5. Nasopharynx; 6. Oropharynx; 7. Fauces
B. 1. Conchae; 2. Vestibule; 3. Naris (nostril); 4. Hard palate; 5. Larynx; 6. Trachea; 7. Laryngopharynx; 8. Oropharynx; 9. Nasopharynx; 10. Uvula; 11. Soft palate; 12. Choana; 13. Meatus

Larynx
A. 1. Thyroid cartilage; 2. Cricoid cartilage; 3. Epiglottis; 4. Vestibular folds; 5. Vocal folds; 6. Laryngitis
B. 1. Epiglottis 2. Hyoid bone; 3. Thyroid cartilage; 4. Cricoid cartilage; 5. Tracheal cartilage; 6. Arytenoid cartilage; 7. Corniculate cartilage; 8. Cuneiform cartilage

Trachea
1. C-shaped cartilage; 2. Trachealis muscle; 3. Main (primary) bronchi; 4. Carina

Tracheobronchial Tree
A. 1. Lobar (secondary) bronchi; 2. Segmental (tertiary) bronchi; 3. Bronchioles; 4. Terminal bronchioles; 5. Respiratory bronchioles; 6. Alveoli; 7. Alveolar ducts; 8. Alveolar sac
B. 1. Trachea; 2. Main (primary) bronchi; 3. Terminal bronchiole; 4. Respiratory bronchiole; 5. Alveolar duct; 6. Alveolar sac; 7. Alveoli
C. 1. Main (primary) bronchi; 2. Lobar (secondary) bronchi; 3. Terminal bronchioles
D. 1. Trachea and bronchi; 2. Bronchioles; 3. Terminal bronchioles; 4. Respiratory bronchioles; 5. Alveolar ducts and alveoli
E. 1. Respiratory membrane; 2. Type I pneumocytes; 3. Type II pneumocytes
F. 1. Alveolar fluid; 2. Alveolar epithelium; 3. Basement membrane of alveolar epithelium; 4. Interstitial space; 5. Basement membrane of capillary epithelium; 6. Capillary endothelium

Lungs
1. Hilum; 2. Root of lung; 3. Lobes; 4. Lobes; 5. Bronchopulmonary segments; 6. Bronchopulmonary segments; 7. Lobules

Thoracic Wall and Muscles of Respiration
A. 1. Muscles of inspiration; 2. Muscles of expiration; 3. Diaphragm; 4. Lateral; 5. Anterior
B. 1. Contract; 2. Relax; 3. Relax; 4. Contract; 5. More; 6. Greater; 7. Greater

Pleura
1. Pleural cavity; 2. Parietal pleura; 3. Visceral pleura; 4; Pleural fluid

Blood Supply
1. Pulmonary arteries; 2. Pulmonary veins; 3. Bronchial arteries; 4. Bronchial veins

Lymphatic Supply
1. Superficial lymphatic vessels; 2. Deep lymphatic vessels

Pressure, Airflow, and Volume Changes
1. Higher; 2. Lower; 3. Radius; 4. Increase; 5. Decrease; 6. Boyle's law; 7. Decrease

Airflow Into and Out of Alveoli
1. Barometric air pressure; 2. Alveolar pressure; 3. End of expiration and end of inspiration; 4. During inspiration; 5. During expiration

Changing Alveolar Volume
A. 1. Elastic recoil and surface tension of alveolar fluid; 2. Surfactant; 3. Pleural pressure; 4. Surfactant and pleural pressure
B. 1. Decreases; 2. Decreases; 3. Increases; 4. Decreases; 5. Increases

Compliance of the Lungs and Thorax
1. Increases; 2. Decreases; 3. Increases

Pulmonary Volumes and Capacities
A. 1. Spirometry; 2. Spirometer; 3. Pulmonary volume; 4. Pulmonary capacity; 5. Pulmonary capacity
B. 1. Decreases; 2. Decreases
C. 1. Inspiratory reserve volume; 2. Tidal volume; 3. Expiratory reserve volume; 4. Residual volume; 5. Inspiratory capacity; 6. Functional residual capacity; 7. Vital capacity; 8. Total lung capacity

Minute Ventilation and Alveolar Ventilation
1. Minute ventilation; 2. Respiratory rate; 3. Anatomical dead space; 4. Physiological dead space; 5. Alveolar ventilation

Partial Pressure
A. 1. Dalton's law; 2. Partial pressure; 3. Water vapor pressure
B. 1. Smaller; 2. Greater; 3. Greater

Diffusion of Gases Into and Out of Liquids
1. Henry's law; 2. Solubility coefficient

Diffusion of Gases Through the Respiratory Membrane
1. Decreases; 2. Increases; 3. Decreases; 4. Increases

Relationship Between Alveolar Ventilation and Pulmonary Capillary Perfusion
A. 1. Inadequate cardiac output; 2. Constriction of bronchioles; 3. Anatomical shunt; 4. Physiological shunt
B. 1. More; 2. More; 3. More; 4. Less; 5. Decreases

Oxygen and Carbon Dioxide Partial Pressure Gradients
1. 40 mm Hg; 2. 45 mm Hg; 3. 104 mm Hg; 4. 40 mm Hg; 5. 95 mm Hg; 6. 40 mm Hg; 7. 95 mm Hg; 8. 45 mm Hg; 9. 40 mm Hg

Hemoglobin and Oxygen Transport
A. 1. Saturated hemoglobin; 2. Decreases; 3. 98%, 4. 23%; 5. 73%
B. 1. Decreases; 2. Bohr effect; 3. Decreases; 4. Shift to the right; 5. Shift to the left; 6. Decreases
C. 1. Increases; 2. Increases; 3. Greater

Transport of Carbon Dioxide
A. 1. HCO_3^-; 2. Hemoglobin and blood proteins; 3. Dissolved in plasma; 4. Haldane effect
B. 1. Carbonic acid; 2. Carbonic anhydrase; 3. HCO_3^-; 4. Cl^-; 5. Chloride shift; 6. Hemoglobin; 7. Buffer; 8. Increases; 9. Bohr effect
C. 1. Increases; 2. Decreases; 3. Increases

Regulation of Rhythmic Ventilation
- A. 1. Medullary respiratory center; 2. Dorsal respiratory groups; 3. Ventral respiratory groups; 4. Pontine respiratory group
- B. 1. Threshold; 2. More; 3. Pontine; 4. Stretch receptors
- C. 1. Cerebral cortex; 2. Limbic system; 3. Apnea
- D. 1. Central; 2. Peripheral; 3. Carbon dioxide; 4. Hypocapnia; 5. Increases; 6. Central; 7. Hypoxia; 8. Increases; 9. Hering-Breuer; 10. Touch, pain, and thermal
- E. 1. Increases; 2. Increases; 3. No significant change; 4. Anaerobic threshold

Respiratory Adaptations to Exercise
1. No change; 2. Increases; 3. Slight decrease; 4. Increases; 5. No change or slight decrease; 6. Increases; 7. Increases

Effect of Aging on the Respiratory System
1. Decreases; 2. Decreases; 3. Increases; 4. Increases; 5. Decreases

QUICK RECALL

1. Gas exchange, regulation of blood pH, voice production, olfaction, and protection
2. Passageway for air, cleans the air, humidifies and warms the air, contains the olfactory epithelium for the sense of smell, and resonating chambers (nasal cavity and paranasal sinuses) for speech
3. Maintains an open passageway for air, prevents swallowed materials from entering the larynx, and sound production
4. Trachea, main (primary) bronchus, lobar (secondary) bronchus, segmental (tertiary) bronchus, bronchioles, terminal bronchiole, respiratory bronchioles, alveolar duct, alveolar sac, alveolus
5. Main (primary) bronchi supply each lung, lobar (secondary) bronchi supply the lobes, segmental (tertiary) bronchi supply the bronchopulmonary segments, and bronchioles supply the lobules.
6. An increased pressure gradient and a decreased resistance to flow cause an increase in airflow; increased volume decreases pressure (Boyle's law).
7. Lung recoil results from elastic fibers and water surface tension; surfactant and pleural pressure prevent the collapse of the alveoli.
8. The pulmonary volumes are tidal volume, inspiratory reserve volume, expiratory reserve volume, and residual volume. Pulmonary capacities are the sum of two or more pulmonary volumes.
9. Alveolar ventilation = respiratory rate times the difference between the tidal volume and the dead space—i.e., $V_A = f(V_T - V_D)$.
10. Thin layer of fluid, alveolar epithelium, alveolar epithelium basement membrane, interstitial space, basement membrane of capillary endothelium, and capillary endothelium
11. Thickness of the respiratory membrane, diffusion coefficient, surface area of the respiratory membrane, and partial pressure gradient
12. Hemoglobin 98.5%, dissolved in plasma 1.5%
13. Bicarbonate ions 70%, blood proteins 23%, and dissolved in plasma 7%
14. Carbon dioxide and water combine to form carbonic acid, which dissociates into H^+ and HCO_3^-
15. At rest: changes in pH, which can be caused by changes in carbon dioxide, are detected by the central chemoreceptors; during exercise: input from the motor cortex and proprioceptors

24 Digestive System

CONTENT LEARNING ACTIVITY

Anatomy of the Digestive System

A. Using the terms provided, complete these statements:

Accessory organs
Alimentary tract (canal)
Gastrointestinal tract
Tube

The digestive system consists of the digestive tract, a(n) (1) extending from the mouth to the anus, and its associated (2) (primarily glands), which secrete fluids into the digestive tract. The digestive tract is also called the (3). The (4) technically consists only of the stomach and intestines but the term is often used as a synonym for *digestive tract*.

1. _____
2. _____
3. _____
4. _____

B. Match these terms with the correct statement or definition:

Anus
Esophagus
Large intestine
Oral cavity
Pharynx
Small intestine

_____ 1. First region of the digestive tract; has salivary glands and tonsils as accessory organs.

_____ 2. Throat; contains tubular mucous glands.

_____ 3. Region of the digestive tract between the throat and stomach; contains tubular mucous glands.

_____ 4. Consists of the duodenum, jejunum, and ileum, with the liver, gallbladder, and pancreas as major accessory organs.

_____ 5. Includes the cecum, colon, rectum, and anal canal; contains mucous glands.

_____ 6. Final region of the digestive tract.

C. Match these terms with the correct parts of the diagram labeled in figure 24.1:

Anus
Appendix
Esophagus
Gallbladder
Large intestine
Liver
Oral cavity (mouth)
Pancreas
Pharynx (throat)
Rectum
Salivary glands
Small intestine
Stomach

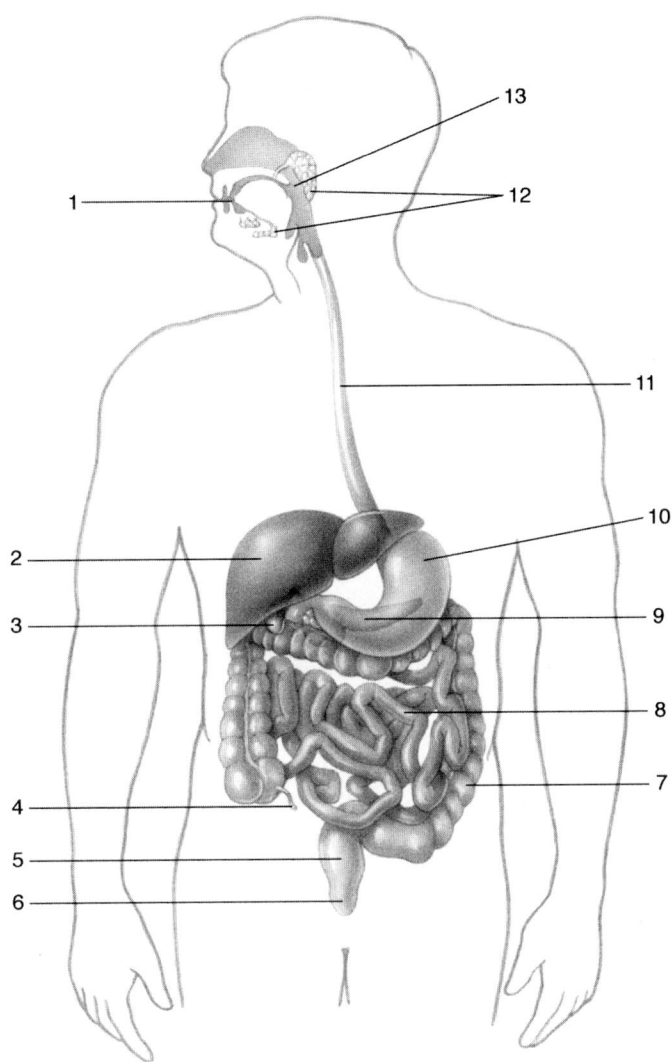

Figure 24.1

1. _____
2. _____
3. _____
4. _____
5. _____
6. _____
7. _____
8. _____
9. _____
10. _____
11. _____
12. _____
13. _____

Chapter 24

Functions of the Digestive System

Match these terms with the correct statement or definition:

Absorption
Defecation
Deglutition
Digestion
Ingestion
Mass movements
Mastication
Peristalsis
Propulsion
Secretion
Segmental contractions

_____ 1. Introduction of solid or liquid food into the stomach.

_____ 2. Process by which the food taken into the mouth is chewed by the teeth.

_____ 3. Movement of food from one end of the digestive tract to the other.

_____ 4. Swallowing; moves a bolus from the mouth into the esophagus.

_____ 5. Contraction of circular and longitudinal muscles in waves; moves food through the digestive tract.

_____ 6. Contractions that extend over large parts of the large intestine.

_____ 7. Mixing contractions that occur in the small intestine.

_____ 8. Adding mucus, water, and enzymes to digestive tract contents.

_____ 9. Breakdown, mechanically or chemically, of large organic molecules into their component parts.

_____ 10. Movement of molecules out of the digestive tract into the circulation or into the lymphatic system.

_____ 11. Elimination of semisolid waste from the digestive tract.

Histology of the Digestive Tract

A. Match these terms with the correct statement or definition:

Mucosa
Muscularis
Serosa (adventitia)
Submucosa

_____ 1. Innermost tunic, consisting of three layers.

_____ 2. Thick connective tissue tunic between the mucosa and the muscularis tunics.

_____ 3. Tunic that consists of two layers of smooth muscle, located between the submucosa and serosa tunics.

_____ 4. Connective tissue tunic that forms the outermost layer of the digestive tract.

B. Match these parts of the tunics with the correct statement or definition:

- Enteric (intramural) plexus
- Interstitial cells
- Lamina propria
- Mucous epithelium
- Muscularis mucosae
- Myenteric plexus
- Submucosal plexus
- Visceral peritoneum

1. _____ Mucosa layer composed of squamous or columnar epithelial cells.

2. _____ Layer of mucosa composed of loose connective tissue.

3. _____ Layer of mucosa composed of a thin, smooth muscle layer.

4. _____ Nerve plexus in the muscularis layer.

5. _____ Collective name for the submucosal and myenteric plexuses.

6. _____ Outermost serosa layer; present on parts of the digestive tract that protrude into the peritoneal cavity.

7. _____ Specialized cells within the myenteric plexus that form a network of "pacemakers."

Regulation of the Digestive Tract

Using the terms provided, complete these statements:

- Enteric
- Enteric interneurons
- Enteric motor neurons
- Enteric sensory neurons
- Hormones
- Inhibits
- Local reflexes
- Paracrine
- Parasympathetic
- Stimulates
- Sympathetic

Nervous and chemical mechanisms regulate the movement, secretion, absorption, and elimination processes. Local neuronal control of the digestive tract occurs within the _(1)_ nervous system (ENS). The ENS coordinates peristalsis and regulates _(2)_, which control activities within specific, short regions of the digestive tract. There are three major types of enteric neurons: _(3)_ detect chemical changes within the digestive tract or stretch of the digestive tract wall, _(4)_ stimulate or inhibit smooth muscle contraction and glandular secretion in the digestive tract, and _(5)_ connect enteric sensory and motor neurons. _(6)_ neurons extend to the digestive tract through the vagus nerves to control responses or alter the activity of the ENS and local reflexes. Alternatively, some _(7)_ neurons inhibit muscle contraction and secretion and decrease blood flow to the digestive system. The digestive tract also produces _(8)_, such as gastrin and secretin, which help regulate many gastrointestinal tract functions. In addition to hormones, other _(9)_ chemicals, such as histamine, are released locally and influence the activity of nearby cells. Two major ENS neurotransmitters are acetylcholine and norepinephrine. In general, acetylcholine _(10)_, and norepinephrine _(11)_ GI tract motility and secretions. Serotonin, another GI tract neurotransmitter, _(12)_ GI tract motility.

1. _____
2. _____
3. _____
4. _____
5. _____
6. _____
7. _____
8. _____
9. _____
10. _____
11. _____
12. _____

Peritoneum

A. Match these terms with the correct statement or definition:

Mesenteries   Retroperitoneal
Parietal peritoneum   Visceral peritoneum

_____ 1. Serous membrane that covers the organs of the abdominal cavity.

_____ 2. Serous membrane that covers the interior surface of the abdominal body wall.

_____ 3. Connective tissue sheets that hold many of the organs in place within the abdominal cavity; consist of two layers of serous membrane with loose connective tissue between them.

_____ 4. Abdominal organs that lie against the body wall and have no mesenteries—e.g., duodenum, pancreas, and kidneys.

B. Match these terms with the correct statement or definition:

Coronary ligament   Mesoappendix
Falciform ligament   Omental bursa
Greater omentum   Sigmoid mesocolon
Lesser omentum   Transverse mesocolon
Mesentery proper

_____ 1. Mesentery connecting the lesser curvature of the stomach and the proximal end of the duodenum to the liver and diaphragm.

_____ 2. Mesentery that extends as a fold from the greater curvature of the stomach to the transverse colon.

_____ 3. Cavity, or pocket, formed between the long, double folds of mesentery in the greater omentum.

_____ 4. Mesentery that attaches the liver to the diaphragm.

_____ 5. Mesentery that attaches the liver to the anterior abdominal wall.

_____ 6. Mesentery associated with the small intestine.

_____ 7. Mesentery that extends from the transverse colon to the posterior body wall.

_____ 8. Mesentery that extends from the sigmoid colon to the posterior body wall.

_____ 9. Mesentery attached to the vermiform appendix.

Oral Cavity

A. Match these terms with the correct statement or definition:

Buccal fat pad Oral cavity proper
Buccinator muscle Palatine tonsils
Fauces Soft palate
Frenula Uvula
Hard palate Vestibule

_____ 1. Posterior boundary of the oral cavity; opening into the pharynx.

_____ 2. Space between the lips or cheeks and the alveolar processes.

_____ 3. Region of the oral cavity medial to the alveolar processes.

_____ 4. Mucosal folds that extend from the alveolar processes of the maxilla and mandible to the lips.

_____ 5. Muscle that flattens the cheek against the teeth.

_____ 6. Structure that rounds out the profile of the side of the face.

_____ 7. Anterior bony part of the palate.

_____ 8. Projection from the posterior edge of the soft palate.

_____ 9. Tonsils located in the lateral wall of the fauces.

B. Match these terms with the correct statement or definition:

Extrinsic muscles Lingual tonsil
Frenulum Terminal sulcus
Intrinsic muscles

_____ 1. Thin fold of tissue that attaches the tongue to the floor of the mouth.

_____ 2. Muscles within the tongue; responsible for changing the shape of the tongue.

_____ 3. Muscles attached to the tongue but located outside it; responsible for protruding, retracting, and moving the tongue from side to side.

_____ 4. Groove that divides the tongue into an anterior two-thirds and a posterior one-third.

_____ 5. Lymphoid tissue on the posterior one-third of the tongue.

Chapter 24

C. Match these numbers with the correct description:

    One   Two   Three

_____ 1. Number of incisors in each quadrant of the mouth.

_____ 2. Number of canines in each quadrant of the mouth.

_____ 3. Number of premolars in each quadrant of the mouth.

_____ 4. Number of molars in each quadrant of the mouth.

D. Match these terms with the correct statement or definition:

    Maxillary and mandibular arches   Primary teeth
    Permanent teeth   Wisdom tooth

_____ 1. Location of the 32 teeth found in a normal adult.

_____ 2. Third molar.

_____ 3. Teeth of the adult mouth; secondary teeth.

_____ 4. Teeth lost during childhood; deciduous or milk teeth.

E. Match these terms with the correct statement or definition:

    Anatomical crown   Neck
    Apical foramen   Pulp
    Cementum   Pulp cavity
    Clinical crown   Root
    Dentin   Root canal
    Enamel

_____ 1. Part of the tooth exposed in the oral cavity.

_____ 2. Entire enamel-covered part of the tooth.

_____ 3. Blood vessels, nerves, and connective tissue in the center of a tooth.

_____ 4. Pulp cavity within the root of the tooth.

_____ 5. Where blood vessels and nerves enter and exit the tooth.

_____ 6. Living, calcified tissue; surrounds the pulp cavity of the tooth.

_____ 7. Extremely hard, nonliving, acellular substance; covers dentin and protects the tooth from abrasion and acids.

_____ 8. Substance that covers the root; helps anchor teeth in the jaw.

F. Match these terms with the correct statement or definition:

Alveoli   Periodontal ligaments
Gingiva

_____ 1. Sockets in the alveolar processes of the maxilla and mandible.

_____ 2. Dense fibrous connective tissue and stratified squamous epithelium that covers the alveolar ridges.

_____ 3. Connective tissue that holds the teeth in the alveoli.

G. Match these terms with the correct parts of the diagram labeled in figure 24.2:

Apical foramen   Gingiva
Cementum   Neck
Crown   Periodontal ligaments
Cusp   Pulp cavity
Dentin   Root
Enamel   Root canal

1. _____
2. _____
3. _____
4. _____
5. _____
6. _____
7. _____
8. _____
9. _____
10. _____
11. _____
12. _____

Figure 24.2

Chapter 24   375

Mastication

Match these terms with the correct statement or definition:

Chewing (mastication) reflex   Medial pterygoid muscles
Incisors and canines   Premolars and molars
Lateral pterygoid muscles   Temporalis muscles
Masseter muscles

_____ 1. Teeth that cut and tear food.

_____ 2. Teeth that crush and grind food.

_____
_____
_____ 3. Muscles of mastication that close the jaw.

_____ 4. Muscles of mastication that open the jaw.

_____ 5. Reflex integrated in the medulla oblongata; controls the basic movements involved in chewing.

Salivary Glands

A. Match these terms with the correct statement or definition:

Parotid glands   Submandibular glands
Sublingual glands

_____ 1. Largest salivary glands, located just anterior to the ear; produce mostly serous fluid.

_____ 2. Salivary glands that become infected with mumps.

_____ 3. Salivary glands located along the inferior border of the posterior half of the mandible; produce more serous fluid than mucus.

_____ 4. Salivary glands located immediately below the mucous membrane of the floor of the mouth; produce mostly mucus.

B. Using the terms provided, complete these statements:

Antibacterial Odors
Immunoglobulin A Parasympathetic
Mucin Starch

The amylase in saliva starts the digestive process by breaking the covalent bonds between glucose molecules in (1) and other polysaccharides. In addition, saliva prevents bacterial infection in the mouth by washing the oral cavity, and it contains substances (e.g., lysozyme) with weak (2) action. Saliva also contains (3), which helps prevent bacterial infection. The mucous secretions of the submandibular and sublingual glands contain a large amount of (4), a proteoglycan that gives a lubricating quality to salivary gland secretion. Salivary gland secretion is stimulated mainly by the (5) nervous system but also by the sympathetic nervous system. Tactile simulation, certain tastes, (6), and higher brain centers also affect the activity of the salivary glands.

1. _____
2. _____
3. _____
4. _____
5. _____
6. _____

Pharynx and Esophagus

Match these terms with the correct statement or definition:

Esophagus Pharyngeal constrictors
Laryngopharynx Upper and lower esophageal
Nasopharynx sphincters
Oropharynx

_____ 1. Portions of the pharynx that normally transmit food.

_____ 2. Portion of the pharynx superior to the oropharynx.

_____ 3. Three muscles located in the posterior walls of the oropharynx and laryngopharynx.

_____ 4. Portion of the digestive tract that extends between the pharynx and stomach.

_____ 5. Muscles that regulate the movement of materials into and out of the esophagus.

Swallowing

A. Match these phases of deglutition with the correct statement or definition:

Esophageal phase
Pharyngeal phase
Voluntary phase

_____ 1. Phase of swallowing that involves forcing a bolus of food into the oropharynx.

_____ 2. Phase of swallowing that involves closing the nasopharynx, forcing food through the pharynx, and covering the opening into the larynx.

_____ 3. Phase of swallowing that is responsible for moving food from the pharynx to the stomach.

B. Match these terms with the correct statement or definition:

Enteric plexus   Pharyngeal constrictor muscles
Epiglottis   Swallowing center
Peristaltic waves

_____ 1. Area in the medulla oblongata that controls swallowing.

_____ 2. Muscles that contract and force food through the pharynx.

_____ 3. Part of the larynx that covers the opening into the larynx.

_____ 4. Muscular contractions of the esophagus.

_____ 5. Stimulated by food in the esophagus; controls peristaltic waves.

Stomach

A. Match these terms with the correct statement or definition:

Body   Lower esophageal sphincter
Cardiac part   Muscularis
Fundus   Pyloric part
Greater curvature Pyloric sphincter
Lesser curvature Rugae

_____ 1. Surrounds the gastroesophageal (cardiac) opening; cardiac sphincter.

_____ 2. Region of the stomach around the gastroesophageal opening.

_____ 3. Part of the stomach to the left of the cardiac part; superior to the gastroesophageal opening.

_____ 4. Largest part of the stomach.

_____ 5. Curvature on the left side of the body of the stomach.

_____ 6. Region of the stomach that joins the small intestine.

_____ 7. Relatively thick ring of smooth muscle that surrounds the opening between the stomach and the small intestine.

_____ 8. Layer of the stomach wall that consists of longitudinal, circular, and oblique layers of smooth muscle.

_____ 9. Large folds of the mucosa and submucosa layers formed when the stomach is empty.

B. Match these terms with the correct parts of the diagram labeled in figure 24.3:

Body
Cardiac part
Fundus
Gastroesophageal opening
Lower esophageal sphincter
Pyloric orifice
Pyloric part
Pyloric sphincter
Rugae

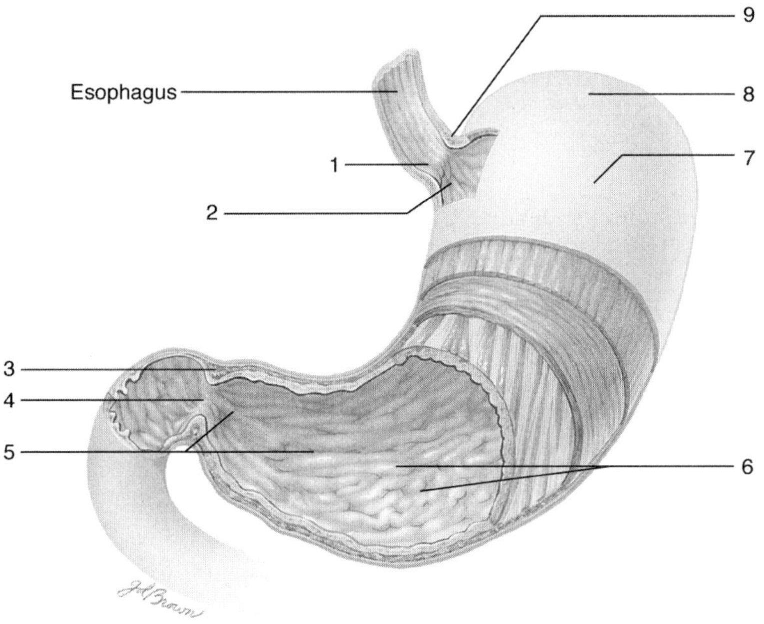

Figure 24.3

1. _____
2. _____
3. _____
4. _____
5. _____
6. _____
7. _____
8. _____
9. _____

Chapter 24

C. Match these terms with the correct statement or definition:

Chief cells
Endocrine cells
Gastric glands
Gastric pits
Mucous neck cells
Parietal cells
Surface mucous cells

_____ 1. Tubelike openings in the mucosal surface of the stomach.

_____ 2. Glands in the stomach that open into the gastric pits.

_____ 3. Epithelial cells in the stomach that secrete mucus.

_____

_____ 4. Epithelial cells in the gastric glands that produce hydrochloric acid and intrinsic factor.

_____ 5. Epithelial cells in the gastric glands that produce pepsinogen.

_____ 6. Epithelial cells in the gastric glands that secrete regulatory hormones.

D. Match these terms with the correct statement or definition:

Gastrin
Hydrochloric acid
Intrinsic factor
Mucus
Pepsin
Pepsinogen

_____ 1. Viscous, alkaline substance that covers the surface of epithelial cells.

_____ 2. Glycoprotein secreted by parietal cells that binds with vitamin B_{12} and makes it more readily absorbed in the ileum.

_____ 3. Stomach secretion that produces a low pH; kills ingested bacteria.

_____ 4. Inactive form of pepsin, packaged in zymogen granules; secreted by chief cells.

_____ 5. Stomach enzyme that catalyzes the cleavage of some covalent bonds in proteins.

_____ 6. Hormone that increases gastric secretion and increases stomach emptying.

E. Match these phases of stomach secretion with the correct statement:

Cephalic phase
Gastric phase
Intestinal phase

_____ 1. Phase of gastric secretion that responds to taste, smell, and the sensations of chewing and swallowing.

_____ 2. Phase of gastric secretion that is stimulated by food in the stomach.

_____ 3. Phase of gastric secretion that is stimulated by the entrance of acidic chyme into the duodenum.

F. Using the terms provided, complete these statements:

Decrease(s)   Increase(s)

Several mechanisms regulate gastric secretions. Through the medulla, the smell, taste, or thought of food can __(1)__ parasympathetic stimulation of parietal cells, chief cells, and endocrine cells. The endocrine cells secrete gastrin, which travels in the blood back to the stomach mucosa and causes a(n) __(2)__ in hydrochloric acid secretion. In addition, gastrin stimulates endocrine cells to secrete histamine, which causes a(n) __(3)__ in the secretion of hydrochloric acid by parietal cells. Distension of the stomach activates CNS and enteric reflexes that __(4)__ gastric secretions, but, if the pH of the stomach contents falls below 2, increased stomach secretions stimulated by distension of the stomach is blocked. Partially digested proteins, caffeine, and alcohol can also __(5)__ secretions of gastrin. Amino acids and peptides released by the digestive action of pepsin __(6)__ hydrochloric acid secretion. Increased acidity in the duodenum stimulates the secretion of secretin, which acts to __(7)__ secretion from both parietal and chief cells; fatty acids in the duodenum stimulate the secretion of cholecystokinin, which __(8)__ gastric secretions. Distension and increased acidic chyme in the duodenum can activate the enterogastric reflex and cause a(n) __(9)__ in gastric secretions. In general, hormonal and neural mechanisms that increase stomach secretion also __(10)__ stomach motility and __(11)__ stomach emptying.

1. _____
2. _____
3. _____
4. _____
5. _____
6. _____
7. _____
8. _____
9. _____
10. _____
11. _____

G. Match these terms with the correct statement or definition:

Chyme   Peristaltic waves
Mixing waves   Pyloric pump

_____ 1. Semifluid material formed from ingested food mixed with stomach gland secretions.

_____ 2. Peristaltic-like contractions that occur every 20 seconds to mix ingested material with the secretions of the stomach.

_____ 3. Strong waves of contraction that force chyme toward the pyloric sphincter.

_____ 4. Movement of chyme through the partially closed pylorus by the force of peristaltic contraction.

Small Intestine

A. Match these terms with the correct statement or definition:

Hepatopancreatic ampulla   Major duodenal papilla
Hepatopancreatic ampullar sphincter   Minor duodenal papilla

_____ 1. Larger mound where the hepatopancreatic ampulla empties into the duodenum.

_____ 2. Smaller mound where the accessory pancreatic duct opens into the duodenum.

_____ 3. Formed from the junction of the common bile duct and the pancreatic duct.

_____ 4. Smooth muscle that keeps the hepatopancreatic ampulla closed.

B. Match these terms with the correct statement or definition:

Absorptive cells   Granular cells
Circular folds (plicae circulares)   Intestinal glands
Duodenal glands   Lacteal
Endocrine cells   Microvilli
Goblet cells   Villi

_____ 1. Folds formed from the mucosa and submucosa that run perpendicular to the long axis of the duodenum.

_____ 2. Tiny, fingerlike projections of the mucosa of the duodenum.

_____ 3. Lymphatic capillary found in a villus.

_____ 4. Cytoplasmic extensions of villi; collectively they form the brush border.

_____ 5. Simple columnar epithelial cells in the duodenum that are specialized to produce digestive enzymes and absorb food.

_____ 6. Simple columnar epithelial cells in the duodenum that produce a protective mucus.

_____ 7. Cells that may help protect the intestinal epithelium from bacteria.

_____ 8. Cells that produce regulatory hormones.

_____ 9. Tubular invaginations of the mucosa at the base of the villi.

_____ 10. Coiled, tubular mucous glands located in the submucosa of the duodenum.

C. Match these terms with the correct statement or definition:

Duodenum Ileum
Ileocecal sphincter Jejunum
Ileocecal valve Peyer's patches

_____ 1. Two parts of the small intestine that are the major sites of nutrient absorption.

_____ 2. Aggregations of lymph nodules in the ileum.

_____ 3. Ring of smooth muscle located at the junction of the ileum and the large intestine (ileocecal junction).

_____ 4. One-way valve at the ileocecal junction.

D. Match these terms with the correct description:

Disaccharidases Peptidases
Mucus Secretin and cholecystokinin
Nucleases

_____ 1. Secreted by duodenal glands, by goblet cells, and within intestinal glands.

_____ 2. Hormones released from the intestinal mucosa that stimulate hepatic and pancreatic secretions.

_____ 3. Enzymes on the intestinal microvilli that break down disaccharides to monosaccharides.

_____ 4. Enzymes on the intestinal microvilli that break peptide bonds between small amino acid chains.

_____ 5. Enzymes on the intestinal microvilli that break down nucleic acids.

E. Match these terms with the correct statement or definition:

Decreases Peristaltic contractions
Increases Segmental contractions

_____ 1. Propagated for only short distances, they mix the intestinal contents.

_____ 2. Propel the intestinal contents along the digestive tract; some may proceed the entire length of the intestine.

_____ 3. Effect of small intestine distension on intestinal smooth muscle contraction.

_____ 4. Effect of low pH, amino acids, and peptides on small intestine contraction.

_____ 5. Effect of parasympathetic stimulation on small intestine contraction.

_____ 6. Effect of cecal distension on constriction of the ileocecal sphincter.

Liver and Gallbladder

A. Match these terms with the correct statement or definition:

Caudate and quadrate
Common bile duct
Common hepatic duct
Cystic duct
Gallbladder
Left and right
Porta

_____ 1. Two minor lobes of the liver.

_____ 2. Located on the inferior surface of the liver where various vessels, ducts, and nerves enter and exit the liver.

_____ 3. Duct formed from the junction of right and left hepatic ducts.

_____ 4. Duct from the gallbladder.

_____ 5. Duct formed from the junction of the common hepatic duct and the cystic duct.

_____ 6. Small sac on the inferior surface of the liver that stores bile.

B. Using the terms provided, complete these statements:

Bile
Bile canaliculi
Central vein
Hepatic ducts
Hepatic portal vein
Hepatic sinusoids
Hepatic veins
Hepatocytes

In the liver, blood from the _(1)_ and the hepatic artery flows into the _(2)_ and becomes mixed. The mixed blood then flows to the _(3)_, which exits the lobule. Central veins unite to form _(4)_, which exit the liver. The liver secretes _(5)_. Bile is produced by _(6)_ in the liver. Bile flows through the _(7)_ toward the triad and exits the liver through the _(8)_.

1. _____
2. _____
3. _____
4. _____
5. _____
6. _____
7. _____
8. _____

C. Match these terms with the correct statement or definition:

Bare area
Endothelial and phagocytic cells
Hepatic cords
Hepatic sinusoids
Hepatocytes
Lobules
Portal triad

_____ 1. Small area on the diaphragmatic surface of the liver not covered by visceral peritoneum.

_____ 2. Portions of the liver divided by connective tissue septa.

_____ 3. Corner of a liver lobule where three vessels are commonly located.

_____ 4. Structures located between the central vein and the septa of each lobule; consist of hepatocytes.

_____ 5. Functional cells of the liver; produce bile.

_____ 6. Blood channels that separate the hepatic cords.

_____ 7. Cells that line the liver sinusoids.

D. Using the terms provided, complete these statements:

Bile pigments
Bile salts
Detoxify
Glycogen
Interconversion
Phagocytize
Store
Synthesize
Urea

Although bile does not contain digestive enzymes, it does have (1) , which emulsify fats. Bile also contains excretory products, such as (2) and cholesterol. Hepatocytes can (3) fat, vitamins (A, D, E, and K), copper, and iron. The liver is an important regulator of blood sugar levels because hepatocytes can remove sugar from the blood and store it as (4) , or they can break down glycogen into sugar that is released into the blood. Another function that hepatocytes perform is (5) of nutrients, in which the proportion of nutrients is controlled by changing one type of nutrient into another (e.g., amino acids into glucose). Substances not readily usable by cells are transformed within the liver. For example, vitamin D is hydroxylated by hepatocytes. The hepatocytes also (6) many harmful substances by altering their structure, such as converting ammonia to (7) . Hepatic phagocytic cells (8) worn-out and dying red and white blood cells, bacteria, and other debris. The liver can also (9) its own unique new compounds, including albumins and other blood proteins.

1. _____
2. _____
3. _____
4. _____
5. _____
6. _____
7. _____
8. _____
9. _____

E. Match these terms with the correct statement:

Inhibits
Stimulates

_____
_____
_____
_____

1. Effect of secretin on bile secretion.
2. Effect of bile salts on bile secretion.
3. Effect of cholecystokinin on contraction of the gallbladder.
4. Effect of vagal stimulation on contraction of the gallbladder.

Chapter 24

Pancreas

A. Using the terms provided, complete these statements:

Acini
Head
Intercalated ducts
Interlobular ducts
Intralobular ducts
Pancreatic duct
Pancreatic islets
Tail

The pancreas consists of a(n) __(1)__ located within the curvature of the duodenum, a body, and a(n) __(2)__, which extends to the spleen. The exocrine portion of the pancreas consists of __(3)__, which produce digestive enzymes. Clusters of acini are connected by small __(4)__ to __(5)__, which leave the lobules to join __(6)__ between the lobules. The interlobular ducts attach to the __(7)__, which joins the common bile duct at the hepatopancreatic ampulla. Insulin and glucagon are produced by cells within the __(8)__.

1. _____
2. _____
3. _____
4. _____
5. _____
6. _____
7. _____
8. _____

B. Match these terms with the correct parts of the diagram labeled in figure 24.4:

Common bile duct
Common hepatic duct
Cystic duct
Hepatic ducts
Hepatopancreatic ampulla
Pancreatic duct

1. _____
2. _____
3. _____
4. _____
5. _____
6. _____

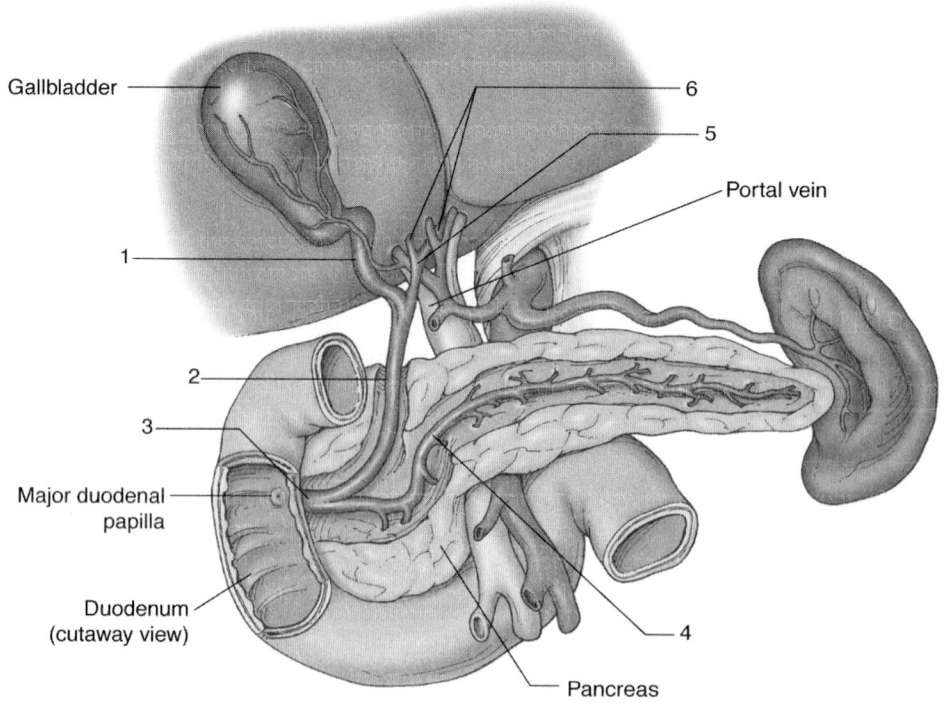

Figure 24.4

386   Chapter 24

C. Match these components of pancreatic juice with the correct statement or definition:

Aqueous component
Enzymatic component

_____ 1. Portion of pancreatic juice that contains Na^+, K^+, and HCO_3^-.

_____ 2. Portion of pancreatic juice produced by the epithelial cells of the smaller pancreatic ducts.

_____ 3. Portion of pancreatic juice that neutralizes acidic chyme.

_____ 4. Portion of pancreatic juice that is important for food digestion.

_____ 5. Portion of pancreatic juice produced by the acinar cells of the pancreas.

_____ 6. Portion of pancreatic juice that has an increased secretion rate because of secretin.

_____ 7. Portion of pancreatic juice that has an increased secretion rate because of cholecystokinin.

_____ 8. Portion of pancreatic juice that has an increased secretion rate because of parasympathetic stimulation.

D. Match these terms with the correct statement or definition:

Carboxypeptidase   Pancreatic amylase
Chymotrypsin   Pancreatic lipase
Deoxyribonuclease   Ribonuclease
Enterokinase   Trypsin

_____
_____
_____ 1. Major proteolytic enzymes in pancreatic juice.

_____ 2. Proteolytic enzyme that cleaves trypsinogen to trypsin; produced by the brush border of the small intestine.

_____ 3. Enzyme that digests polysaccharides.

_____ 4. Enzyme that digests lipids.

_____ 5. Enzyme that digests DNA to its component nucleotides.

_____ 6. Enzyme that digests ribonucleic acids to its component nucleotides.

Chapter 24

Large Intestine

A. Match these terms with the correct statement or definition:

Anal canal
Anus
Ascending colon
Cecum
Crypts
Descending colon
Epiploic appendages
External anal sphincter
Haustra
Internal anal sphincter
Rectum
Sigmoid colon
Teniae coli
Transverse colon
Vermiform appendix

_____ 1. Blind sac that is the proximal end of the large intestine.

_____ 2. Small blind tube that contains many lymph nodules; attached to the cecum.

_____ 3. Portion of the colon that extends from the right colic flexure to the left colic flexure.

_____ 4. Portion of the colon that forms an S-shaped tube that ends at the rectum.

_____ 5. Three bands of longitudinal smooth muscle that run the length of the colon.

_____ 6. Pouches formed in the colon when the teniae coli contract.

_____ 7. Small, fat-filled connective tissue pouches attached to the outer surface of the colon.

_____ 8. Straight, tubular glands in the epithelium of the large intestine.

_____ 9. Straight, muscular tube between the sigmoid colon and anal canal.

_____ 10. Last 2–3 cm of the digestive tube.

_____ 11. Thick layer of smooth muscle at the superior end of the anal canal.

B. Match these terms with the correct statement or definition:

Bicarbonate ions
Flatus
Microorganisms
Mucus

_____ 1. Substance secreted by goblet cells in the colon.

_____ 2. Neutralize the acid produced by bacteria in the colon.

_____ 3. Source of vitamin K synthesis; 30% of the dry weight of feces.

_____ 4. Gases produced by bacterial action in the colon.

C. Match these terms with the correct statement or definition:

Defecation reflex Gastrocolic reflex
Duodenocolic reflex Mass movements

_____ 1. Strong peristaltic contractions of the transverse and descending colon.

_____ 2. Strong peristaltic contractions of the colon initiated by the stomach.

_____ 3. Distension of the rectal wall by feces initiates this reflex.

_____ 4. Results in reinforcement of peristaltic contractions in the lower colon and rectum and relaxation of the internal anal sphincter.

Digestion, Absorption, and Transport

Match these terms with the correct statement or definition:

Absorption Lipids and lipid-soluble
Digestion   substances
Ions and water- Transport
   soluble substances

_____ 1. Begins in the oral cavity.

_____ 2. Most occurs in the duodenum and jejunum, although some occurs in the ileum.

_____ 3. Transported through the hepatic portal system to the liver.

_____ 4. Transported into lacteals, through the lymphatic system to the left subclavian vein, and then to the liver or adipose tissue.

Carbohydrates

Match these terms with the correct statement or definition:

Disaccharidases Pancreatic amylase
Glucose   Salivary amylase
Insulin

_____ 1. Enzyme that digests starch and is secreted into the oral cavity.

_____ 2. Enzymes bound to the microvilli of the intestinal epithelium.

_____ 3. Sugar transported by the circulatory system to cells that need energy.

_____ 4. Hormone that greatly increases the rate of glucose transport into most types of cells.

Chapter 24

Lipids

A. Using the terms provided, complete these statements:

Bile salts
Chylomicrons
Emulsification
Lacteal
Lipase
Liver
Micelles
Triglycerides

Lipids include triglycerides, phospholipids, steroids, and fat-soluble vitamins. __(1)__ consist of one glycerol molecule and three fatty acids covalently bound together. The first step in lipid digestion is __(2)__, which is the transformation of large lipid droplets into much smaller droplets. This process is accomplished by __(3)__ secreted by the liver. __(4)__ secreted by the pancreas digests lipid molecules. Once lipids are digested in the intestine, bile salts aggregate around the small droplets to form __(5)__. When these structures come into contact with epithelial cells of the small intestine, their contents pass through the cell membrane of the epithelial cell by the process of simple diffusion. Within the smooth endoplasmic reticulum of the intestinal epithelial cells, free fatty acids combine with glycerol to form triglyceride. Proteins in the epithelial cells coat droplets of triglycerides, phospholipids, and cholesterol to form __(6)__, which leave the epithelial cell to enter a(n) __(7)__. From there, the chylomicrons are transported to the blood and are carried to adipose tissue or the __(8)__.

1. _____
2. _____
3. _____
4. _____
5. _____
6. _____
7. _____
8. _____

B. Match these terms with the correct statement or definition:

High-density lipoproteins (HDLs)
Low-density lipoproteins (LDLs)
Very low-density lipoproteins (VLDLs)

1. Molecules that are 92% lipid and 8% protein.

2. Molecules that are 75% lipid and 25% protein.

3. Molecules that are 55% lipid and 45% protein.

4. Molecules produced from VLDL when triglycerides are removed.

5. Molecules delivered to cells; attach to receptors in "pits" on cell surface.

6. Molecules that transport excess lipids back to the liver for recycling or dispersal.

7. Aerobic exercise increases the level of these molecules in the blood.

Proteins

Match these terms with the correct statement or definition:

Growth hormone and insulin   Peptidase
Pepsin   Trypsin

_____ 1. Enzyme in the stomach that catalyzes the cleavage of covalent bonds in proteins.

_____ 2. Enzyme produced by the pancreas that continues the digestion of proteins started in the stomach; produces small peptide chains.

_____ 3. Enzyme bound to the microvilli and located inside intestinal epithelial cells; completes the breakdown of small peptide chains.

_____ 4. Hormones that stimulate the transport of amino acids.

Water and Ions

Match these terms with the correct statement or definition:

Active transport   Into lumen of intestine
Diffusion   Osmosis
Into circulation

_____ 1. Mechanism responsible for water movement across the wall of the small intestine.

_____ 2. Mechanism that moves sodium, potassium, magnesium, calcium, and phosphate into the epithelial cells of the small intestine.

_____ 3. Passive movement of negative ions (e.g., Cl^-.) as they follow positive ions (e.g., NA^+.) into intestinal epithelial cells.

_____ 4. Direction of water movement when the chyme is very concentrated.

_____ 5. Direction of water movement as nutrients are absorbed.

Effects of Aging on the Digestive System

Match these terms with the correct statement or definition:

Decreases   Increases

_____ 1. Effect of aging on blood flow to the digestive tract.

_____ 2. Effect of aging on mucous secretions from digestive cells and glands and motility of the digestive tract.

_____ 3. Effect of aging on the liver's ability to detoxify certain chemicals.

_____ 4. Effect of aging on susceptibility to infections, ulcerations, and cancer and effects of toxic agents.

_____ 5. Effect of aging on the function of the teeth and muscles of mastication.

Quick Recall

1. Name the four layers, or tunics, of the digestive tract.

2. Group the four types of teeth in humans according to their function.

3. List the three large pairs of multicellular salivary glands, and name the digestive enzyme in saliva.

4. Name six sphincters that control the movement of materials through the digestive tract.

5. Name the five types of epithelial cells in the stomach, and list their secretions.

6. List the three structural modifications that increase surface area in the small intestine.

7. List the three major types of cells in the intestinal mucosa.

8. Name the three phases of swallowing.

9. List the types of contraction (movement) that occur in the stomach, small intestine, and large intestine.

10. List the three phases of gastric secretion.

11. Name a substance in pancreatic juice that is responsible for each of these activities: neutralizes acid, digests proteins, digests fats, digests carbohydrates.

12. List three functions of bile.

13. List four major functions of the liver in addition to the production of bile.

14. List three major functions of the colon.

15. In this table, indicate if the control mechanism stimulates (S), inhibits (I), or has no effect (O) on the activity:

	GASTRIN	CHOLE-CYSTO-KININ	SECRETIN	PARA-SYMPA-THETIC
Stomach secretion	___	___	___	___
Bile secretion	___	___	___	___
Pancreas secretion	___	___	___	___
Contraction of gallbladder	___	___	___	___
Gastric motility	___	___	___	___

16. List the breakdown products of carbohydrates, proteins, and triglycerides.

17. List the locations in the digestive tract where carbohydrate digestion, lipid digestion, and protein digestion occur.

18. Name the routes by which water-soluble and lipid-soluble molecules leave the intestinal epithelial cells.

ANSWERS TO CHAPTER 24

CONTENT LEARNING ACTIVITY

Anatomy of the Digestive System
A. 1. Tube; 2. Accessory organs; 3. Alimentary tract (canal); 4. Gastrointestinal tract
B. 1. Mouth (oral cavity); 2. Pharynx; 3. Esophagus; 4. Small intestine; 5. Large intestine; 6. Anus
C. 1. Mouth; 2. Liver; 3. Gallbladder; 4. Appendix; 5. Rectum; 6. Anus; 7. Large intestine; 8. Small intestine; 9. Pancreas; 10. Stomach; 11. Esophagus; 12. Salivary glands; 13. Pharynx (throat)

Functions of the Digestive System
1. Ingestion; 2. Mastication; 3. Propulsion; 4. Deglutition; 5. Peristalsis; 6. Mass movements; 7. Segmental contractions; 8. Secretion; 9. Digestion; 10. Absorption; 11. Defecation

Histology of the Digestive Tract
A. 1. Mucosa; 2. Submucosa; 3. Muscularis; 4. Serosa (adventitia)
B. 1. Mucous epithelium; 2. Lamina propria; 3. Muscularis mucosae; 4. Myenteric plexus; 5. Enteric (intramural) plexus; 6. Visceral peritoneum; 7. Interstitial cells

Regulation of the Digestive Tract
1. Enteric; 2. Local reflexes; 3. Enteric sensory neurons; 4. Enteric motor neurons; 5. Enteric interneurons; 6. Parasympathetic; 7. Sympathetic; 8. Hormones; 9. Paracrine; 10. Stimulates; 11. Inhibits; 12. Stimulates

Peritoneum
- A. 1. Visceral peritoneum; 2. Parietal peritoneum; 3. Mesenteries; 4. Retroperitoneal
- B. 1. Lesser omentum; 2. Greater omentum; 3. Omental bursa; 4. Coronary ligament; 5. Falciform ligament; 6. Mesentery proper; 7. Transverse mesocolon; 8. Sigmoid mesocolon; 9. Mesoappendix

Oral Cavity
- A. 1. Fauces; 2. Vestibule; 3. Oral cavity proper; 4. Frenula; 5. Buccinator muscle; 6. Buccal fat pad; 7. Hard palate; 8. Uvula; 9. Palatine tonsils
- B. 1. Frenulum; 2. Intrinsic muscles; 3. Extrinsic muscles; 4. Terminal sulcus; 5. Lingual tonsil
- C. 1. Two; 2. One; 3. Two; 4. Three
- D. 1. Maxillary and mandibular arches; 2. Wisdom tooth; 3. Permanent teeth; 4. Primary teeth
- E. 1. Clinical crown; 2. Anatomical crown; 3. Pulp; 4. Root canal; 5. Apical foramen; 6. Dentin; 7. Enamel; 8. Cementum
- F. 1. Alveoli; 2. Gingiva; 3. Periodontal ligaments
- G. 1. Cusp; 2. Enamel; 3. Gingiva; 4. Dentin; 5. Pulp cavity; 6. Periodontal ligaments; 7. Root canal; 8. Cementum; 9. Apical foramen; 10. Root; 11. Neck; 12. Crown

Mastication
1. Incisors and canines; 2. Premolars and molars; 3. Masseter muscles, Temporalis muscles, and Medial pterygoid muscles; 4. Lateral pterygoid muscles; 5. Chewing (mastication) reflex

Salivary Glands
- A. 1. Parotid glands; 2. Parotid glands; 3. Submandibular glands; 4. Sublingual glands
- B. 1. Starch; 2. Antibacterial; 3. Immunoglobulin A; 4. Mucin; 5. Parasympathetic; 6. Odors

Pharynx and Esophagus
1. Oropharynx and laryngopharynx; 2. Nasopharynx; 3. Pharyngeal constrictors; 4. Esophagus; 5. Upper and lower esophageal sphincter

Swallowing
- A. 1. Voluntary phase; 2. Pharyngeal phase; 3. Esophageal phase
- B. 1. Swallowing center; 2. Pharyngeal constrictor muscles; 3. Epiglottis; 4. Peristaltic waves; 5. Enteric plexus

Stomach
- A. 1. Lower esophageal sphincter; 2. Cardiac part; 3. Fundus; 4. Body; 5. Greater curvature; 6. Pyloric part; 7. Pyloric sphincter; 8. Muscularis; 9. Rugae
- B. 1. Gastroesophageal opening; 2. Cardiac part; 3. Pyloric sphincter; 4. Pyloric orifice; 5. Pyloric part; 6. Rugae; 7. Body; 8. Fundus; 9. Lower esophageal sphincter
- C. 1. Gastric pits; 2. Gastric glands; 3. Surface mucous cells and mucous neck cells; 4. Parietal cells; 5. Chief cells; 6. Endocrine cells
- D. 1. Mucus; 2. Intrinsic factor; 3. Hydrochloric acid; 4. Pepsinogen; 5. Pepsin; 6. Gastrin
- E. 1. Cephalic phase; 2. Gastric phase; 3. Intestinal phase
- F. 1. Increase; 2. Increase; 3. Increase; 4. Increase; 5. Increase; 6. Increase; 7. Decrease; 8. Decreases; 9. Decrease; 10. Increase; 11. Increase
- G. 1. Chyme; 2. Mixing waves; 3. Peristaltic waves; 4. Pyloric pump

Small Intestine
- A. 1. Major duodenal papilla; 2. Minor duodenal papilla; 3. Hepatopancreatic ampulla; 4. Hepatopancreatic ampullar sphincter
- B. 1. Circular folds (plicae circulares); 2. Villi; 3. Lacteal; 4. Microvilli; 5. Absorptive cells; 6. Goblet cells; 7. Granular cells; 8. Endocrine cells; 9. Intestinal glands; 10. Duodenal glands
- C. 1. Duodenum and jejunum; 2. Peyer's patches; 3. Ileocecal sphincter; 4. Ileocecal valve
- D. 1. Mucus; 2. Secretin and cholecystokinin; 3. Disaccharidases; 4. Peptidases; 5. Nucleases
- E. 1. Segmental contractions; 2. Peristaltic contractions; 3. Increases; 4. Increases; 5. Increases; 6. Increases

Liver and Gallbladder
- A. 1. Caudate and quadrate; 2. Porta; 3. Common hepatic duct; 4. Cystic duct; 5. Common bile duct; 6. Gallbladder
- B. 1. Hepatic portal vein; 2. Hepatic sinusoid; 3. Central vein; 4. Hepatic veins; 5. Bile; 6. Hepatocytes; 7. Bile canaliculi; 8. Hepatic ducts
- C. 1. Bare area; 2. Lobules; 3. Portal triad; 4. Hepatic cords; 5. Hepatocytes; 6. Hepatic sinusoids; 7. Endothelial and phagocytic cells
- D. 1. Bile salts; 2. Bile pigments; 3. Store; 4. Glycogen; 5. Interconversion; 6. Detoxify; 7. Urea; 8. Phagocytize; 9. Synthesize
- E. 1. Stimulates; 2. Stimulates; 3. Stimulates; 4. Stimulates

Pancreas
- A. 1. Head; 2. Tail; 3. Acini; 4. Intercalated ducts; 5. Intralobular ducts; 6. Interlobular ducts; 7. Pancreatic duct; 8. Pancreatic islets
- B. 1. Cystic duct; 2. Common bile duct; 3. Hepatopancreatic ampulla; 4. Pancreatic duct; 5. Common hepatic duct; 6. Hepatic ducts

C. 1. Aqueous component; 2. Aqueous component; 3. Aqueous component; 4. Enzymatic component; 5. Enzymatic component; 6. Aqueous component; 7. Enzymatic component; 8. Enzymatic component
D. 1. Chymotrypsin, trypsin, and carboxypeptidase; 2. Enterokinase; 3. Amylase; 4. Lipase; 5. Deoxyribonuclease; 6. Ribonuclease

Large Intestine
A. 1. Cecum; 2. Vermiform appendix; 3. Transverse colon; 4. Sigmoid colon; 5. Teniae coli; 6. Haustra; 7. Epiploic appendages; 8. Crypts; 9. Rectum; 10. Anal canal; 11. Internal anal sphincter
B. 1. Mucus; 2. Bicarbonate ions; 3. Microorganisms; 4. Flatus
C. 1. Mass movements; 2. Gastrocolic reflex; 3. Defecation reflex; 4. Defecation reflex

Digestion, Absorption, and Transport
1. Digestion; 2. Absorption; 3. Ions and water-soluble substances; 4. Lipids and lipid-soluble substances

Carbohydrates
1. Salivary amylase; 2. Disaccharidases; 3. Glucose; 4. Insulin

Lipids
A. 1. Triglycerides; 2. Emulsification; 3. Bile salts; 4. Lipase; 5. Micelles; 6. Chylomicrons; 7. Lacteal; 8. Liver
B. 1. Very-low density lipoproteins (VLDLs); 2. Low-density lipoproteins (LDLs); 3. High density lipoproteins (HDLs); 4. Low-density lipoproteins (LDLs); 5. Low-density lipoproteins (LDLs); 6. High-density lipoproteins (HDLs); 7. High-density lipoproteins (HDLs)

Proteins
1. Pepsin; 2. Trypsin; 3. Peptidase; 4. Growth hormone and insulin

Water and Ions
1. Osmosis; 2. Active transport; 3. Diffusion; 4. Into lumen of intestine; 5. Into circulatory system

Effects of Aging on the Digestive System
1. Decreases; 2. Decreases; 3. Decreases; 4. Increases; 5. Decreases

QUICK RECALL

1. Mucosa, submucosa, muscularis, and serosa or adventitia
2. Incisors and canines: cutting and tearing food; molars and premolars: crushing and grinding food
3. Parotid, submandibular, and sublingual glands; amylase is the enzyme in saliva
4. Upper esophageal sphincter, lower esophageal sphincter, pyloric sphincter, ileocecal sphincter, internal anal sphincter, external anal sphincter
5. Surface mucous cells: mucus; mucous neck cells: mucus; parietal cells: hydrochloric acid and intrinsic factor; chief cells: pepsinogen; endocrine cells: gastrin
6. Plicae circulares, villi, and microvilli
7. Absorptive cells, goblet cells, and endocrine cells
8. Voluntary, pharyngeal, and esophageal
9. Stomach: mixing waves and peristaltic waves; small intestine: segmental contractions and peristaltic contractions; large intestine: segmental movements and mass movements
10. Cephalic, gastric, and intestinal
11. Bicarbonate ions neutralize acid; trypsin, chymotrypsin, and carboxypeptidase digest protein; lipase digests fats; and amylase digests starch.
12. Neutralizes stomach acids, emulsifies fats, carries out excretory products
13. Storage, nutrient interconversion, detoxification, phagocytosis, and synthesis
14. Reabsorb water and salts, secretion of mucus, absorption of vitamins produced by microorganisms, storage of feces
15. Stomach secretion: S, I, I, S
    Bile secretion: O, O, S, S
    Pancreas secretion: O, S, S, S
    Contraction of gallbladder: O, S, O, S
    Gastric motility: S, I, I, S
16. Carbohydrates: monosaccharides; proteins: amino acids; triglycerides: fatty acids and glycerol
17. Carbohydrate digestion: mouth, small intestine; lipid digestion: small intestine; protein digestion: stomach, small intestine
18. Water-soluble molecules enter the hepatic portal system; lipid-soluble molecules enter the lacteals.

25 Nutrition, Metabolism, and Temperature Regulation

CONTENT LEARNING ACTIVITY

Nutrition

Match these terms with the correct statement or definition:

Calorie
Carbohydrate
Essential nutrients
Fat
Kilocalorie (Calorie)
MyPyramid
Nutrition
Protein

_____ 1. Process by which food is obtained and used by the body.

_____ 2. Certain amino acids and fatty acids, most vitamins, minerals, water, and a minimum amount of carbohydrates must be ingested.

_____ 3. Amount of energy required to raise the temperature of 1 g of water 1°C.

_____ 4. Used to express the amount of energy in food; 1000 calories.

_____ 5. Contains approximately 9 kcal per gram.

_____ 6. The U.S. government's dietary guidelines, which take into account a person's age, sex, and activity level.

Carbohydrates

Match these terms with the correct statement or definition:

Cellulose
Complex carbohydrates
Disaccharides
Fructose
Glucose
Glycogen
Starch
Sucrose

_____ 1. Category to which sucrose, lactose, and maltose belong.

_____ 2. Table sugar; a disaccharide of glucose and fructose.

_____ 3. Category to which starch, glycogen, and cellulose belong; polysaccharides.

Chapter 25   397

_____ 4. Not digestible by humans; provides "roughage."

_____ 5. Monosaccharide converted into glucose by the liver.

_____ 6. Primary energy source for most cells.

_____ 7. Energy-storage molecule produced from glucose in animals.

_____ 8. Energy-storage molecule produced from glucose in plants.

Lipids

A. Match these terms with the correct statement or definition:

Adipose　　　　　　　Phospholipid
Cholesterol　　　　　Saturated fat
Eicosanoids　　　　　Triglycerides
Essential fatty acids　Unsaturated fat

_____ 1. About 95% of the lipids in the human diet.

_____ 2. Have only single covalent bonds between their carbon atoms.

_____ 3. Have one or more double covalent bonds between carbon atoms.

_____ 4. Excess triglycerides are stored in this tissue.

_____ 5. Part of the plasma membrane, modified to form bile salts and steroid hormones.

_____ 6. Derived from fatty acids—e.g., prostaglandins and leukotrienes.

_____ 7. Lecithin is an example.

B. Match these terms with the correct statement or definition:

Alpha-linolenic acid　Linoleic acid
Cis form　　　　　　*Trans* form
Essential fatty acids

_____ 1. Omega-3 fatty acid.

_____ 2. Omega-6 fatty acid.

_____ 3. Alpha-linolenic acid and linoleic acid; humans lack the enzymes necessary to synthesize them.

_____ 4. Type of fatty acid in which the hydrogen atoms are on the same side of the carbon–carbon double bond.

_____ 5. Processed foods and oil account for most of this type of fatty acid in the American diet; it is associated with a greater risk for cardiovascular disease.

Proteins

Match these terms with the correct statement or definition:

Antibody
Collagen
Complete
Enzyme
Essential
Hemoglobin
Incomplete
Nonessential

_____ 1. Type of amino acids that can be manufactured by the body.

_____ 2. Protein that contains all eight essential amino acids.

_____ 3. Provides structural strength.

_____ 4. Regulates the rate of chemical reactions.

_____ 5. Transports oxygen and carbon dioxide.

_____ 6. Protects against microorganisms and other foreign substances.

Sources and Recommended Amounts

A. Match these terms with the correct statement or definition:

Carbohydrate
Cholesterol
Fat (all)
Fat (saturated)
Fat (unsaturated)
Protein

_____ 1. Should account for 45%–65% of the daily intake of kilocalories; otherwise, acidosis or breakdown of muscle tissue occurs.

_____ 2. Should account for 20%–35% or less of the total kilocaloric intake.

_____ 3. Should contribute no more than 10% of total fat intake.

_____ 4. Should be limited to 300 mg or less per day.

_____ 5. A person is in nitrogen balance if the amount of nitrogen in this ingested food is equal to the amount of nitrogen excreted in the urine or feces.

B. Match these terms with the correct statement or definition:

Carbohydrate
Cholesterol
Monounsaturated fat
Polyunsaturated fat
Protein (complete)
Protein (incomplete)
Saturated fat

_____ 1. Fruit, cereal, lactose (in milk).

_____ 2. Lipid in fats of meat, whole milk, cheese, butter, coconut oil, and palm oil.

_____ 3. Lipid in olive and peanut oil.

_____ 4. Lipid in fish, safflower, sunflower, and corn oils.

_____ 5. Lipid in high concentration in liver and egg yolks.

Chapter 25

_____ 6. Proteins in meat, fish, poultry, milk, cheese, and eggs.

_____ 7. Proteins in leafy, green vegetables, grains, peas, and beans.

Vitamins

A. Match these terms with the correct statement or definition:

Essential vitamins Coenzyme
Fat-soluble vitamins Provitamins
Carotene Water-soluble vitamins

_____ 1. Vitamins that must be obtained from the diet.

_____ 2. Portions of vitamins that can be assembled or modified by the body into functional vitamins.

_____ 3. Combine with enzymes to make them functional.

_____ 4. Can be modified to form vitamin A.

_____ 5. B-complex vitamins and vitamin C.

_____ 6. Vitamins A, D, E, and K.

_____ 7. Vitamins that can be stored in the body.

B. Match these vitamins with the correct deficiency symptom:

A (retinol) Folate
B_{12} (cobalamin) Hypervitaminosis
C (ascorbic acid) K (phylloquinone)
D (cholecalciferol)

_____ 1. Scurvy; defective collagen formation and poor wound healing.

_____ 2. Night blindness, retarded growth, and skin disorders.

_____ 3. Excessive bleeding resulting from retarded blood clotting.

_____ 4. Neural tube defects.

_____ 5. Pernicious anemia and nervous system disorders.

_____ 6. Accumulation of the vitamin to the point of toxicity.

Minerals

Match these minerals with the correct function:

Calcium Minor mineral
Chlorine Phosphorus
Iodine Potassium
Iron Sodium
Major mineral

_____ 1. The daily requirement for this type of mineral is 100 mg or more.

_____ 2. Bone and teeth formation, blood clotting, muscle activity, and nerve function.

_____ 3. Blood acid–base balance; hydrochloric acid production in the stomach.

_____ 4. Thyroid hormone production; maintenance of normal metabolic rate.

_____ 5. Component of hemoglobin; ATP production in electron-transport system.

_____ 6. Bone and teeth formation; important in ATP formation; a component of nucleic acids.

_____ 7. Muscle and nerve function.

_____ 8. Osmotic pressure regulation; nerve and muscle function.

Daily Values

Match these terms with the correct statement or definition:

Daily Values
Daily Reference Values (DRVs)
Percent Daily Value (% Daily Value)
Recommended Dietary Allowance (RDA)
Reference Daily Intakes (RDIs)

_____ 1. Nutrient intake sufficient to meet the needs of nearly all people in certain age and gender groups; first established in 1941.

_____ 2. Dietary reference values now appearing on food labels; a combination of Reference Daily Intakes (RDIs) and Daily Reference Values (DRVs).

_____ 3. Based on the 1968 RDAs for certain vitamins and minerals.

_____ 4. Set for total fat, saturated fat, cholesterol, total carbohydrate, dietary fiber, sodium, potassium, and protein.

_____ 5. Amount of the Daily Value of a nutrient in a particular food.

Metabolism

Match these terms with the correct statement or definition:

Anabolism   Oxidation–reduction
ATP   Oxidized
Catabolism   Reduced
Metabolism

_____ 1. Total of all the chemical changes that occur in the body.

_____ 2. Energy-requiring process by which small molecules are joined to form larger molecules.

_____ 3. Energy-releasing process by which large molecules are broken down into smaller molecules.

_____ 4. Energy currency of the cell; used to drive cell activities.

_____ 5. Type of chemical reaction responsible for the transfer of energy from the chemical bonds of nutrient molecules to ATP molecules.

_____ 6. Molecule that has gained electrons, H$^+$, and energy.

Glycolysis

Match these terms with the correct statement or definition:

ATP
Four
NAD$^+$
NADH
One
Phosphorylation
Two

_____ 1. Process of attaching a phosphate group to a molecule.

_____ 2. Reduced form of nicotinamide adenine dinucleotide.

_____ 3. Number of ATP molecules required to start glycolysis for one glucose molecule.

_____ 4. Net number of ATP molecules produced from one glucose molecule by glycolysis.

_____ 5. Number of NADH molecules produced from one glucose molecule by glycolysis.

_____ 6. Number of pyruvic acid molecules produced from one glucose molecule by glycolysis.

Anaerobic Respiration

Match these terms with the correct statement or definition:

ATP
Cori cycle
Lactic acid
NADH
Oxygen deficit

_____ 1. The net energy gain from the anaerobic respiration of one molecule of glucose is two of these molecules.

_____ 2. Formed by the reduction of pyruvic acid.

_____ 3. Two of these molecules are produced in glycolysis and used (oxidized) when pyruvic acid is reduced.

_____ 4. Lactic acid released from cells is transported to the liver; the lactic acid is converted to glucose that is transported back to cells.

_____ 5. The oxygen necessary for the synthesis of the ATP used to convert lactic acid to glucose is part of this.

Aerobic Respiration

A. Match these terms with the correct parts of the diagram labeled in figure 25.1:

Acetyl-CoA
ADP
Aerobic respiration
Anaerobic respiration
ATP

Citric acid cycle
Glycolysis
H₂O
Lactic acid
NADH

1. _____
2. _____
3. _____
4. _____
5. _____
6. _____
7. _____
8. _____
9. _____
10. _____

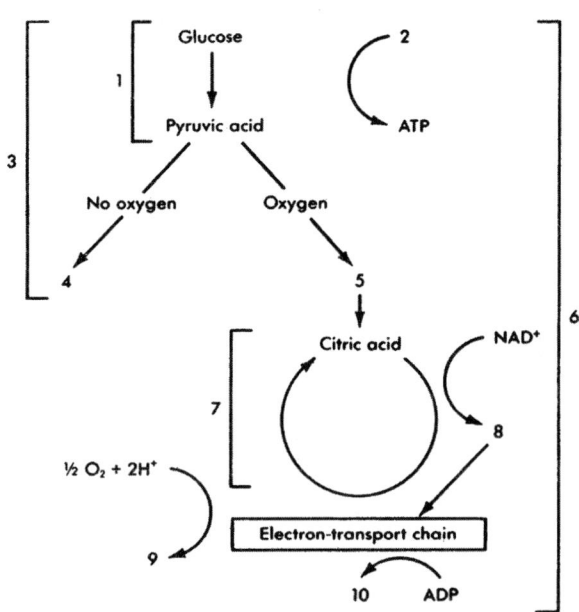

Figure 25.1

B. Match these terms with the correct statement or definition:

Acetyl-CoA formation
Aerobic respiration
Chemiosmotic model

Citric acid (Krebs) cycle
Electron-transport chain
Glycolysis

_____ 1. First phase of aerobic respiration; produces two ATP and two NADH molecules per glucose molecule.

_____ 2. Second phase of aerobic respiration in which pyruvic acid is modified to form acetyl-CoA; produces two NADH and two carbon dioxide molecules per glucose molecule.

_____ 3. Third phase of aerobic respiration, in which acetyl-CoA is combined with oxaloacetic acid to form citric acid; citric acid is then converted by a series of reactions into oxaloacetic acid; produces six NADH, two FADH₂, four carbon dioxide, and two ATP molecules per glucose molecule.

_____ 4. Produces 38 (or 36) ATP molecules for each glucose molecule broken down.

_____ 5. Process produces three ATP molecules for every NADH oxidized and two ATP molecules for every FADH₂ oxidized; occurs within the mitochondria.

_____ 6. Process uses oxygen as a final electron acceptor, producing water.

_____ 7. Hydrogen ions from NADH and FADH$_2$ are actively pumped out of the inner mitochondrial compartment; diffusion of the H$^+$ back into the inner mitochondrial compartment provides energy for ATP production.

Lipid Metabolism

Match these terms with the correct statement or definition:

Beta-oxidation Ketone bodies
Free fatty acids Triglycerides
Ketogenesis

_____ 1. Primary storage form of lipids in adipose tissue.

_____ 2. Fatty acids released into the blood from the breakdown of triglycerides; used as an energy source by muscle and liver cells.

_____ 3. Series of reactions in which two carbons are removed from the end of a fatty acid chain to form acetyl-CoA.

_____ 4. Formation of ketone bodies from acetyl-CoA.

_____ 5. Acetoacetic acid, beta-hydroxybutyric acid, and acetone; an energy source, especially for skeletal muscle.

Protein Metabolism

Using the terms provided, complete the following statements:

Citric acid cycle Proteins
Keto acid Stored
NADH Transamination
Oxidative deamination Urea

Amino acids can be used to synthesize _(1)_ or as a source of energy but, unlike carbohydrates and lipids, amino acids are not _(2)_ in the body. The synthesis of a nonessential amino acid usually begins with a _(3)_, which is usually converted to an amino acid. This process, called _(4)_, involves the transfer of an amine group from an amino acid to the keto acid. Amino acids can be used as a source of energy in an _(5)_ reaction. In this reaction, an amine group is removed from an amino acid, leaving ammonia, a keto acid, and _(6)_ that can be used to produce ATP. The ammonia is converted to _(7)_, which is eliminated by the kidney. Keto acids can also enter the _(8)_ or be converted into pyruvic acid or acetyl-CoA.

1. _____
2. _____
3. _____
4. _____
5. _____
6. _____
7. _____
8. _____

Interconversion of Nutrient Molecules

A. Using the terms provided, complete the following statements:

Gluconeogenesis   Glycogenolysis
Glycogenesis   Lipogenesis

If there is excess glucose, it can be used to form glycogen through a process called __(1)__. Once glycogen stores are filled, glucose and amino acids are used to synthesize lipids, a process called __(2)__. When glucose is needed, glycogen can be broken down into glucose-6-phosphate through a set of reactions called __(3)__. When liver glycogen levels are inadequate to supply glucose, amino acids from proteins and glycerol from triglycerides are used to produce glucose in a process called __(4)__.

1. _____
2. _____
3. _____
4. _____

B. Match these terms with the correct parts of the diagram labeled in figure 25.2:

Gluconeogenesis   Glycogenolysis
Glycogenesis   Glycolysis

1. _____
2. _____
3. _____
4. _____

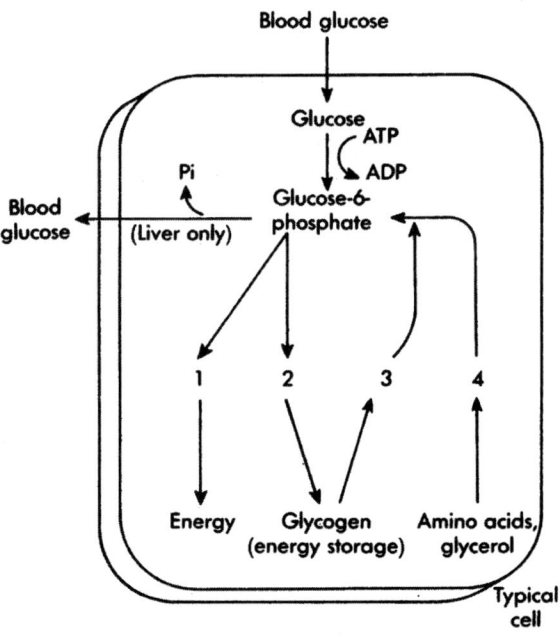

Figure 25.2

Metabolic States

Match these terms with the correct statement or definition:

Absorptive state
Postabsorptive state

_____ 1. During this state, most of the glucose entering the circulation is used by the cell; the remainder is converted into glycogen or fat.

Chapter 25   405

_____ 2. During this state, blood glucose levels are maintained by converting other molecules to glucose.

_____ 3. During this state, glycogen is used preferentially, then fats and ketones, and then proteins.

Metabolic Rate

Match these terms with the correct statement or definition:

    Basal metabolic rate (BMR)   Increases
    Calorie   Kilocalorie
    Decreases

_____ 1. Energy produced and used by the body at rest; calculated in kilocalories per square meter of body surface per hour.

_____ 2. Effect of increased muscle tissue on BMR.

_____ 3. Effect of increasing age on BMR.

_____ 4. Effect of dieting or fasting on BMR.

_____ 5. Effect of physical activity on expenditure of energy.

_____ 6. Effect of the thermic effect of food on expenditure of energy.

Body Temperature Regulation

Match these terms with the correct statement or definition:

    Conduction   Free energy
    Convection   Homeotherms
    Decrease   Increase
    Evaporation   Radiation

_____ 1. Animals that can regulate their body temperatures.

_____ 2. Total amount of energy that can be liberated by the complete catabolism of food.

_____ 3. Exchange of heat between objects in direct contact with each other.

_____ 4. Exchange of heat between the body and the air.

_____ 5. Loss of water, which carries heat with it, from the body.

_____ 6. Effect of vasodilation on skin temperature.

QUICK RECALL

1. List the six major classes of nutrients.

2. List the main functions of carbohydrates in the body.

3. List the main function of lipids in the body.

4. Name the main functions of proteins in the body.

5. Give the total energy gain to the cell from the breakdown of one molecule of glucose during anaerobic respiration and aerobic respiration.

6. Name three energy-storing compounds produced in aerobic respiration.

7. List the end products formed by anaerobic and aerobic respiration.

8. Name the three phases of aerobic respiration. Assuming the electron-transport chain is operating, give the number of ATP molecules produced from a glucose molecule by each phase.

9. Name the chemical reactions by which fatty acids and amino acids are used as a source of energy.

10. Name four processes that involve the interconversion of nutrient molecules.

11. List three ways that metabolic energy can be used.

ANSWERS TO CHAPTER 25

CONTENT LEARNING ACTIVITY

Nutrition
1. Nutrition; 2. Essential nutrients; 3. Calorie; 4. Kilocalorie (Calorie); 5. Fat; 6. MyPyramid

Carbohydrates
1. Disaccharides; 2. Sucrose; 3. Complex carbohydrates; 4. Cellulose; 5. Fructose; 6. Glucose; 7. Glycogen; 8. Starch

Lipids
A. 1. Triglycerides; 2. Saturated fat; 3. Unsaturated fat; 4. Adipose; 5. Cholesterol; 6. Eicosanoids; 7. Phospholipid
B. 1. Alpha-linolenic acid; 2. Linoleic acid; 3. Essential fatty acids; 4. *Cis* form; 5. *Trans* form

Proteins
1. Nonessential; 2. Complete; 3. Collagen; 4. Enzyme; 5. Hemoglobin; 6. Antibody

Sources and Recommended Amounts
A. 1. Carbohydrate; 2. Fat (all); 3. Fat (saturated); 4. Cholesterol; 5. Protein
B. 1. Carbohydrate; 2. Saturated fat; 3. Monounsaturated fat; 4. Polyunsaturated fat; 5. Cholesterol; 6. Protein (complete); 7. Protein (incomplete)

Vitamins
A. 1. Essential vitamins; 2. Provitamins; 3. Coenzyme; 4. Carotene; 5. Water-soluble vitamins; 6. Fat-soluble vitamins; 7. Fat-soluble vitamins
B. 1. C (ascorbic acid); 2. A (retinol); 3. K (phylloquinone); 4. Folate; 5. B_{12} (cobalamin); 6. Hypervitaminosis

Minerals
1. Major mineral; 2. Calcium; 3. Chlorine; 4. Iodine; 5. Iron; 6. Phosphorus; 7. Potassium; 8. Sodium

Daily Values
1. Recommended Dietary Allowance (RDA); 2. Daily Values; 3. Reference Daily Intakes (RDIs); 4. Daily Reference values (DRVs); 5. Percent Daily Value (% Daily Value)

Metabolism
1. Metabolism; 2. Anabolism; 3. Catabolism; 4. ATP; 5. Oxidation–reduction; 6. Reduced

Glycolysis
1. Phosphorylation; 2. NADH; 3. Two; 4. Two; 5. Two; 6. Two

Anaerobic Respiration
1. ATP; 2. Lactic acid; 3. NADH; 4. Cori cycle; 5. Oxygen deficit

Aerobic Respiration
A. 1. Glycolysis; 2. ADP; 3. Anaerobic respiration; 4. Lactic acid; 5. Acetyl-CoA; 6. Aerobic respiration; 7. Citric acid cycle; 8. NADH; 9. H_2O; 10. ATP
B. 1. Glycolysis; 2. Acetyl-CoA formation; 3. Citric acid (Krebs) cycle; 4. Aerobic respiration; 5. Electron-transport chain; 6. Electron-transport chain; 7. Chemiosmotic model

Lipid Metabolism
   1. Triglycerides; 2. Free fatty acids; 3. Beta-oxidation; 4. Ketogenesis; 5. Ketone bodies

Protein Metabolism
   1. Proteins; 2. Stored; 3. Keto acid; 4. Transamination; 5. Oxidative deamination; 6. NADH; 7. Urea; 8. Citric acid cycle

Interconversion of Nutrient Molecules
   A. 1. Glycogenesis; 2. Lipogenesis; 3. Glycogenolysis; 4. Gluconeogenesis
   B. 1. Glycolysis; 2. Glycogenesis; 3. Glycogenolysis; 4. Gluconeogenesis

Metabolic States
   1. Absorptive state; 2. Postabsorptive state; 3. Postabsorptive state

Metabolic Rate
   1. Basal metabolic rate (BMR); 2 Increases; 3. Decreases; 4. Decreases; 5. Increases; 6. Increases

Body Temperature Regulation
   1. Homeotherms; 2. Free energy; 3. Conduction; 4. Convection; 5. Evaporation; 6. Increase

QUICK RECALL

1. Carbohydrates, lipids, proteins, vitamins, minerals, and water
2. Main energy source of body, minor energy storage (e.g., glycogen), and structural (e.g., DNA, RNA, ATP)
3. Energy source and major energy storage (e.g., adipose tissue)
4. Structural (e.g., collagen), regulatory (e.g., enzymes, hormones, buffers), transport (e.g., hemoglobin, carrier molecules), protection (e.g., antibodies), and energy source
5. Anaerobic respiration: two ATP molecules; aerobic respiration: 38 ATP molecules
6. NADH, $FADH_2$, and ATP
7. Anaerobic respiration: lactic acid; aerobic respiration: carbon dioxide and water
8. Glycolysis: 8 ATP ; acetyl-CoA formation: 6 ATP; citric acid cycle: 24 ATP
9. Fatty acids: beta-oxidation; amino acids: oxidative deamination
10. Glycogenesis, glycogenolysis, lipogenesis, and gluconeogenesis
11. Basal metabolic rate, muscular energy, and thermic effect of food

26 Urinary System

CONTENT LEARNING ACTIVITY

Functions of the Urinary System

Using the terms provided, complete these statements:

Blood
Filter
pH
Reabsorbed
Red blood cells
Solutes
Toxic
Vitamin D

The kidneys (1) the blood, retaining proteins and blood cells and producing a large volume of filtrate. Most of the filtrate is (2) into the blood, but a small volume of metabolic wastes, (3) molecules, and excess ions are retained in the urine. The kidneys play a major role in regulating (4) volume, the concentration of (5) in the blood, and the (6) of the blood. The kidneys secrete erythropoietin, which regulates the synthesis of (7). The kidneys also regulate the synthesis of (8), which is important in regulating blood levels of Ca^{2+}.

1. _____
2. _____
3. _____
4. _____
5. _____
6. _____
7. _____
8. _____

Kidney Anatomy and Histology

A. Match these terms with the correct statement or definition:

Cortex
Hilum
Medulla
Medullary rays
Perirenal fat
Renal capsule
Renal columns
Renal fascia
Renal pyramids
Renal sinus

_____ 1. Fibrous connective tissue that surrounds each kidney.

_____ 2. Dense deposit of adipose tissue surrounding the renal capsule.

_____ 3. Thin layer of connective tissue that attaches the kidneys and surrounding adipose tissue to the abdominal wall.

_____ 4. Site where the renal artery and nerves enter the kidney and the renal vein and ureter exit the kidney.

_____ 5. Cavity filled with fat and connective tissue into which the hilum opens.

410 Chapter 26

_____ 6. Portion of the kidney outside of the bases of the renal pyramids.

_____ 7. Inner portion of the kidney that surrounds the renal sinus.

_____ 8. Cone-shaped structures located in the medulla of the kidney.

_____ 9. Extensions from the base of the renal pyramid that are located in the cortex.

_____ 10. Extensions of the cortex that project between the pyramids.

B. Match these terms with the correct statement or definition:

Major calyces   Renal pelvis
Minor calyces   Ureter
Renal papilla

_____ 1. Apex of the renal pyramid; located in the medulla.

_____ 2. Funnel-shaped structures that surround the renal papillae.

_____ 3. Larger funnels that converge to form the renal pelvis.

_____ 4. Enlarged urinary channel in the center of the renal sinus.

_____ 5. Tube that extends from the renal pelvis to the urinary bladder.

C. Match these terms with the correct parts of the diagram labeled in figure 26.1:

Cortex   Renal column
Major calyx   Renal papilla
Medulla   Renal pelvis
Minor calyx   Renal pyramid
Renal capsule   Ureter

1. _____
2. _____
3. _____
4. _____
5. _____
6. _____
7. _____
8. _____
9. _____
10. _____

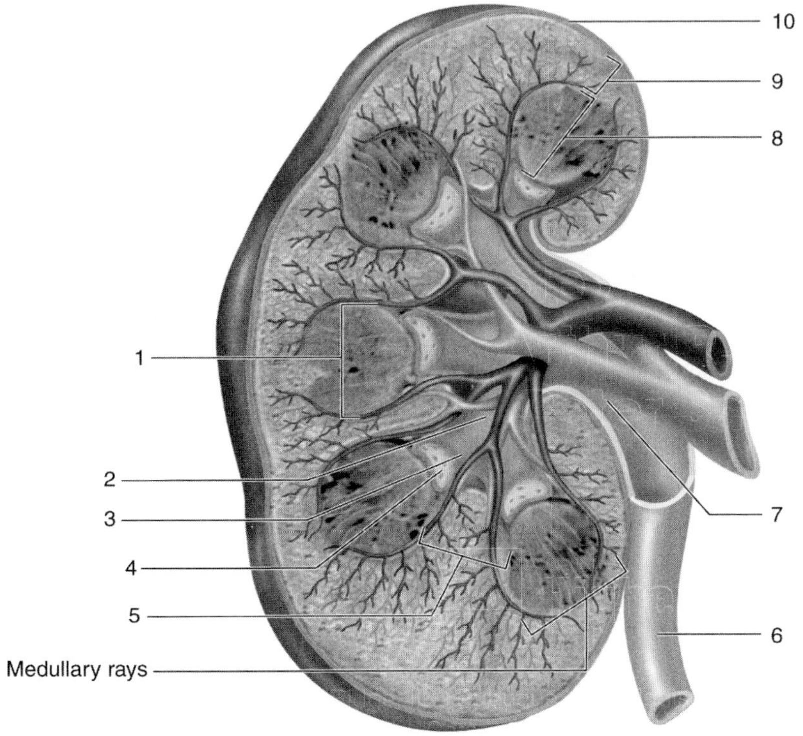

Figure 26.1

Chapter 26

D. Match these terms with the correct statement or definition:

Bowman's capsule
Fenestrae
Filtration membrane
Glomerulus
Juxtamedullary nephrons
Nephron
Papillary duct
Podocytes
Renal corpuscle

_____ 1. Basic histological and functional unit of the kidney.

_____ 2. Formed when several collecting ducts merge to form a larger-diameter tubule near the tip of the renal papilla.

_____ 3. Nephrons that lie near the medulla.

_____ 4. Capillary portion of the renal corpuscle.

_____ 5. Tubule portion of the renal corpuscle, composed of a parietal and visceral layer; proximal end of the nephron.

_____ 6. Specialized cells found in the visceral layer of Bowman's capsule; gaps between their processes are the filtration slits.

_____ 7. Openings in endothelial cells that line glomerular capillaries.

_____ 8. Collective name for the capillary endothelium, basement membrane, and podocytes of Bowman's capsule.

E. Match these terms with the correct statement or definition:

Ascending limb
Collecting duct
Descending limb
Distal convoluted tubule
Juxtaglomerular apparatus
Juxtaglomerular cells
Macula densa
Proximal convoluted tubule

_____ 1. Smooth muscle cells; form a cuff around the afferent arteriole.

_____ 2. Specialized tubule cells found in the distal convoluted tubule, where it is adjacent to the afferent and efferent arterioles.

_____ 3. Collectively, the juxtaglomerular cells and the macula densa where they contact each other.

_____ 4. Part of the nephron between Bowman's capsule and the loop of Henle.

_____ 5. Part of the nephron between the loop of Henle and the collecting duct.

_____ 6. Portion of the loop of Henle connected to the distal convoluted tubule.

_____ 7. Duct to which distal convoluted tubules of many nephrons join.

F. Match these terms with the correct parts of the diagram labeled in figure 26.2:

Afferent arteriole
Collecting duct
Distal convoluted tubule
Efferent arteriole
Loop of Henle
Proximal convoluted tubule
Renal corpuscle

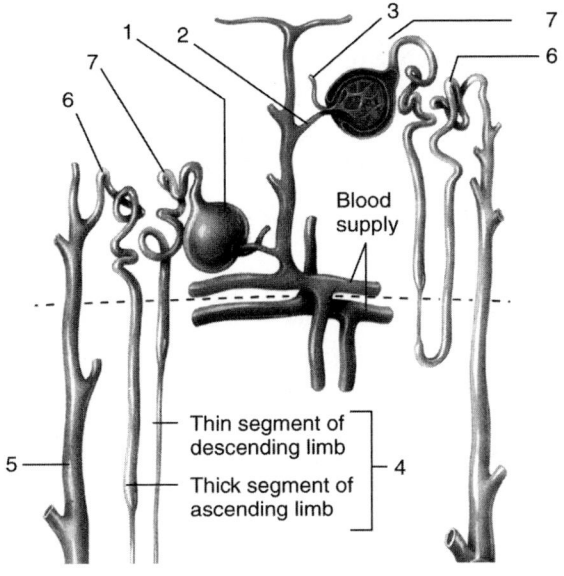

Figure 26.2

1. _____
2. _____
3. _____
4. _____
5. _____
6. _____
7. _____

G. Place these vessels in the correct sequence that blood passes through them, from the abdominal aorta to the interlobular veins:

Afferent arteriole
Arcuate artery
Efferent arteriole
Glomerulus
Interlobar artery
Interlobular artery
Peritubular capillaries
Renal artery
Segmental artery
Vasa recta

1. Abdominal aorta
2. _____
3. _____
4. _____
5. _____
6. _____
7. _____
8. _____
9. _____
10. _____
11. _____
12. Interlobular veins

Urine Production

Match these terms with the correct statement or definition:

Filtration
Tubular reabsorption
Tubular secretion

_____

_____

_____

1. Movement of water and small solutes across the filtration membrane because of a pressure difference.

2. Movement of water and solutes from the filtrate back into the blood.

3. Active transport of solutes into the nephron.

Filtration

A. Match these terms with the correct statement or definition:

Filtration fraction   Renal blood flow rate
Glomerular filtration rate   Renal fraction
(GFR)   Renal plasma flow rate

_____ 1. Part of the total cardiac output that passes through the kidneys; used to calculate renal blood flow rate.

_____ 2. Cardiac output multiplied by the renal fraction.

_____ 3. Renal blood flow rate multiplied by the portion of the blood that is made up of plasma.

_____ 4. Part of the plasma volume that is filtered through the filtration membrane to become filtrate.

_____ 5. Amount of filtrate produced per minute.

B. Using the terms provided, complete these statements:

Capillary
Endocytosis
Filtration barrier
Hormones
Podocyte
Proximal convoluted tubule
Urine

The filtration membrane is a _(1)_, which prevents the entry of blood cells and proteins into the lumen of Bowman's capsule but allows other blood components to enter. The filtration membrane is many times more permeable than a typical _(2)_. The fenestrae of the glomerular capillary, the basement membrane, and the _(3)_ cells prevent molecules larger than 7 nm in diameter from passing through. Most plasma proteins are larger than 7 nm and are retained in the glomerular capillaries. However, albumin and protein _(4)_ are small enough to pass through the filtration barrier. Proteins that do pass through the filtration membrane are actively reabsorbed by _(5)_ and metabolized by cells in the _(6)_. Consequently, normal _(7)_ contains little protein.

1. _____
2. _____
3. _____
4. _____
5. _____
6. _____
7. _____

C. Match these terms with the correct statement or definition:

Capsule pressure   Filtration pressure
Colloid osmotic pressure   Glomerular capillary pressure

_____ 1. Blood pressure within the glomerulus.

_____ 2. Pressure of filtrate already in Bowman's capsule; opposes the glomerular capillary pressure.

_____ 3. Pressure caused by unfiltered plasma proteins remaining within the glomerular capillary.

_____ 4. Net pressure gradient that forces fluid through the filtration membrane into Bowman's capsule.

D. Using the terms provided, complete these statements:

Bowman's capsule
Decrease(s)
Increase(s)
Peritubular capillaries

As the diameter of a vessel decreases, the resistance to flow through the vessel _(1)_. Because the efferent arteriole has a small diameter, resistance to blood flow _(2)_ and blood pressure within the glomerulus is high. Consequently, filtration pressure is high, and filtrate moves from the glomerulus into _(3)_. After blood passes through the efferent arteriole, blood pressure _(4)_; therefore, fluid moves out of the interstitial spaces and into the _(5)_. Changing the diameter of the efferent arteriole can alter filtration pressure. For example, dilation of the afferent arterioles or constriction of the efferent arteriole causes filtration pressure to _(6)_ and results in an _(7)_ in urine production.

1. _____
2. _____
3. _____
4. _____
5. _____
6. _____
7. _____

Regulation of Glomerular Filtration Rate (GFR)

Match these terms with the correct statement:

Autoregulation
Constrict
Decreases
Dilate
Increases

_____ 1. Maintenance of a relatively stable GFR over a wide range of systemic blood pressures.

_____ 2. During autoregulation, the response of the afferent arterioles to increased systemic blood pressure.

_____ 3. Effect of constriction of the afferent arterioles on renal blood flow, filtration pressure, and GFR.

_____ 4. During autoregulation, the response of the afferent arterioles to increased rate of filtrate flow past the macula densa.

_____ 5. Effect of sympathetic stimulation of the afferent arteriole on renal blood flow, filtration pressure, and GFR.

Tubular Reabsorption

A. Using the terms provided, complete these statements:

    Antiport   Interstitial
    Apical membrane   Into
    Basal membrane   Osmosis
    Diffusion   Out of
    Facilitated diffusion   Symport

In the proximal convoluted tubule, amino acids, glucose, and fructose, as well as Na^+, K^+, Ca^{2+}, HCO_3^- and Cl^- are reabsorbed from the lumen of the nephron to the __(1)__ fluid. The __(2)__ of sodium ions across the __(3)__ to the interstitial fluid produces a low concentration of Na^+ inside nephron cells. Within the __(4)__ of the nephron cells, there are carrier molecules that bind to a substance to be transported and to sodium ions. The concentration gradient for Na^+ then provides the energy for movement of both Na^+ and other molecules or ions __(5)__ the nephron cells by the process of __(6)__. Once the molecules are inside the cell, they cross the basal membrane of the cell mainly by __(7)__. Some solutes also move by __(8)__ *between* the cells from the lumen of the nephron into the interstitial fluid. Because the proximal convoluted tubule is permeable to water, as solute particles are transported and diffuse to the interstitial spaces, water follows by __(9)__.

1. _____
2. _____
3. _____
4. _____
5. _____
6. _____
7. _____
8. _____
9. _____

B. Using the terms provided, complete these statements:

    Actiport   Into
    Decreased   Out of
    Diffusion   Permeable
    Increased   Symport
    Interstitial

As filtrate moves through the thin segment of the descending limb of the loop of Henle, water moves __(1)__ the nephron, and some solutes diffuse __(2)__ the nephron. In the ascending limb of the loop of Henle, Na^+ are moved out of tubule cells into the interstitial fluid by __(3)__, resulting in a low concentration of Na^+ inside the tubule cells. K^+ and Cl^- move with sodium ions from the filtrate across the apical membrane of the tubule cells by the process of __(4)__. Then K^+ and Cl^- move across the basal cell membrane into the interstitial fluid by __(5)__. Water does not follow the Na^+, K^+, and Cl^- into the interstitial fluid because the ascending limb of the loop of Henle is not __(6)__ to water. Because the ascending limb is impermeable to water and because ions are transported out of the ascending limb, the concentration of solutes is __(7)__ to about 100 mOsm/kg, and the filtrate entering the distal tubule is much more dilute than the __(8)__ fluid surrounding it.

1. _____
2. _____
3. _____
4. _____
5. _____
6. _____
7. _____
8. _____

C. Match these terms with the correct statement or definition:

Antiport Symport
Osmosis Urea

_____ 1. Process by which Na⁺ are moved across the basal membrane of the distal convoluted tubule and collecting duct cells.

_____ 2. Process by which Cl⁻ are moved across the apical membrane into the distal convoluted tubule and collecting duct cells.

_____ 3. Method by which water moves out of the distal convoluted tubules and collecting ducts when they are permeable to water.

_____ 4. Toxic substance that becomes concentrated in filtrate because renal tubules are not as permeable to it as they are to water.

Tubular Secretion

Match these terms with the correct statements as they pertain to tubular secretion:

Antiport
Diffusion

_____ 1. Method of movement of ammonia into lumen of the nephron.

_____ 2. Method of movement of H⁺, K⁺, and penicillin into the distal convoluted tubules and collecting ducts.

Urine Concentration Mechanism

A. Using the terms provided, complete these statements:

Active transport Symport
Diffusion Vasa recta
Impermeability

The major mechanisms that create and maintain the high solute concentration in the renal medulla include the active transport of Na⁺ and the __(1)__ of K⁺ and Cl⁻ out of the thick portion of the ascending limb of the loop of Henle; the __(2)__ of the thin and thick parts of the ascending limb of the loop of Henle to water; the __(3)__, which remove excess water and solutes that enter the medulla without destroying the high concentration of solutes in the interstitial fluid of the medulla; __(4)__ of ions from the collecting ducts into the interstitial fluid of the medulla; and the urea cycling due to __(5)__ of urea from the collecting ducts into the interstitial fluid of the medulla and into the descending limb of the loop of Henle.

1. _____
2. _____
3. _____
4. _____
5. _____

B. Match these terms with the correct statement or definition:

Loops of Henle   Vasa recta
Urea

_____ 1. Countercurrent mechanism that supplies blood to the kidney medulla; removes excess water and solutes from the medulla without changing the high concentration of solutes in medullary interstitial fluid.

_____ 2. Together, these maintain a high concentration of solutes in interstitial fluid.

_____ 3. Molecules responsible for a substantial portion of the high osmolality in the kidney medulla.

C. Match these terms with the correct statement or definition:

Ascending limb of the   Distal convoluted tubule
  loop of Henle   Proximal convoluted tubule
Collecting duct
Descending limb of the
  loop of Henle

_____ 1. Sixty-five percent of the filtrate is reabsorbed at this location.

_____ 2. As the filtrate moves through these, water moves out and solutes diffuse into the nephron; osmolality increases up to about 1200 mOsm/L, and 15% more filtrate is reabsorbed.

_____ 3. This part is not permeable to water, but Na^+, Cl^-, and K^+ are transported into the interstitial fluid.

_____ 4. From this structure and the collecting ducts, if ADH is present, water diffuses into the interstitial fluid and 19% of the filtrate is reabsorbed, resulting in a concentrated urine.

_____ 5. Urea diffuses out of this part of the nephron into the interstitial fluid of the medulla.

_____ 6. Urea diffuses into this part of the nephron from the interstitial fluid.

Regulation of Urine Concentration and Volume

A. Using the terms provided, complete these statements:

Decrease(s)   Increase(s)

The rate of renin secretion by the juxtaglomerular apparatus increases when blood pressure _(1)_ or Na^+ concentration in the filtrate _(2)_. Renin converts angiotensinogen to angiotensin I, and angiotensin-converting enzyme converts angiotensin I to angiotensin II. Increased angiotensin II _(3)_ blood pressure in two major ways. First, angiotensin II _(4)_ peripheral resistance. Second, it _(5)_ the rate of aldosterone secretion, which leads to a(n) _(6)_ in the reabsorption of Na^+ from the filtrate into the blood. As water follows the Na^+ by osmosis, urine volume _(7)_ and blood volume _(8)_, which results in a(n) _(9)_ in blood pressure.

1. _____
2. _____
3. _____
4. _____
5. _____
6. _____
7. _____
8. _____
9. _____

B. Match these terms with the correct statement:

Decreases
Increases

_____ 1. Effect of increased blood osmolality on ADH secretion.

_____ 2. Effect of increased ADH secretion on blood osmolality.

_____ 3. Effect of increased blood pressure on ADH secretion.

_____ 4. Effect of decreased ADH secretion on blood pressure.

C. Match these terms with the correct statement:

Decreases
Increases

_____ 1. Effect of increased blood pressure in the right atrium on atrial natriuretic hormone secretion.

_____ 2. Effect of atrial natriuretic hormone on Na^+ reabsorption.

_____ 3. Effect of atrial natriuretic hormone on ADH secretion.

_____ 4. Effect of atrial natriuretic hormone on urine production.

_____ 5. Effect of atrial natriuretic hormone on blood volume and blood pressure.

Plasma Clearance and Tubular Maximum

Match these terms with the correct statement or definition:

Plasma clearance   Tubular maximum
Tubular load

_____ 1. Volume of plasma that is cleared of a specific substance each minute.

_____ 2. Total amount of a substance that filters through the filtration membrane into the nephrons each minute.

_____ 3. Maximum rate at which a substance can be actively reabsorbed.

Anatomy and Histology of the Ureters and Urinary Bladder

Match these terms with the correct statement or definition:

Detrusor muscle   Transitional epithelium
External urinary sphincter   Trigone
Internal urinary sphincter

_____ 1. Triangular area of the bladder wall between the two ureters posteriorly and the urethra anteriorly.

_____ 2. Multiple layers of smooth muscle external to the epithelium in the bladder wall.

Chapter 26

_____ 3. This internal lining permits changes in the size of the urinary bladder and ureter.

_____ 4. In males, elastic tissue and smooth muscle that surround the urethra where it exits the urinary bladder.

_____ 5. Skeletal muscle that surrounds the urethra as it extends through the pelvic floor.

Urine Movement

A. Match these terms with the correct statement or definition:

Hydrostatic pressure
Peristaltic contractions

_____ 1. Mechanism responsible for urine flow from the nephron into the renal pelvis.

_____ 2. Mechanism responsible for urine flow through the ureters into the urinary bladder.

B. Match these terms with the correct statement or definition:

External urinary sphincter   Micturition reflex
Higher brain centers

_____ 1. Reflex initiated by stretching of the bladder wall, which results in contraction of the bladder and inhibition of the external urinary sphincter.

_____ 2. Part of the nervous system responsible for inhibition or stimulation of the micturition reflex.

_____ 3. This structure is kept tonically contracted by the higher brain centers.

Effects of Aging on the Kidneys

Match these terms with the correct statement or definition:

Decrease(s)   No effect
Increase(s)

_____ 1. Effect of aging on kidney size.

_____ 2. Effect of aging on blood flow through the kidneys.

_____ 3. Effect of aging on number of functional glomeruli.

_____ 4. Effect of aging on risk for dehydration.

_____ 5. Effect of aging on the ability to eliminate uric acid, urea, creatine, and toxins from the blood.

_____ 6. Effect of aging on the responsiveness of the kidneys to ADH and aldosterone.

_____ 7. Effect of aging on the likelihood of Ca^{2+} deficiency, osteoporosis, and bone fractures due to decreased vitamin D synthesis.

QUICK RECALL

1. List the five major parts of a nephron.

2. Name the two parts of the juxtaglomerular apparatus.

3. List the three steps in urine formation.

4. Write the formula for filtration pressure.

5. Complete these table by placing a "+" under the location where the condition exists and a "—" under the location where the condition does not exist.

CONDITION	PROXIMAL CONVOLUTED TUBULE	DESCENDING LIMB	ASCENDING LIMB	DISTAL CONVOLUTED TUBULE
Na^+ are reabsorbed from tubule (active transport/ symport).	_____	_____	_____	_____
Na^+ passively diffuse into tubule.	_____	_____	_____	_____
Water moves out of tubule by osmosis.	_____	_____	_____	_____

Chapter 26

6. Complete this table by placing a "+" in each column where the filtrate concentration exists and a "—" where the condition does not exist.

FILTRATE	PROXIMAL CONVOLUTED TUBULE	TIP OF LOOP OF HENLE	DISTAL CONVOLUTED TUBULE
100 mOsm	_____	_____	_____
300 mOsm (if ADH is present)	_____	_____	_____
1200 mOsm	_____	_____	_____

7. Name two ions actively secreted into the distal tubule.

8. Complete this table by placing a "+" in each column where the correct percentage of filtrate volume reduction occurs.

FILTRATE VOLUME REDUCED BY	DESCENDING LIMB	ASCENDING LIMB	PROXIMAL CONVOLUTED TUBULE	DISTAL CONVOLUTED TUBULE	COLLECTING DUCT
65%	_____	_____	_____	_____	_____
15%	_____	_____	_____	_____	_____
19% (if ADH is present)	_____	_____	_____	_____	_____

9. List three hormones that affect urine production, and give the major effect of each on urine production. List the corresponding effects these hormones have on extracellular fluid concentration.

10. List the events that result in micturition following stretch of the bladder.

ANSWERS TO CHAPTER 26

CONTENT LEARNING ACTIVITY

Functions of the Urinary System
1. Filter; 2. Reabsorbed; 3. Toxic; 4. Blood; 5. Solutes; 6. pH; 7. Red blood cells; 8. Vitamin D

Kidney Anatomy and Histology
A. 1. Renal capsule; 2. Perirenal fat; 3. Renal fascia; 4. Hilum; 5. Renal sinus; 6. Cortex; 7. Medulla; 8. Renal pyramids; 9. Medullary rays; 10. Renal columns
B. 1. Renal papilla; 2. Minor calyces; 3. Major calyces; 4. Renal pelvis; 5. Ureter
C. 1. Renal pyramid; 2. Renal papilla; 3. Minor calyx; 4. Major calyx; 5. Renal column; 6. Ureter; 7. Renal pelvis; 8. Medulla; 9. Cortex; 10. Renal capsule
D. 1. Nephron; 2. Papillary duct; 3. Juxtamedullary nephrons; 4. Glomerulus; 5. Bowman's capsule; 6. Podocytes; 7. Fenestrae; 8. Filtration membrane
E. 1. Juxtaglomerular cells; 2. Macula densa; 3. Juxtaglomerular apparatus; 4. Proximal convoluted tubule; 5. Distal convoluted tubule; 6. Ascending limb; 7. Collecting duct
F. 1. Renal corpuscle; 2. Afferent arteriole; 3. Efferent arteriole; 4. Loop of Henle; 5. Collecting duct; 6. Distal convoluted tubule; 7. Proximal convoluted tubule
G. 2. Renal artery; 3. Segmental artery; 4. Interlobar artery; 5. Arcuate artery; 6. Interlobular artery; 7. Afferent arteriole; 8. Glomerulus; 9. Efferent arteriole; 10. Peritubular capillaries; 11. Vasa recta

Urine Production
1. Filtration; 2. Tubular reabsorption; 3. Tubular secretion

Filtration
A. 1. Renal fraction; 2. Renal blood flow rate; 3. Renal plasma flow rate; 4. Filtration fraction; 5. Glomerular filtration rate (GFR)
B. 1. Filtration barrier; 2. Capillary; 3. Podocyte; 4. Hormones; 5. Endocytosis; 6. Proximal convoluted tubule; 7. Urine
C. 1. Glomerular capillary pressure; 2. Capsule pressure; 3. Colloid osmotic pressure; 4. Filtration pressure
D. 1. Increases; 2. Increases; 3. Bowman's capsule; 4. Decreases; 5. Peritubular capillaries; 6. Increase; 7. Increase

Regulation of Glomerular Filtration Rate (GFR)
1. Autoregulation; 2. Constrict; 3. Decreases; 4. Constrict; 5. Decreases

Tubular Reabsorption
A. 1. Interstitial; 2. Antiport; 3. Basal membrane; 4. Apical membrane; 5. Into; 6. Symport; 7. Facilitated diffusion; 8. Diffusion; 9. Osmosis
B. 1. Out of; 2. Into; 3. Antiport; 4. Symport; 5. Diffusion; 6. Permeable; 7. Decreased; 8. Interstitial
C. 1. Active transport; 2. Symport; 3. Osmosis; 4. Urea

Tubular Secretion
1. Diffusion; 2. Antiport

Urine Concentration Mechanism
A. 1. Symport; 2. Impermeability; 3. Vasa recta; 4. Active transport; 5. Diffusion
B. 1. Vasa recta; 2. Loops of Henle and vasa recta; 3. Urea
C. 1. Proximal convoluted tubule; 2. Descending limb of the loop of Henle; 3. Ascending limb of the loop of Henle; 4. Distal convoluted tubule; 5. Collecting duct; 6. Descending limb of the loop of Henle

Regulation of Urine Concentration and Volume
A. 1. Decreases; 2. Decreases; 3. Increases; 4. Increases; 5. Increases; 6. Increase; 7. Decreases; 8. Increases; 9. Increase
B. 1. Increases; 2. Decreases; 3. Decreases; 4. Decreases
C. 1. Increases; 2. Decreases; 3. Decreases; 4. Increases; 5. Decreases

Plasma Clearance and Tubular Maximum
1. Plasma clearance; 2. Tubular load; 3. Tubular maximum

Anatomy and Histology of the Ureters and Urinary Bladder
1. Trigone; 2. Detrusor muscle; 3. Transitional epithelium; 4. Internal urinary sphincter; 5. External urinary sphincter

Urine Movement
A. 1. Hydrostatic pressure; 2. Peristaltic contractions
B. 1. Micturition reflex; 2. Higher brain centers; 3. External urinary sphincter

Effects of Aging on the Urinary System
1. Decreases; 2. Decreases; 3. Decreases; 4. Increases; 5. Decreases; 6. Decreases; 7. Increases

Quick Recall

1. Glomerulus, Bowman's capsule, proximal convoluted tubule, loop of Henle, and distal convoluted tubule

2. Macula densa and juxtaglomerular cells of the afferent arterioles

3. Filtration, reabsorption, and secretion

4. Filtration pressure equals glomerular capillary pressure minus capsule pressure minus colloid osmotic pressure.

5.
	PCT	DL	AL	DCT
Na$^+$ are reabsorbed.	+	-	+	+
Na$^+$ passively diffuse in.	-	+	-	-
Water moves out by osmosis.	+	+	-	+

6.
	PCT	TLH	DCT
100 mOsm	-	-	+
300 mOsm (if ADH present)	+	-	-
1200 mOsm	-	+	-

7. Potassium ions and hydrogen ions

8.
	PCT	DL	AL	DCT	CD
65%	+	-	-	-	-
15%	-	+	-	-	-
19% (if ADH present)	-	-	-	+	+

9. Aldosterone: increased sodium ion reabsorption, resulting in decreased urine concentration and volume, increased extracellular fluid volume; ADH: decreased urine volume, increased extracellular fluid volume; atrial natriuretic factor: inhibits ADH production, resulting in increased urine volume, and decreased extracellular fluid volume

10. Stretch of bladder, reflex stimulated, bladder contracts, and external urinary sphincter inhibited

Water, Electrolytes, and Acid–Base Balance

CONTENT LEARNING ACTIVITY

Body Fluids

Match these terms with the correct statement or definition:
Extracellular fluid Intracellular fluid
Interstitial fluid Plasma

_____ 1. Accounts for about 40% of the total body weight and includes the small amount of fluid in trillions of cells.

_____ 2. Accounts for about 20% of the total body weight and includes plasma, interstitial fluid, lymph, cerebrospinal fluid, and synovial fluid.

_____ 3. Extracellular fluid that occupies spaces outside blood vessels.

_____ 4. Extracellular fluid that occupies space within blood vessels.

Regulation of Water Content

A. Match these terms with the correct statement or definition:
10% Decreases
90% Increases

_____ 1. Effect of increased osmolality of extracellular fluids on thirst; mediated through the hypothalamus.

_____ 2. Effect of decreased blood pressure on thirst; mediated through baroreceptors.

_____ 3. Effect of decreased blood pressure on thirst; mediated through increased angiotensin II.

_____ 4. Effect of wetting the oral mucosa on thirst.

_____ 5. Effect of stretch of the gastrointestinal wall on thirst.

_____ 6. Percentage of water ingested.

_____ 7. Percentage of water produced by metabolism.

B. Match these terms with the correct statement or definition:

4%
35%
61%

Decreases
Increases

_____ 1. Water lost through evaporation from the skin.

_____ 2. Secreted by sweat glands; contains solutes and is usually hyposmotic to plasma.

_____ 3. Percentage of water loss in the urine.

_____ 4. Percentage of water loss by evaporation from respiratory passageways and perspiration.

_____ 5. Percentage of water loss in the feces.

Regulation of Extracellular Fluid Osmolality

Match these terms with the correct statement or definition:

Decreases
Increases

_____ 1. Effect of increased blood osmolality on ADH production.

_____ 2. Effect of increased ADH on water reabsorption from the kidneys.

_____ 3. Effect of increased water reabsorption from the kidneys and increased thirst (drinking) on blood osmolality.

Regulation of Extracellular Fluid Volume

A. Match these terms with the correct statement or definition:

Decreases
Increases

_____ 1. Effect of decreased blood pressure on stimulation of the afferent arterioles of the kidneys; mediated through baroreceptors and the sympathetic division.

_____ 2. Effect of increased sympathetic stimulation of the afferent arterioles on glomerular filtration rate and urine volume.

_____ 3. Effect of decreased urine production on blood volume and blood pressure.

B. Match these terms with the correct statement or definition:

Decreases
Increases

_____ 1. Effect of a decrease in blood pressure in the afferent arterioles on renin secretion by the juxtaglomerular apparati.

_____ 2. Effect of an increased renin on angiotensin II formation.

_____ 3. Effect of increased angiotensin II on aldosterone secretion by the adrenal cortex.

_____ 4. Effect of increased aldosterone on the rate of Na⁺ and water reabsorption from the filtrate into the blood.

_____ 5. Effect of an increase in water reabsorption on blood volume and pressure.

C. Match these terms with the correct statement or definition:
Decreases
Increases

_____ 1. Effect of an increase in pressure in the atria of the heart on the secretion of atrial natriuretic hormone.

_____ 2. Effect of atrial natriuretic hormone on Na⁺ reabsorption in the distal convoluted tubules and collecting ducts.

_____ 3. Effect of atrial natriuretic hormone on urine volume.

_____ 4. Effect of atrial natriuretic hormone on blood volume and pressure.

D. Match these terms with the correct statement or definition:
Decreases
Increases

_____ 1. Effect of a large decrease in blood pressure on ADH secretion.

_____ 2. Effect of increased ADH on water reabsorption.

_____ 3. Effect of increased water reabsorption on blood volume and pressure.

Regulation of Specific Electrolytes in the Extracellular Fluid

A. Match these terms with the correct statement or definition:
Atrial natriuretic hormone Kidneys
Aldosterone Na⁺
Hypernatremia Sweat
Hyponatremia

_____ 1. Dominant extracellular positively charged ion; 90%–95% of the osmotic pressure of extracellular fluid results from these ions and the negatively charged ions associated with them.

_____ 2. Major route by which Na⁺ are secreted.

_____ 3. Hormone that increases Na⁺ reabsorption from the distal convoluted tubule and collecting duct.

_____ 4. Hormone that decreases Na⁺ reabsorption from the distal convoluted tubule and collecting duct.

_____ 5. Elevated plasma Na⁺ concentration.

Chapter 27

B. Using the terms provided, complete these statements:

Decrease(s)   Increase(s)

Hormones regulating Na$^+$ and water reabsorption work together to maintain blood pressure and blood osmolality. When blood pressure decreases, the renin-angiotensin-aldosterone mechanism causes Na$^+$ reabsorption to _(1)_ and blood osmolality to _(2)_. Consequently, ADH secretion _(3)_, water reabsorption _(4)_, and blood osmolality _(5)_. Thus, there is increased reabsorption of Na$^+$ and water, which _(6)_ blood volume but maintains blood osmolality. The increased blood volume _(7)_ blood pressure.

1. _____
2. _____
3. _____
4. _____
5. _____
6. _____
7. _____

C. Match these terms with the correct statement or definition:

Decreases
Depolarization
Distal convoluted tubule
Hyperkalemia
Hypokalemia
Hypopolarization
Increases
Proximal convoluted tubule

1. Effect of an increase in extracellular K$^+$ on the resting membrane potential.
2. Part of the nephron where K$^+$ are actively reabsorbed.
3. Part of the nephron where K$^+$ are actively secreted.
4. Effect of increased aldosterone on K$^+$ secretion.
5. Effect of increased aldosterone on blood K$^+$ levels.
6. Effect of increased blood K$^+$ on aldosterone secretion.
7. Abnormally high levels of K$^+$ in the blood.

D. Match these terms with the correct statement or definition:

Decreases
Increases
Hypercalcemia
Hypocalcemia

1. Effect of reduced extracellular Ca^{2+} concentration on spontaneous action potential generation.
2. Effect of increased parathyroid hormone on extracellular Ca^{2+} levels.
3. Effect of decreased Ca^{2+} levels on the secretion of parathyroid hormone.
4. Effect of increased parathyroid hormone on Ca^{2+} reabsorption from the kidneys.
5. Effect of increased parathyroid hormone on the production of active vitamin D.

_____ 6. Effect of vitamin D on Ca^{2+} absorption in the gastrointestinal tract.

_____ 7. Effect of increased calcitonin on extracellular Ca^{2+} concentration.

_____ 8. Effect of elevated extracellular Ca^{2+} levels on the secretion of calcitonin.

_____ 9. Above normal levels of Ca^{2+} in the extracellular fluid.

Regulation of Acid–Base Balance

A. Match these terms with the correct statement or definition:

Acid   Neutral
Base   pH scale

_____ 1. Measurement of the acidity of a solution; numbers above 7 are basic, whereas those below 7 are acidic.

_____ 2. Substance with a pH of 7.

_____ 3. Substance that releases H^+ into a solution.

B. Using the terms provided, complete these statements:

Acidosis
Alkalosis
Bicarbonate ions
Decreased
Depression
Hyperexcitability
Increased
Lactic acid
Urine

When body fluid pH is below 7.35, the condition is referred to as _(1)_; when the pH is above 7.45, the condition is called _(2)_. The major effect of acidosis is _(3)_ of the central nervous system, whereas a major effect of alkalosis is _(4)_ of the nervous system. Respiratory acidosis may occur with _(5)_ elimination of carbon dioxide from the body fluids through the respiratory system. Metabolic acidosis may occur through a loss of _(6)_ through diarrhea, ingestion of acidic drugs, untreated diabetes mellitus, or _(7)_ buildup from severe exercise, heart failure, or shock. Respiratory alkalosis can result with _(8)_ elimination of carbon dioxide from the body fluids through the respiratory system. Metabolic alkalosis can result when loss of large amounts of acidic stomach contents occurs, when alkaline substances are ingested, or when there is a higher than normal loss of H^+ in the _(9)_.

1. _____
2. _____
3. _____
4. _____
5. _____
6. _____
7. _____
8. _____
9. _____

C. **Match these terms with the correct statement or definition:**

Bicarbonate buffer system   Protein buffer system
Phosphate buffer system

_____ 1. Provides three-fourths of the buffer system of the body; includes plasma proteins and hemoglobin.

_____ 2. Plays an exceptionally important role in controlling the pH of extracellular fluid; involves carbonic acid.

_____ 3. Important intracellular buffer system; extracellular concentrations of this system are low, compared with those of the other two buffer systems.

D. **Match these terms with the correct statement or definition:**

Respiratory system
Urinary system

_____ 1. System that responds most rapidly to pH change.

_____ 2. System with the greatest regulatory capacity for acid–base balance but with a slower response to pH change.

E. **Using the terms provided, complete these statements:**

Carbonic acid–bicarbonate
Carbonic anhydrase
Decreases
Increases

The respiratory system regulates acid–base balance through the (1) buffer system. The reaction between CO_2 and H_2O is catalyzed by (2), which is located on the surface of capillary epithelial cells. As CO_2 levels increase, pH of the body fluids (3), neurons in the medullary respiratory center of the brain are stimulated, and the rate and depth of ventilation (4). Carbon dioxide elimination (5), and the concentration of CO_2 in the body fluids (6). This causes H^+ to combine with HCO_3^- to form H_2CO_3, which then forms CO_2 and H_2O. As a result of these reactions, the concentration of H^+ (7) and pH increases to its normal range.

1. _____
2. _____
3. _____
4. _____
5. _____
6. _____
7. _____

F. **Using the terms provided, complete these statements:**

Buffers
Decrease(s)
H^+
HCO_3^-
H_2CO_3
Increase(s)
Na^+
NH_3

When blood pH increases, the movement of CO_2 into nephron cells (1). Within these cells, CO_2 and H_2O combine to form (2), which dissociates into H^+ and HCO_3^-. The H^+ are secreted into the filtrate by an active transport pump that exchanges (3) for H^+. The Na^+ and (4) pass into the extracellular fluid, where the HCO_3^- combine with H^+, causing blood pH to (5). In the filtrate, H^+ combine with (6) and pass out in the urine. An example of a filtrate buffer is (7), which is produced from amino acids. Ammonia combines with H^+ to form NH_4^+ (ammonium ions).

1. _____
2. _____
3. _____
4. _____
5. _____
6. _____
7. _____

QUICK RECALL

1. List three hormones that influence Na^+ concentration, and state if they cause an increase or a decrease in blood Na^+ concentration.

2. Name the major hormone that regulates K^+ concentration, and describe its effect.

3. List three compounds and their influence on extracellular Ca^{2+} levels.

4. List three ways the sensation of thirst is increased and two ways it is decreased.

5. Name two stimuli that increase ADH and renin secretion; state the effect of these substances on urine production and concentration.

6. Name three important buffer systems in the body.

7. State the effects on blood pH when respiration rate increases above normal or decreases below normal.

8. State the effects on blood pH when the acidity of the urine increases or decreases.

Chapter 27

ANSWERS TO CHAPTER 27

CONTENT LEARNING ACTIVITY

Body Fluids
1. Intracellular fluid; 2. Extracellular fluid; 3. Interstitial fluid; 4. Plasma

Regulation of Water Content
- A. 1. Increases; 2. Increases; 3. Increases; 4. Decreases; 5. Decreases; 6. 90%; 7. 10%
- B. 1. Insensible perspiration; 2. Sweat (sensible perspiration; 3. 61%; 4. 35%; 5. 4%

Regulation of Extracellular Fluid Osmolality
1. Increases; 2. Increases; 3. Decreases

Regulation of Extracellular Fluid Volume
- A. 1. Increases; 2. Decreases; 3. Increases
- B. 1. Increases; 2. Increases; 3. Increases; 4. Increases; 5. Increases
- C. 1. Increases; 2. Decreases; 3. Increases; 4. Decreases
- D. 1. Increases; 2. Increases; 3. Increases

Regulation of Specific Electrolytes in the Extracellular Fluid
- A. 1. Na^+; 2. Kidneys; 3. Aldosterone; 4. Atrial natriuretic hormone; 5. Hypernatremia
- B. 1. Increase; 2. Increase; 3. Increases; 4. Increases; 5. Decrease; 6. Increases; 7. Increases
- C. 1. Depolarization; 2. Proximal convoluted tubule; 3. Distal convoluted tubule; 4. Increase; 5. Decreases; 6. Increases; 7. Hyperkalemia
- D. 1. Increases; 2. Increases; 3. Increases; 4. Increases; 5. Increases; 6. Increases; 7. Decreases; 8. Increases; 9. Hypercalcemia

Regulation of Acid–Base Balance
- A. 1. pH scale; 2. Neutral; 3. Acid
- B. 1. Acidosis; 2. Alkalosis; 3. Depression; 4. Hyperexcitability; 5. Decreased; 6. Bicarbonate ions; 7. Lactic acid; 8. Increased; 9. Urine
- C. 1. Protein buffer system; 2. Bicarbonate buffer system; 3. Phosphate buffer system
- D. 1. Respiratory system; 2. Urinary system
- E. 1. Carbonic acid–bicarbonate; 2. Carbonic anhydrase; 3. Decreases; 4. Increases; 5. Increases; 6. Decreases; 7. Decreases
- F. 1. Increases; 2. H_2CO_3; 3. Na^+; 4. HCO_3^-; 5. Decrease; 6. Buffers; 7. NH_3

QUICK RECALL

1. ADH: decreases blood Na^+ concentration (increases water reabsorption); aldosterone: increases blood Na^+ concentration (increases Na^+ reabsorption); atrial natriuretic hormone: decreases blood Na^+ concentration (inhibits Na^+ reabsorption)
2. Aldosterone increases K^+ secretion in the distal convoluted tubule.
3. Parathyroid hormone: increases extracellular Ca^{2+} levels; calcitonin: decreases extracellular Ca^{2+} levels; vitamin D: increases extracellular Ca^{2+} by increasing Ca^{2+} uptake in the intestine
4. Increased osmolality of body fluid, reduction in plasma volume, and decrease in blood pressure cause increased thirst. Wetting of the oral mucosa and stretching of the gastrointestinal tract decrease thirst.
5. Decreased blood pressure and increased tissue osmolality cause an increase in ADH and renin production, producing less urine that is more highly concentrated.
6. Plasma protein, carbonic acid–bicarbonate, and phosphate buffer systems
7. Above normal: increase in blood pH; below normal: decrease in blood pH
8. If the acidity of the urine increases, the blood pH increases; if the acidity of the urine decreases, the blood pH decreases.

28 Reproductive System

CONTENT LEARNING ACTIVITY

Scrotum

Match these terms with the
correct statement or definition:

Cremaster muscles   Raphe
Dartos muscle

_____ 1. Irregular ridge on the midline of the scrotum.

_____ 2. Layer of smooth muscle surrounding the scrotum; contracts and causes the skin of the scrotum to become firm and wrinkled.

_____ 3. Extensions of abdominal muscles; contract and pull the testes closer to the body.

Perineum

Match these terms with the
correct statement or definition:

Anal triangle
Urogenital triangle

_____ 1. Anterior portion of the perineum; contains the base of the penis and the scrotum.

_____ 2. Smaller, posterior portion of the perineum; contains the anal opening.

Testes

A. Match these terms with the
correct statement or definition:

Efferent ductules   Seminiferous tubules
Interstitial cells   Septa
  (Leydig cells)   Tubuli recti
Rete testis   Tunica albuginea

_____ 1. Outer, thick, white capsule of the testis.

_____ 2. Connective tissue of the tunica albuginea that divides the testis into lobules.

Chapter 28   433

_____ 3. Structures in which sperm cells develop.

_____ 4. Endocrine cells in the loose connective tissue around the seminiferous tubules.

_____ 5. Short, straight tubules between the seminiferous tubules and the rete testis.

_____ 6. Tubules, connected to the rete testis, that pierce the tunica albuginea and exit the testis.

B. Match these terms with the correct statement or definition:

Cryptorchidism   Process vaginalis
Deep inguinal rings   Superficial inguinal rings
Gubernaculum   Tunica vaginalis
Inguinal canal

_____ 1. Fibromuscular cord that connects the testes to the scrotum.

_____ 2. Bilateral oblique passageways in the anterior abdominal wall through which the testes descend.

_____ 3. Outpocketing of peritoneum that precedes the testis as it moves to the scrotum during fetal development.

_____ 4. Small, closed sac that covers most of the testis.

_____ 5. Origin of the inguinal canal; opens through the aponeurosis of the transversus abdominis muscle.

_____ 6. End of the inguinal canal; opens in the aponeurosis of the external abdominal oblique muscle.

_____ 7. Failure of one or both of the testes to descend into the scrotum.

C. Match these terms with the correct statement or definition:

Androgen-binding protein   Germ cells
Dihydrotestosterone   Spermatogenesis
Estrogen   Sustentacular (Sertoli) cells

_____ 1. Production of sperm cells.

_____ 2. Large cells that nourish the germ cells, produce hormones, and form the blood–testis barrier; nurse cells.

_____ 3. Cells from which sperm cells are derived.

_____ 4. Two hormones produced from testosterone by Sertoli cells; may be the active hormones that promote sperm cell development.

_____ 5. Protein secreted by sustentacular cells; binds to testosterone and dihydrotestosterone.

D. Match these terms with the correct statement or definition:

Acrosome　　　　　　　　Spermatids
Primary spermatocytes　　Spermatogenesis
Secondary spermatocytes　Spermatogonia
Sperm cells (spermatozoa)　Spermiogenesis

_____ 1. Most peripheral germ cells; divide by mitosis.

_____ 2. Germ cells produced from spermatogonia that are passing through the first meiotic division.

_____ 3. Germ cells produced from primary spermatocytes by the first meiotic division.

_____ 4. Germ cells produced from secondary spermatocytes by the second meiotic division.

_____ 5. Mature male reproductive cell.

_____ 6. Process during which a spermatid becomes a sperm cell.

_____ 7. Cap found on the head of a sperm cell, containing enzymes necessary for penetrating the female oocyte.

Ducts

Match these terms with the correct statement or definition:

Duct of the epididymis　　Membranous urethra
Ductus (vas) deferens　　Prostatic urethra
Efferent ductules　　　　Spermatic cord
Ejaculatory duct　　　　　Spongy (penile) urethra
Epididymis　　　　　　　Urethral glands

_____ 1. Tubules that become convoluted and empty into the duct of the epididymis.

_____ 2. Comma-shaped structure on the posterior testis in which maturation of sperm cells occurs.

_____ 3. Single convoluted ductule located primarily in the body of the epididymis; contains elongated microvilli called stereocilia.

_____ 4. Duct running from the tail of the epididymis to the ejaculatory duct; the end enlarges to form the ampulla.

_____ 5. The ductus deferens, the blood vessels and nerves that supply the testis, fascia, and the cremaster muscle compose this structure.

_____ 6. Duct formed when the duct of the seminal vesicle joins the ductus deferens.

_____ 7. Portion of the urethra that passes through the prostate gland and into which the prostatic ducts and ejaculatory ducts empty.

_____ 8. Shortest portion of the urethra; extends from the prostate through the urogenital diaphragm.

_____ 9. Several minute mucus-secreting glands that empty into the urethra.

Penis

Match these terms with the correct statement or definition:

Bulb of the penis
Corpora cavernosa
Corpus spongiosum
Crus of the penis
Erection
External urethral orifice
Glans penis
Prepuce (foreskin)
Root of the penis

_____ 1. Engorgement of erectile tissue with blood, which causes the penis to enlarge and become firm.

_____ 2. Two erectile columns that form the dorsum and sides of the penis.

_____ 3. Smallest erectile column, which occupies the ventral portion of the penis; the spongy urethra passes through it.

_____ 4. Cap, formed from the corpus spongiosum, over the distal end of the penis.

_____ 5. Expansion of the corpus spongiosum at the base of the penis.

_____ 6. Expansion of the corpora cavernosa at the base of the penis.

_____ 7. Collective name for the crus of the penis and the bulb of the penis; attaches the penis to the coxal bones.

_____ 8. Loose fold of skin that covers the glans penis.

Accessory Glands

A. Match these terms with the correct statement or definition:

Bulbourethral glands
Ejaculation
Emission
Prostate gland
Semen
Seminal vesicles

_____ 1. Sac-shaped glands located near the ampullae of the ductus deferentia.

_____ 2. Gland the size and shape of a walnut located dorsal to the symphysis pubis at the base of the bladder; surrounds part of the urethra and the ejaculatory ducts.

_____ 3. Pair of small mucous glands near the membranous portion of the urethra.

_____ 4. Discharge of semen into the prostatic urethra; stimulated by sympathetic impulses.

_____ 5. Forceful expulsion of semen from the urethra.

_____ 6. Glands producing a mucous secretion that lubricates the urethra, neutralizes the contents of the spongy urethra, provides lubrication during intercourse, and reduces the acidity of the vagina.

_____ 7. Glands producing a thick, mucuslike secretion containing fructose, fibrinogen, and prostaglandins.

_____ 8. Gland that produces a thin, milky, alkaline secretion that neutralizes the urethra and vagina.

_____ 9. Gland that produces a secretion containing clotting factors, which convert fibrinogen to fibrin, resulting in coagulation and a sticky mass of semen.

_____ 10. Gland that produces a secretion containing fibrinolysin, which dissolves fibrin, resulting in the release of sperm cells.

B. Match these terms with the correct parts of the diagram labeled in figure 28.1:

Bulbourethral gland
Ductus deferens
Ejaculatory duct
Epididymis
External urethral orifice
Penis
Prostate gland
Prostatic urethra
Scrotum
Seminal vesicle
Spongy urethra

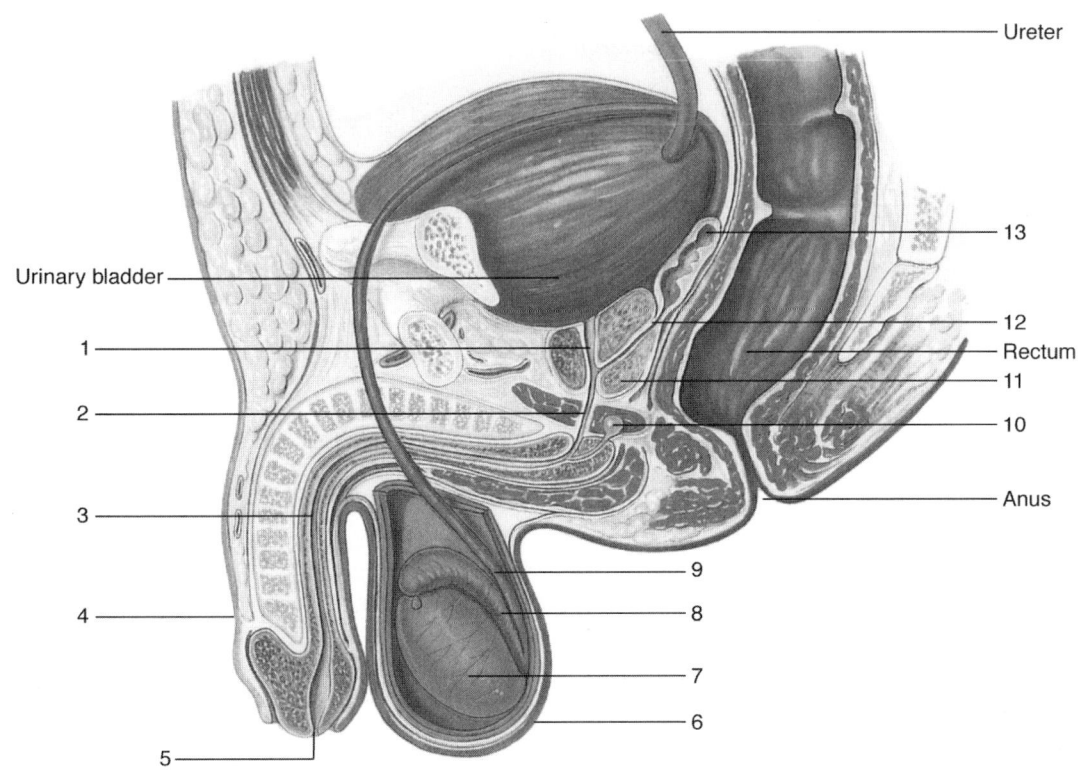

Figure 28.1

Chapter 28

1. _____   6. _____   10. _____
2. _____   7. _____   11. _____
3. _____   8. _____   12. _____
4. _____   9. _____   13. _____
5. _____

Regulation of Sex Hormone Secretion

Match these terms with the correct statement or definition:

Androgen
Follicle-stimulating hormone (FSH)
Gonadotropin-releasing hormone (GnRH)
Gonadotropins
Inhibin
Luteinizing hormone (LH)
Testosterone

_____ 1. Small peptide hormone released from neurons in the hypothalamus; stimulates cells in the anterior pituitary; luteinizing hormone–releasing hormone (LHRH).

_____ 2. General term for hormones that affect the gonads; secreted by the anterior pituitary.

_____ 3. Hormone that binds to the interstitial cells in the testes and stimulates testosterone synthesis and secretion; interstitial cell-stimulating hormone (ICSH).

_____ 4. Hormone that binds to the sustentacular cells in the seminiferous tubules of the testes and promotes sperm cell development.

_____ 5. Major male hormone secreted by the testes.

_____ 6. General name for male hormones that stimulate the development of male secondary sexual characteristics.

_____ 7. Secreted by the sustentacular cells of the testes, this hormone inhibits FSH secretion from the anterior pituitary.

Puberty

Using the terms provided, complete these statements:

Androgens
GnRH
Human chorionic gonadotropin (HCG)
Interstitial cells
LH and FSH
Sperm cell

_(1)_, which is secreted by the placenta, stimulates the synthesis and secretion of testosterone before birth. After birth, only small amounts of testosterone are secreted until puberty, when the hypothalamus becomes less sensitive to the inhibitory effect of _(2)_ and the rate of _(3)_ secretion from the hypothalamus increases, leading to increased _(4)_ release from the anterior pituitary. Elevated FSH levels promote _(5)_ formation, and elevated LH levels cause the _(6)_ to secrete testosterone.

1. _____
2. _____
3. _____
4. _____
5. _____
6. _____

Effects of Testosterone

Using the terms provided, complete these statements:

- Body fluids
- Coarser
- Hypertrophy
- Melanin
- Protein synthesis
- Rapid bone growth
- Sebaceous glands
- Testes

Testosterone causes the enlargement and differentiation of the male reproductive system, is necessary for spermatogenesis, and is required for the descent of the (1) during fetal development. Testosterone stimulates hair growth in several regions and causes the texture of the hair and skin to become (2) . Testosterone increases the quantity of (3) in the skin, causing it to become darker, and increases the rate of secretion from (4) , frequently resulting in acne. Testosterone also causes (5) of the larynx, deepening the voice. Testosterone stimulates red blood cell production, stimulates metabolism, and causes the retention of sodium in the body, resulting in an increase in the volume of (6) . Testosterone promotes (7) , resulting in an increased skeletal muscle mass, and (8) , resulting in an increase in height.

1. _____
2. _____
3. _____
4. _____
5. _____
6. _____
7. _____
8. _____

Male Sexual Behavior and the Male Sex Act

Match these terms with the correct statement or definition:

- Acetylcholine and nitric oxide (NO)
- Ejaculation
- Emission
- Erectile dysfunction
- Erection
- Orgasm
- Resolution

1. Pleasurable climax sensation associated with ejaculation.
2. Period after ejaculation when the penis becomes flaccid and the male is unable to achieve erection and a second ejaculation.
3. Inability to accomplish the male sex act because of psychic or physical factors.
4. Occurs when parasympathetic or sympathetic impulses cause the dilation of the arteries that supply blood to the erectile tissues in the penis.
5. Neurotransmitters that cause the dilation of the arteries that supply blood to the erectile tissues of the penis.
6. Stimulated by sympathetic centers in the spinal cord as the level of sexual tension increases; secretions of the prostate and seminal vesicles are released.
7. Rhythmic contractions that force semen out of the urethra; triggered by somatic motor action potentials.

Ovaries

A. Match these terms with the correct statement or definition:

Cortex
Medulla
Mesovarium
Oocyte
Ovarian follicles
Ovarian (germinal) epithelium
Ovarian ligament
Suspensory ligament
Tunica albuginea

_____ 1. Peritoneal fold that attaches the ovaries to the broad ligament.

_____ 2. Ligament that extends from the mesovarium to the body wall.

_____ 3. Ligament that attaches the ovary to the uterus.

_____ 4. Peritoneum covering the surface of the ovary.

_____ 5. Layer of dense fibrous connective tissue; surrounds the ovary.

_____ 6. Looser, inner portion of the ovary.

_____ 7. Small vesicles, each of which contains an oocyte, distributed throughout the cortex of the ovary.

_____ 8. Egg cell.

B. Match these terms with the correct statement or definition:

Oogenesis
Oogonium
Polar body
Primary oocyte
Secondary oocyte

_____ 1. Production of a secondary oocyte within the ovaries.

_____ 2. Cell from which an oocyte develops.

_____ 3. Oogonium that has started meiosis but has stopped at prophase I.

_____ 4. Two structures produced from a primary oocyte by the first meiotic division.

C. Match these terms with the correct statement or definition:

Antrum
Corona radiata
Cumulus mass
Mature (graafian) follicle
Polar body
Primary follicle
Primordial follicle
Secondary follicle
Theca
Vesicles
Zona pellucida

_____ 1. Primary oocyte with a surrounding layer of flat granulosa cells.

_____ 2. Layer of clear material that is deposited around a primary oocyte.

_____ 3. Primary oocyte surrounded by one or more layers of cuboidal granulosa cells and the zona pellucida.

_____ 4. Follicle that is not yet mature but contains several fluid-filled spaces and is surrounded by the theca.

_____ 5. Two layers of cells (interna and externa) molded around the secondary follicle or graafian follicle to form a capsule.

_____ 6. Completely developed follicle with an antrum.

_____ 7. Mass of follicular cells that surrounds the oocyte in a secondary follicle.

_____ 8. Irregular, fluid-filled spaces among the granulosa cells.

_____ 9. Fluid-filled chamber formed from the fusion of vesicles; indicates a mature follicle.

_____ 10. Innermost cells of the cumulus mass.

_____ 11. Structure with little cytoplasm produced when the first meiotic division is completed.

D. Using the terms provided, complete these statements:

Corpus albicans
Corpus luteum
Corpus luteum of pregnancy
Fertilization
Luteal cells
Ovulation
Polar body
Progesterone and estrogen
Secondary oocyte
Zygote

The release of the secondary oocyte from the follicle is called _(1)_. During ovulation, the development of the _(2)_ has stopped at metaphase II. Continuation of the second meiotic division is triggered by _(3)_, the entry of the sperm cell into the secondary oocyte. If fertilization does not occur, meiosis is not completed; if fertilization occurs, meiosis II is completed and a second _(4)_ is formed. The fertilized oocyte is now called a _(5)_. After ovulation, the follicle is transformed into a glandular structure called the _(6)_. The granulosa cells and the theca interna, now called _(7)_, enlarge and begin to secrete _(8)_. If pregnancy occurs, the corpus luteum enlarges and remains throughout pregnancy as the _(9)_. If pregnancy does not occur, the connective tissue cells in the corpus luteum become enlarged and clear and give the whole structure a whitish color; therefore, it is called the _(10)_.

1. _____
2. _____
3. _____
4. _____
5. _____
6. _____
7. _____
8. _____
9. _____
10. _____

Uterine Tubes

Match these terms with the correct statement or definition:

Ampulla Mucosa
Fimbriae Muscular layer
Infundibulum Serosa
Isthmus Uterine (intramural) part
Mesosalpinx

_____ 1. Portion of the broad ligament most directly associated with the uterine tube.

_____ 2. Funnel-shaped end of the uterine tube.

_____ 3. Long, thin processes that surround the opening into the uterine tube.

_____ 4. Portion of the uterine tube closest to the infundibulum; the widest and longest part of the tube.

_____ 5. Narrow, thick-walled portion of the uterine tube.

_____ 6. Part of the tube that passes through the uterine wall.

_____ 7. Middle layer of the uterine tube wall.

_____ 8. Inner layer of the uterine tube wall.

Uterus

A. Match these terms with the correct statement or definition:

Body Isthmus
Broad ligament Ostium
Cervical canal Round ligaments
Cervix Uterine cavity
Fundus Uterosacral ligaments

_____ 1. Larger, rounded portion of the uterus, directed superiorly.

_____ 2. Narrower portion of the uterus, directed inferiorly.

_____ 3. Part of the uterus between the fundus and cervix.

_____ 4. Slight constriction that marks the junction between the body of the uterus and the cervix.

_____ 5. Cavity inside the cervix of the uterus.

_____ 6. Opening of the cervix into the vagina.

_____ 7. Major ligament that spreads on both sides of the uterus and to which the ovaries and uterine tubes are attached.

_____ 8. Major ligaments that extend from the uterus through the inguinal canals to the external genitalia.

B. Match these terms with the correct statement or definition:

Basal layer
Cervical mucous glands
Endometrium (mucous membrane)
Functional layer
Myometrium (muscular layer)
Perimetrium (serous layer)

_____ 1. Outer coat of the uterus; peritoneum.

_____ 2. Thickest layer of the uterine wall; thickest layer of smooth muscle in the body.

_____ 3. Innermost layer of the uterus.

_____ 4. Thin, deep layer of the endometrium that is continuous with the myometrium.

_____ 5. Thicker, superficial layer of the endometrium; undergoes the greatest change during the menstrual cycle.

_____ 6. Structures that produce mucus, which fills the cervical canal.

C. Match these terms with the correct parts of the diagram labeled in figure 28.2:

Body of uterus
Cervix
Endometrium
Fundus of uterus
Myometrium
Ovarian ligament
Ovary
Perimetrium
Round ligament
Uterine tube
Vagina

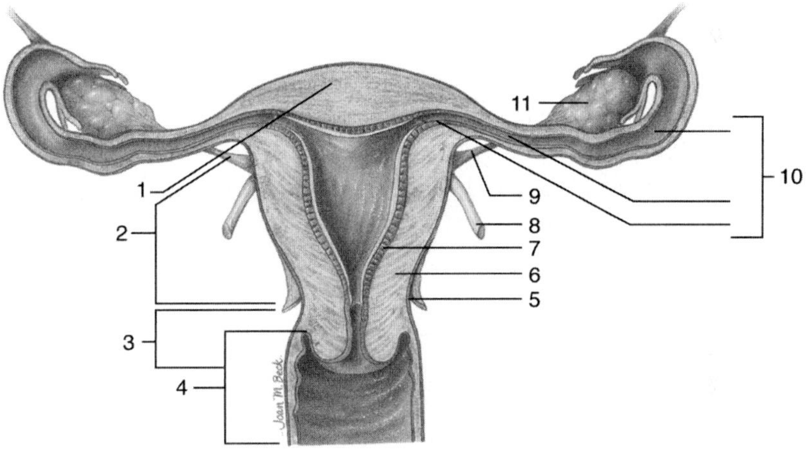

Figure 28.2

1. _____   5. _____   9. _____

2. _____   6. _____   10. _____

3. _____   7. _____   11. _____

4. _____   8. _____

Chapter 28   443

Vagina

Match these terms with the correct statement or definition:

Columns Hymen
Fornix Rugae

_____ 1. Longitudinal ridges that extend the length of the vaginal walls.

_____ 2. Transverse ridges in the vagina.

_____ 3. Superior, domed portion of the vagina.

_____ 4. Thin mucous membrane that may cover the vaginal opening (orifice).

External Genitalia

A. Match these terms with the correct statement or definition:

Bulb of the vestibule Lesser vestibular glands
Clitoris Mons pubis
Crus of the clitoris Prepuce
Greater vestibular glands Pudendal cleft
Labia majora Vestibule
Labia minora Vulva (pudendum)

_____ 1. External female genitalia.

_____ 2. Space into which the vagina and urethra open.

_____ 3. Thin, longitudinal skin folds bordering the vestibule.

_____ 4. Erectile structure that contains two corpora cavernosa; located in the anterior margin of the vestibule.

_____ 5. Expansions at the base end of the corpora cavernosa; attach the clitoris to the coxal bones.

_____ 6. Fold of skin over the clitoris formed by the two labia minora.

_____ 7. Erectile tissue; corresponds to corpus spongiosum in the male.

_____ 8. Ducts of these glands open on each side of the vestibule between the vaginal opening and the labia minora.

_____ 9. Small mucous glands near the clitoris and urethral opening.

_____ 10. Round folds of skin lateral to the labia minora.

_____ 11. Mound located over the symphysis pubis.

_____ 12. Space between the labia majora.

B. Match these terms with the correct parts of the diagram labeled in figure 28.3:

Clitoris
Labia majora
Labia minora
Mons pubis
Prepuce
Urethra
Vagina
Vestibule

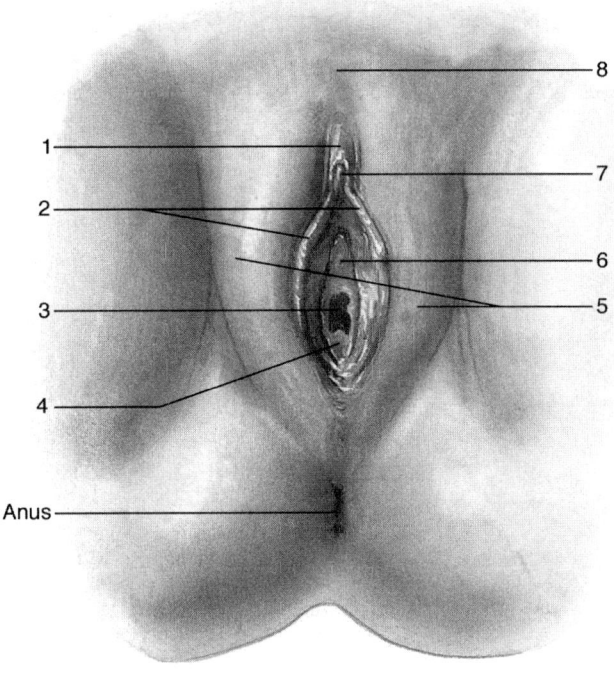

Figure 28.3

1. _____   4. _____   7. _____

2. _____   5. _____   8. _____

3. _____   6. _____

Perineum

Match these terms with the correct statement or definition:

Anal triangle   Episiotomy
Clinical perineum   Urogenital triangle

_____   1. Anterior portion of the perineum; contains the external genitalia.

_____   2. Posterior portion of the perineum; contains the anal opening.

_____   3. Region between the vagina and the anus.

_____   4. Incision in the clinical perineum to prevent tearing during childbirth.

Mammary Glands

A. Match these terms with the correct statement or definition:

Alveoli   Lobes
Areola   Lobules
Areolar glands   Mammary (Cooper's) ligaments
Gynecomastia   Mammary glands
Lactiferous duct   Nipple

_____ 1. Organs of milk production; located within the mammae.

_____ 2. Circular, pigmented area surrounding the nipple.

_____ 3. Rudimentary mammary glands located just below the surface of the areola.

_____ 4. Condition in which male breasts become enlarged.

_____ 5. Compartments of the mammary gland, each of which possesses a single lactiferous duct.

_____ 6. Smaller subcompartments of a lobe; their ducts join a lactiferous duct.

_____ 7. Duct opening to the nipple that carries milk from one lobe; contains a sinus that accumulates milk.

_____ 8. Secretory sacs in the milk-producing mammary gland.

_____ 9. Ligaments that support and hold the breast in place.

B. Match these terms with the correct parts of the diagram labeled in figure 28.4:

Areola   Lobe
Lactiferous duct   Lobule
Lactiferous sinus   Nipple

1. _____
2. _____
3. _____
4. _____
5. _____
6. _____

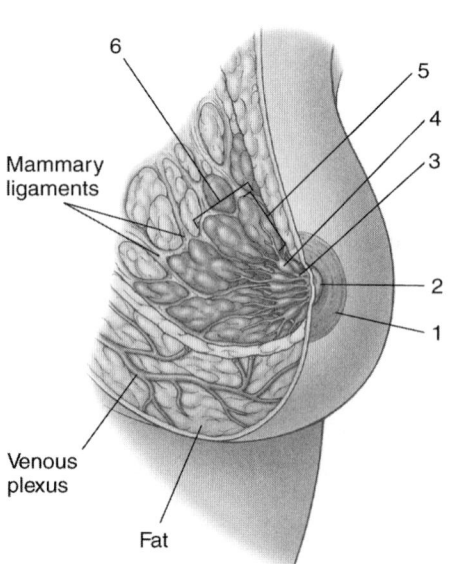

Figure 28.4

446 Chapter 28

Puberty

Using the terms provided, complete these statements:

    Estrogen and progesterone   GnRH
    FSH and LH

Changes associated with puberty result primarily from elevated levels of _(1)_ secreted by the ovaries. These hormones are secreted in response to an increasing and cyclic pattern of _(2)_ secretion by the anterior pituitary. These hormones, in turn, are secreted in response to an increasing _(3)_ secretion by the hypothalamus. The cyclic surge of _(4)_ results in ovulation, and the monthly changes in the secretion of _(5)_ produce the changes in the uterus that characterize the menstrual cycle.

1. _____
2. _____
3. _____
4. _____
5. _____

Menstrual Cycle

Match these terms with the correct statement or definition:

    Follicular (proliferative) phase   Luteal (secretory) phase
    Menses

_____ 1. Period of mild hemorrhage during which the uterine epithelium is sloughed and expelled from the uterus.

_____ 2. Time between the ending of menses and ovulation.

_____ 3. Time between ovulation and the beginning of menses.

Ovarian Cycle

A. Using the terms provided, complete these statements:

    FSH   LH
    GnRH   Ovarian cycle

The _(1)_ is the series of events that occurs in a regular fashion in the ovaries of sexually mature, nonpregnant women. Early in the menstrual cycle, the release of _(2)_ from the hypothalamus stimulates the release of FSH and LH. A number of follicles begin to mature during each menstrual cycle, under the influence of _(3)_, but normally only one is ovulated.

1. _____
2. _____
3. _____

B. Using the terms provided, complete these statements:

Androgens
Decrease(s)
Estrogen
FSH surge
Granulosa
HCG
Increase(s)
LH surge
Negative-feedback
Ovulation
Positive-feedback
Progesterone
Theca interna

Before ovulation, LH stimulates the theca interna cells of the developing follicle to produce (1) , which diffuse to granulosa cells. FSH stimulates the granulosa cells to convert androgens to (2) . In addition, FSH gradually increases LH receptors in the (3) cells, and estrogen produced by the granulosa cells increases LH receptors in the (4) cells. LH stimulates the granulosa cells to produce some (5) , which diffuses to the theca interna cells, where it is converted to (6) . Thus, androgen production by the theca interna cells increases, and the conversion of androgens to estrogen by the granulosa cells increases. As estrogen levels begin to increase in the follicular phase, they have a (7) effect on FSH and LH secretion. Late in the follicular phase, a sustained increase in estrogen begins to have a (8) effect, and LH and FSH secretion increase rapidly. The increase in blood level of LH is called the (9) , and the increase in the blood level of FSH is called the (10) . The LH surge initiates (11) and causes the ovulated follicle to become the corpus luteum. Shortly after ovulation, the production of estrogen by the follicle (12) , and the production of progesterone (13) as granulosa cells are converted to corpus luteum cells. The increased progesterone and estrogen have a (14) effect on GnRH release from the hypothalamus. As a result, LH and FSH release from the anterior pituitary (15) , and, without a fertilized oocyte, the cells of the corpus luteum begin to atrophy, blood levels of estrogen and progesterone (16) rapidly, and menses occurs. If fertilization does occur, the developing embryonic mass begins to secrete (17) , which keeps the corpus luteum from degenerating, estrogen and progesterone do not (18) , and menses does not occur.

1. _____
2. _____
3. _____
4. _____
5. _____
6. _____
7. _____
8. _____
9. _____
10. _____
11. _____
12. _____
13. _____
14. _____
15. _____
16. _____
17. _____
18. _____

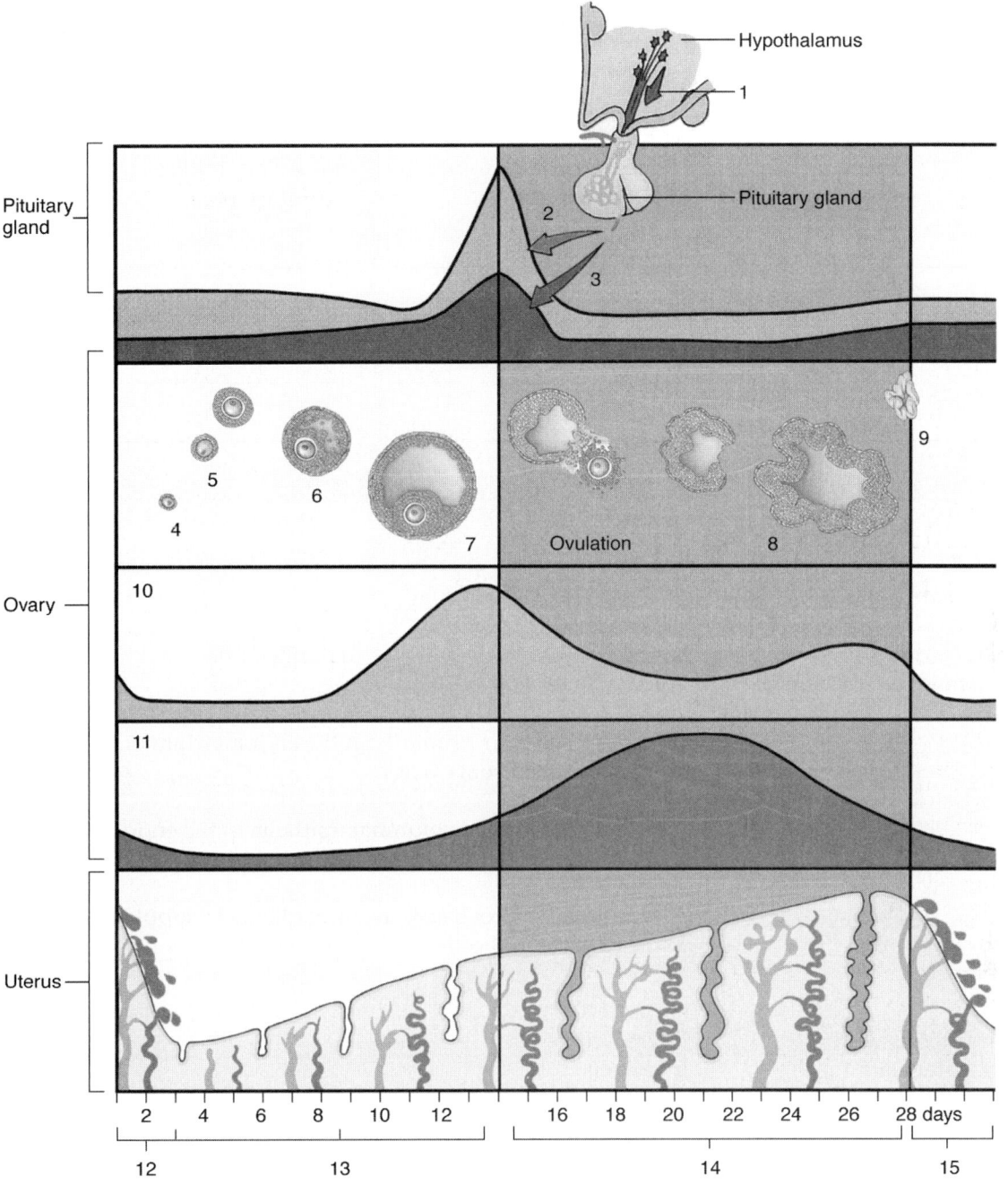

Figure 28.5

C. Match these terms with the correct parts of the diagram labeled in figure 28.5:

Corpus albicans
Corpus luteum
Estrogen level
FSH
GnRH
Mature (graafian) follicle
LH
Menses
Primary follicle
Primordial follicle
Progesterone level
Proliferative phase
Secondary follicle
Secretory phase

1. _____
2. _____
3. _____
4. _____
5. _____
6. _____
7. _____
8. _____
9. _____
10. _____
11. _____
12. _____
13. _____
14. _____
15. _____

Uterine Cycle

A. Match these terms with the correct statement or definition:

Spiral arteries
Spiral glands
Uterine cycle

_____
_____
_____

1. Changes that occur primarily in the endometrium of the uterus during the menstrual cycle.
2. Glands formed when columnar epithelia in the endometrium are thrown into folds; secrete a glycogen-rich fluid.
3. Arteries that project between spiral glands to supply nutrients to the endometrial cells.

B. Match these terms with the correct statement:

Decrease(s)
Increase(s)

_____
_____
_____
_____
_____
_____

1. Effect of estrogen on the thickness of the endometrium; affects cell division.
2. Effect of estrogen on the sensitivity of the endometrial cells to progesterone.
3. Effect of progesterone on the thickness of the endometrium and the myometrium; affects cell size.
4. Effect of progesterone on the secretory ability of endometrial cells.
5. Effect of progesterone on smooth muscle contraction.
6. Effect of declining progesterone levels on blood supply to the endometrium.

Female Sexual Behavior and the Female Sex Act

Using the terms provided, complete these statements:

Clitoris
Fertilization
Orgasm
Psychologic factors
Resolution
Sacral
Vagina and vestibular glands

Androgens and possibly estrogens affect cells in the brain and influence sexual behavior; however, (1) also play a role in sexual behavior. The (2) region of the spinal cord is the area that integrates sexual reflexes, which are modulated by cerebral influences. During sexual excitement, parasympathetic stimulation causes erectile tissue in the (3) and around the vaginal opening to become engorged with blood. Secretions from the (4) provide lubrication for the movement of the penis. Tactile stimulation during intercourse, as well as psychologic stimuli, normally trigger an (5), the female climax. After the sexual act, there is a period of (6), characterized by an overall sense of satisfaction and relaxation. Although orgasm is a pleasurable component of sexual intercourse, it is not required for (7) to occur.

1. _____
2. _____
3. _____
4. _____
5. _____
6. _____
7. _____

Female Fertility and Pregnancy

Match these terms with the correct statement or definition:

Capacitation
HCG
Oxytocin and prostaglandins
Placenta
Swimming
Trophoblast

_____ 1. One of the forces that propel the sperm cells.

_____ 2. Hormones that stimulate smooth muscle contraction, moving the sperm cells toward the ampulla.

_____ 3. Enables the sperm cells to release acrosomal enzymes that allow penetration of the oocyte.

_____ 4. Outer layer of the developing embryonic mass; responsible for implantation.

_____ 5. Hormone, secreted by the trophoblast, that causes the corpus luteum to remain functional.

_____ 6. Organ responsible for secreting estrogen and progesterone during most of pregnancy.

Menopause

Using the terms provided, complete these statements:

Estrogen and progesterone
Female climacteric (perimenopause)
Hot flashes
LH and FSH
Menopause
Ovary

The whole time period from the onset of irregular menstrual cycles to their complete cessation is called the __(1)__. The major cause of menopause is age-related changes in the __(2)__. Follicles become less sensitive to stimulation by __(3)__, and fewer mature follicles and corpora lutea are produced. Gradual morphological changes occur in the female in response to reduced amounts of __(4)__ produced by the ovaries. Symptoms include __(5)__, irritability, fatigue, anxiety, and occasionally severe emotional disturbances.

1. _____
2. _____
3. _____
4. _____
5. _____

Effects of Aging on the Reproductive System

Match these terms with the correct statement or definition:

Decrease
Increase
No effect

_____ 1. Effect of aging on blood flow to the testes and sperm cell production in males.

_____ 2. Effect of aging on the incidence of benign prostatic hypertrophy and the occurrence of prostate cancer in males.

_____ 3. Effect of aging on the occurrence of erectile dysfunction in males.

_____ 4. Effect of aging on the production of estrogen and progesterone by the ovaries in females.

_____ 5. Effect of aging on the size of the uterus in females.

_____ 6. Effect of aging on the incidence of breast cancer in females.

_____ 7. Effect of aging on the occurrence of endometrial, cervical, and ovarian cancer in females.

Quick Recall

1. List the stages of development of sperm cells.

2. List these structures in the correct order from the site of sperm cell production to the exterior of the body: ductus deferens, efferent ductules, ejaculatory duct, epididymis, membranous urethra, prostatic urethra, rete testis, seminiferous tubules, spongy urethra.

3. List three glands involved in reproduction in the male, and describe the secretions they produce.

4. For the male, list the hormones involved with the reproductive system that are secreted by the hypothalamus, the anterior pituitary, and the testes.

5. List six effects that testosterone has in the male.

6. Arrange these stages in the development of follicle cells in the correct order: corpus albicans, corpus luteum, primary follicle, primordial follicle, secondary follicle, mature (graafian) follicle.

7. List the stages of development of the oocyte, and state at what stage fertilization normally takes place.

8. Starting with the site of milk production, name the structures a drop of milk passes through on the way to the outside of the woman's body.

9. Name the three phases of the menstrual cycle.

10. List the effect these hormones have on the ovarian cycle: GnRH, FSH, and LH.

11. List the effects of estrogen and progesterone on the uterine cycle.

ANSWERS TO CHAPTER 28

Content Learning Activity

Scrotum
1. Raphe; 2. Dartos muscle; 3. Cremaster muscles

Perineum
1. Urogenital triangle; 2. Anal triangle

Testes
A. 1. Tunica albuginea; 2. Septa; 3. Seminiferous tubules; 4. Interstitial cells (Leydig cells); 5. Tubuli recti; 6. Efferent ductules
B. 1. Gubernaculum; 2. Inguinal canal; 3. Process vaginalis; 4. Tunica vaginalis; 5. Deep inguinal ring; 6. Superficial inguinal ring; 7. Cryptorchidism
C. 1. Spermatogenesis; 2. Sustentacular (Sertoli) cells; 3. Germ cells; 4. Dihydrotestosterone and estrogen; 5. Androgen-binding protein
D. 1. Spermatogonia; 2. Primary spermatocytes; 3. Secondary spermatocytes; 4. Spermatids; 5. Sperm cells (spermatozoa); 6. Spermiogenesis; 7. Acrosome

Ducts
1. Efferent ductules; 2. Epididymis; 3. Duct of the epididymis; 4. Ductus (vas) deferens; 5. Spermatic cord; 6. Ejaculatory duct; 7. Prostatic urethra; 8. Membranous urethra; 9. Urethral glands

Penis
1. Erection; 2. Corpora cavernosa; 3. Corpus spongiosum; 4. Glans penis; 5. Bulb of the penis; 6. Crus of the penis; 7. Root of the penis; 8. Prepuce (foreskin)

Accessory Glands
A. 1. Seminal vesicles; 2. Prostate gland; 3. Bulbourethral glands; 4. Emission; 5. Ejaculation; 6. Bulbourethral glands; 7. Seminal vesicles; 8. Prostate gland; 9. Prostate gland; 10. Prostate gland
B. 1. Prostatic urethra; 2. Membranous urethra; 3. Spongy urethra; 4. Penis; 5. External urethral orifice; 6. Scrotum; 7. Testis; 8. Epididymis; 9. Ductus deferens; 10. Bulbourethral gland; 11. Prostate gland; 12. Ejaculatory duct; 13. Seminal vesicle

Regulation of Sex Hormone Secretion
1. Gonadotropin-releasing hormone (GnRH); 2. Gonadotropins; 3. Luteinizing hormone (LH); 4. Follicle-stimulating hormone (FSH); 5. Testosterone; 6. Androgen; 7. Inhibin

Puberty
1. HCG; 2. Androgens; 3. GnRH; 4. LH and FSH; 5. Sperm cell; 6. Interstitial cells

Effects of Testosterone
1. Testes; 2. Coarser; 3. Melanin; 4. Sebaceous glands; 5. Hypertrophy; 6. Body fluids; 7. Protein synthesis; 8. Rapid bone growth

Male Sexual Behavior and the Male Sex Act
1. Orgasm; 2. Resolution; 3. Erectile dysfunction; 4. Erection; 5. Acetylcholine and nitric oxide (NO); 6. Emission; 7. Ejaculation

Ovaries
A. 1. Mesovarium; 2. Suspensory ligament; 3. Ovarian ligament; 4. Ovarian (germinal) epithelium; 5. Tunica albuginea; 6. Medulla; 7. Ovarian follicles; 8. Oocyte
B. 1. Oogenesis; 2. Oogonium; 3. Primary oocyte; 4. Secondary oocyte, polar body
C. 1. Primordial follicle; 2. Zona pellucida; 3. Primary follicle; 4. Secondary follicle; 5. Theca; 6. Mature (graafian) follicle; 7. Cumulus mass; 8. Vesicles; 9. Antrum; 10. Corona radiata; 11. Polar body
D. 1. Ovulation; 2. Secondary oocyte; 3. Fertilization; 4. Polar body; 5. Zygote; 6. Corpus luteum; 7. Luteal cells; 8. Progesterone and estrogen; 9. Corpus luteum of pregnancy; 10. Corpus albicans

Uterine Tubes
1. Mesosalpinx; 2. Infundibulum; 3. Fimbriae; 4. Ampulla; 5. Isthmus; 6. Muscular layer; 7. Muscular layer; 8. Mucosa

Uterus
A. 1. Fundus; 2. Cervix; 3. Body; 4. Isthmus; 5. Cervical canal; 6. Ostium; 7. Broad ligament; 8. Round ligaments
B. 1. Perimetrium (serous layer); 2. Myometrium (muscular layer); 3. Endometrium (mucous membrane); 4. Basal layer; 5. Functional layer; 6. Cervical mucous glands
C. 1. Fundus of uterus; 2. Body of uterus; 3. Cervix; 4. Vagina; 5. Perimetrium; 6. Myometrium; 7. Endometrium; 8. Round ligament; 9. Ovarian ligament; 10. Uterine tube; 11. Ovary

Vagina
1. Columns; 2. Rugae; 3. Fornix; 4. Hymen

External Genitalia
A. 1. Vulva (pudendum); 2. Vestibule; 3. Labia minora; 4. Clitoris; 5. Crus of the clitoris; 6. Prepuce; 7. Bulb of the vestibule; 8. Greater vestibular glands; 9. Lesser vestibular glands; 10. Labia majora; 11. Mons pubis; 12. Pudendal cleft
B. 1. Prepuce; 2. Labia minora; 3. Vagina; 4. Vestibule; 5. Labia majora; 6. Urethra; 7. Clitoris; 8. Mons pubis

Perineum
1. Urogenital triangle; 2. Anal triangle; 3. Clinical perineum; 4. Episiotomy

Mammary Glands
A. 1. Mammary gland; 2. Areola; 3. Areolar glands; 4. Gynecomastia; 5. Lobes; 6. Lobules; 7. Lactiferous duct; 8. Alveoli; 9. Mammary (Cooper's) ligaments
B. 1. Areola; 2. Nipple; 3. Lactiferous duct; 4. Lactiferous sinus; 5. Lobule; 6. Lobe

Puberty
1. Estrogen and progesterone; 2. FSH and LH; 3. GnRH; 4. FSH and LH; 5. Estrogen and progesterone

Menstrual Cycle
1. Menses; 2. Follicular (proliferative) stage; 3. Luteal (secretory) stage

Ovarian Cycle
A. 1. Ovarian cycle; 2. GnRH; 3. FSH
B. 1. Androgens; 2. Estrogen; 3. Granulosa; 4. Theca interna; 5. Progesterone; 6. Androgens; 7. Negative-feedback; 8. Positive-feedback; 9. LH surge; 10. FSH surge; 11. Ovulation; 12. Decreases; 13. Increases; 14. Negative-feedback; 15. Decreases; 16. Decrease; 17. HCG; 18. Decrease
C. 1. GnRH; 2. LH; 3. FSH; 4. Primordial follicle; 5. Primary follicle; 6. Secondary follicle; 7. Mature (graafian) follicle; 8. Corpus luteum; 9. Corpus albicans; 10. Estrogen level; 11. Progesterone level; 12. Menses; 13. Proliferative phase; 14. Secretory phase; 15. Menses

Uterine Cycle
A. 1. Uterine cycle; 2. Spiral glands; 3. Spiral arteries
B. 1. Increases; 2. Increases; 3. Increases; 4. Increases; 5. Decreases; 6. Decreases

Female Sexual Behavior and the Female Sex Act
1. Psychologic factors; 2. Sacral; 3. Clitoris; 4. Vagina and vestibular glands; 5. Orgasm; 6. Resolution; 7. Fertilization

Female Fertility and Pregnancy
1. Swimming; 2. Oxytocin and prostaglandins; 3. Capacitation; 4. Trophoblast; 5. HCG; 6. Placenta

Menopause
1. Female climacteric (perimenopause); 2. Ovary; 3. LH and FSH; 4. Estrogen and progesterone; 5. Hot flashes

Effects of Aging on the Reproductive System
1. Decrease; 2. Increase; 3. Increase; 4. Decrease; 5. Decrease; 6. Increase; 7. Increase

QUICK RECALL

1. Spermatogonia, primary spermatocyte, secondary spermatocyte, spermatid, sperm cell (spermatozoon)
2. Seminiferous tubules, rete testis, efferent ductules, epididymis, ductus deferens, ejaculatory duct, prostatic urethra, membranous urethra, spongy urethra
3. Seminal vesicles: thick, mucoid secretions containing nutrients, fibrinogen, and prostaglandins; prostate gland: thin, milky, alkaline secretions containing clotting factors; bulbourethral glands: alkaline mucous secretions
4. Hypothalamus: GnRH; anterior pituitary: FSH and LH; testes: testosterone and inhibin
5. Enlargement and differentiation of the reproductive system, descent of testes, spermatogenesis, hair growth, increased skin pigmentation, increased sebaceous secretions, increased muscle mass, increased body fluids, increased skeletal growth, laryngeal hypertrophy, increased metabolism, and increased red blood cell count
6. Primordial follicle, primary follicle, secondary follicle, vesicular (graafian) follicle, corpus luteum, corpus albicans
7. Oogonia, primary oocyte, secondary oocyte; fertilization occurs in the secondary oocyte
8. Alveoli, lobule, lobe, lactiferous sinus, lactiferous duct, nipple
9. Menses, proliferative (follicular) phase, secretory (luteal) phase
10. GnRH: stimulates release of FSH and LH from the anterior pituitary; FSH: stimulates follicle development; LH: triggers ovulation, development of corpus luteum
11. Estrogen: thickening of endometrium; progesterone: thickening, glandularity, and vascularity of endometrium stimulated, decreased smooth muscle contraction

29 Development, Growth, and Aging

CONTENT LEARNING ACTIVITY

Prenatal Development

Match these terms with the correct statement or definition:

Clinical age
Embryonic period
Fetal period
Germinal period
Postovulatory age
Prenatal period

_____ 1. Period from conception to birth.

_____ 2. Period of development during which primitive germ layers are formed; the first 2 weeks of development.

_____ 3. Period of development during which the major organ systems are formed; the second to the end of the eighth week of development.

_____ 4. Period of development in which the organ systems grow and become more mature; the last 30 weeks of development.

_____ 5. Calculation of developmental age that uses the last menstrual period (LMP) as the starting point.

_____ 6. Calculation of developmental age that is 14 days less than the clinical age; used by developmental biologists.

Fertilization

A. Match these terms with the correct statement or definition:

Female pronucleus
Male pronucleus
Second meiotic division
Zygote

_____ 1. Division triggered by entrance of the sperm cell into the oocyte; a second polar body is formed.

_____ 2. Haploid nucleus of the oocyte, after the second meiotic division.

_____ 3. Haploid nucleus of the sperm cell.

_____ 4. Result of the process of fertilization.

Chapter 29

B. Match these terms with the correct statement or definition:

Acrosomal reaction Perivitelline space
Corona radiata Slow block to polyspermy
Fast block to polyspermy Zona pellucida

_____ 1. Barrier of cells surrounding the oocyte.

_____ 2. Extracellular membrane between the corona radiata and the oocyte; contains a species-specific receptor (ZP3) for sperm cells.

_____ 3. The activation of digestive enzymes in a sperm cell when it attaches to the zona pellucida.

_____ 4. Depolarization of the oocyte plasma membrane when a sperm cell attaches.

_____ 5. Release of fluid from the oocyte and denaturation of the zona pellucida.

_____ 6. Fluid-filled space between the oocyte and the zona pellucida.

Early Cell Division: Morula and Blastocyst

Match these terms with the correct statement or definition:

Blastocele Pluripotent
Blastocyst Totipotent
Inner cell mass Trophoblast
Morula

_____ 1. Ability of embryonic cells to give rise to any tissue type necessary for development.

_____ 2. Ability of embryonic cells to develop into a wide variety of tissues.

_____ 3. Embryo once it has 12 or more cells.

_____ 4. Embryo once a cavity begins to form inside.

_____ 5. Cavity within the blastocyst.

_____ 6. Single layer of cells that surrounds most of the blastocele and develops into the placenta.

_____ 7. Thickened area, several cell layers thick, at one end of the blastocyst, that will develop into the embryo.

Implantation of the Blastocyst and Development of the Placenta

Match these terms with the correct statement or definition:

Cytotrophoblast
Implantation
Lacunae
Placenta
Syncytiotrophoblast

_____ 1. Burrowing of the blastocyst into the uterine wall.

_____ 2. Organ of nutrient and waste product exchange between the fetus and the mother.

_____ 3. Nondividing, multinucleate cell that invades maternal tissues; does not trigger an immune reaction.

_____ 4. Cavities produced when the syncytiotrophoblast surrounds and digests away the walls of maternal blood vessels.

_____ 5. Dividing population of trophoblast cells that protrudes into the lacunae (chorionic villi) and is followed by embryonic blood vessels; later disappears, leaving only a thin layer separating the maternal and fetal blood supplies.

Formation of Germ Layers

Match these terms with the correct statement or definition:

Amniotic cavity
Ectoderm
Embryo
Embryonic disk
Endoderm
Gastrulation
Mesoderm
Notochord
Primitive streak
Yolk sac

_____ 1. New cavity formed inside the inner cell mass after implantation occurs; lined by the amniotic sac.

_____ 2. Flat disk of tissue composed of two cell layers.

_____ 3. Cell layer of the embryonic disk that is adjacent to the amniotic cavity.

_____ 4. Third cavity produced by the endoderm that forms inside the blastocele.

_____ 5. The phase of development in which cell movement results in the formation of three distinct germ layers.

_____ 6. Thickened line, formed by cells of the ectoderm migrating to the center and caudal end of the embryonic disk.

_____ 7. Third germ layer, formed from ectodermal cells that migrate through the primitive streak and emerge between the ectoderm and endoderm.

Chapter 29

Neural Tube and Neural Crest Formation

Match these terms with the correct statement or definition:

Mesenchyme
Neural crest cells
Neural folds
Neural plate
Neural tube
Neuroectoderm

_____ 1. Thickened layer of cells produced when the notochord stimulates overlying ectoderm cells.

_____ 2. Lateral edges of the neural plate that rise and move toward each other.

_____ 3. Structure formed when the neural folds meet at the midline and fuse; consists of cells called neuroectoderm, which become the brain and spinal cord.

_____ 4. Cells that become part of the peripheral nervous system (autonomic ganglia neurons or enteric nervous system neurons), adrenal medulla, melanocytes, or varied structures in the head.

_____ 5. Cells derived from neural crest cells or mesoderm.

Formation of Somites, the Gut, and Body Cavities

Match these terms with the correct statement or definition:

Branchial arches
Celom
Cloacal membrane
Evaginations
Oropharyngeal membrane
Pharyngeal pouches
Somites
Somitomeres

_____ 1. Distinct segments formed from mesoderm adjacent to the neural tube.

_____ 2. First few somites in the head region that never become clearly divided.

_____ 3. Membrane formed in the hindgut, where it is in close contact with ectoderm; this later becomes the urethra and the anus.

_____ 4. Outpocketings from the GI tract; these develop along the head.

_____ 5. Solid bars of tissue that develop along the head.

_____ 6. Pockets of the pharynx that extend between the branchial arches; give rise to the tonsils, thymus, and parathyroid glands.

_____ 7. Collective name for the body cavities, subdivided to form the pericardial, pleural, and peritoneal cavities.

Limb Bud and Face Development

Match these terms with the correct statement or definition:

Apical ectodermal ridge
Frontonasal process
Mandibular processes
Maxillary processes
Nasal placodes
Primary palate
Secondary palate

_____ 1. Specialized thickening of the ectoderm that develops on the lateral margin of each limb bud.

_____ 2. Process that forms the forehead, nose, and midportion of the upper jaw and lip.

_____ 3. Processes that form the lateral portion of the upper jaw and lip.

_____ 4. Develops at the lateral margin of the frontonasal process and become the nose and center of the upper jaw and lip.

_____ 5. Upper jaw and lip, formed by fusion of the nasal placodes and the maxillary processes.

_____ 6. Roof of the mouth; failure to fuse is a cleft palate.

Development of the Organ Systems

A. Match these terms with the correct structure that develops from them:

Mesoderm
Neural crest cells
Neural tube
Neural tube cavity

_____ 1. Dermis of the skin (excluding the face).

_____ 2. Melanocytes and sensory receptors in the skin.

_____ 3. Appendicular skeleton.

_____ 4. Ventricles of the brain and the central canal of the spinal cord.

B. Match these terms with the correct statement or definition:

Anterior pituitary
Myoblasts
Olfactory bulb and nerve
Optic stalk and optic vesicle
Posterior pituitary

_____ 1. Multinucleated cells that develop into skeletal muscle fibers.

_____ 2. Develops as an evagination from the telencephalon.

_____ 3. Develops as an evagination from the diencephalon.

_____ 4. Develops as an evagination of the floor of the diencephalon.

C. Match these terms with the correct statement or definition:

Adrenal cortex
Adrenal medulla
Pancreas
Parathyroid glands
Thyroid gland

_____ 1. Originates as an evagination from the floor of the pharynx in the region of the developing tongue.

_____ 2. Gland part that arises from neural crest cells and consists of specialized postganglionic neurons of the sympathetic division.

_____ 3. Gland part that originates from mesoderm.

_____ 4. Originates as two evaginations from the duodenum, which later fuse.

D. Match these terms with the correct statement or definition:

Blood islands
Bulbus cordis
Foramen ovale
Sinus venosus

_____ 1. The blood vessels are formed from these structures, which are located on the surface of the yolk sac and inside the embryo.

_____ 2. Site of blood entering the embryonic heart; part of this structure later becomes the SA node.

_____ 3. Site of blood exiting the embryonic heart; this structure is later incorporated into the ventricles.

_____ 4. Opening in the interatrial septum that allows blood to flow from the right to the left atrium.

E. Match these terms with the correct statement or definition:

Allantois
Cloaca
Mesonephros
Metanephros
Pronephros
Ureter

_____ 1. Consists of a duct and simple tubules connected to the celomic cavity; it is probably never functional in the embryo.

_____ 2. Consists of tubules that open into the mesonephric duct; the other end forms a glomerulus.

_____ 3. Common junction of the digestive, urinary, and genital systems.

_____ 4. Blind tube that extends into the umbilical cord; develops into the urinary bladder.

_____ 5. Distal end of the ureter enlarges and branches to form the duct system of this structure; becomes the adult kidney.

F. Match these terms with the correct statement or definition:

Gonads
Mesonephric ducts
Paramesonephric ducts
Primordial germ cells

_____ 1. Cells that form on the surface of the yolk sac, migrate into the embryo, and enter the gonadal ridge.

_____ 2. Ducts that form just lateral to the mesonephric ducts; if no testosterone or müllerian-inhibiting hormone is present, these ducts become the uterine tubes, the uterus, and part of the vagina.

_____ 3. Under the influence of testosterone, will develop into the epididymis, ductus deferens, seminal vesicles, and prostate gland.

G. Match these terms with the structures that they later become:

Genital tubercle
Labioscrotal swellings
Urogenital folds

_____ 1. Penis.

_____

_____ 2. Scrotum or labia majora.

_____ 3. Clitoris.

_____ 4. Labia minora.

Growth of the Fetus

Using the terms provided, complete these statements:

Embryonic
Growing
Lanugo
Subcutaneous fat
Vernix caseosa

Most morphological changes occur in the _(1)_ phase of development, whereas the fetal period is primarily a _(2)_ phase. Fine, soft hair called _(3)_ covers the fetus, and a waxy coat of sloughed epithelial cells called _(4)_ protects the fetus from the somewhat toxic nature of the amniotic fluid. _(5)_ accumulates in the late fetus; this provides a nutrient reserve, helps insulate the infant, and assists the infant in sucking by supporting the cheeks.

1. _____
2. _____
3. _____
4. _____
5. _____

Parturition

A. Match these terms with the correct statement or definition:

First stage of labor   Third stage of labor
Second stage of labor

_____ 1. Stage of labor from the onset of regular uterine contractions until maximum cervical dilation occurs.

_____ 2. Stage of labor from maximal cervical dilation until the baby exits the vagina.

_____ 3. Expulsion of the placenta.

B. Match these terms with the correct statement:

Decrease(s)
Increases(s)

_____ 1. Effect of stress on fetal ACTH production.

_____ 2. Effect of ACTH on glucocorticoid production in the fetus.

_____ 3. Effect of fetal glucocorticoids on maternal progesterone production.

_____ 4. Effect of fetal glucocorticoids on maternal estrogen production.

_____ 5. Effect of declining progesterone and increasing estrogen levels on smooth muscle contraction.

_____ 6. Effect of prostaglandins on uterine contractions.

_____ 7. Effect of oxytocin on uterine contractions.

_____ 8. Effect of stretching the uterus on oxytocin production.

_____ 9. Effect of decreased progesterone on oxytocin release.

The Newborn

A. Match these terms with the correct statement or definition:

Ductus arteriosus   Ligamentum teres
Ductus venosus   Ligamentum venosum
Foramen ovale   Medial umbilical ligaments
Fossa ovalis   Umbilical vein
Ligamentum arteriosum

_____ 1. Opening between the right and left atria; allows blood to bypass the lungs.

_____ 2. Closed foramen ovale.

_____ 3. Connection between the pulmonary trunk and the aorta in the fetus; allows blood to bypass the lungs.

_____ 4. Degenerated ductus arteriosus.

_____ 5. Vessel from the umbilical cord that returns blood toward the liver.

_____ 6. Degenerated umbilical vein; also called the round ligament of the liver.

_____ 7. Vessel that receives blood from the umbilical vein; allows blood to bypass the liver.

_____ 8. Degenerated ductus venosus.

_____ 9. Degenerated umbilical arteries.

B. Using the terms provided, complete these statements:

Amylase   Meconium
Bilirubin   Stomach pH
Lactose

Swallowed amniotic fluid, cells sloughed from the mucosal lining, mucus, and bile from the liver pass from the GI tract as a greenish discharge called _(1)_. The pH of the stomach is nearly neutral at birth, but gastric acid secretion occurs, causing _(2)_ to decrease. The newborn digestive system is capable of digesting _(3)_ from the time of birth, but _(4)_ secretion by the salivary glands remains low until after the first year. The neonatal liver also lacks adequate amounts of the enzyme required to produce _(5)_, which may lead to jaundice.

1. _____

2. _____

3. _____

4. _____

5. _____

C. Match these terms with the correct statement or definition:

Congenital disorders
Teratogens

_____ 1. Defects present at birth, regardless of cause.

_____ 2. Agents that cause abnormal fetal development—e.g., alcohol can cause fetal alcohol syndrome.

Chapter 29

Lactation

Match these terms with the correct statement or definition:

Colostrum
Estrogen
Oxytocin
Progesterone
Prolactin
Prolactin-inhibiting factor (PIF)
Prolactin-releasing factor (PRF)

_____ 1. Hormone primarily responsible for breast growth during pregnancy.

_____ 2. Hormone responsible for development of the breast's secretory alveoli.

_____ 3. Hormone responsible for stimulating the mammary glands to produce milk.

_____ 4. Increased release of this substance by the hypothalamus occurs in response to mechanical stimulation of the breast.

_____ 5. Substance produced by the breasts for the first few days after parturition that contains little fat and less lactose than milk.

_____ 6. Hormone that causes cells around the alveoli to contract; responsible for milk letdown.

_____ 7. Substance secreted by the posterior pituitary in response to mechanical stimulation of the breast.

Life Stages

Match these life stages with the correct description:

Adolescent
Adult
Child
Embryo
Fetus
Germinal
Infant
Neonate

_____ 1. Period from fertilization to 14 days.

_____ 2. Period from 14 to 56 days after fertilization.

_____ 3. Period from 56 days after fertilization to birth.

_____ 4. Period from birth to 1 month after birth.

_____ 5. Period from 1 month after birth to 1 or 2 years of age.

_____ 6. Period from 1 or 2 years after birth to puberty.

_____ 7. Period from puberty to 20 years.

Aging

Using the terms provided, complete these statements:

Atherosclerosis
Arteriosclerosis
Autoimmunity
Collagen
Cytologic aging
Decrease
Embolus
Filtration
Genetic
Heart
Increase
Thrombus

With age, more cross-links are formed between __(1)__ molecules, rendering the tissues more rigid. Death of nondividing cell produces irreversible damage; as a result, the number of muscle cells and neurons __(2)__ with age. The __(3)__ loses elastic recoil ability and muscular contractility, causing a decline in cardiac output, which may also cause a decreased blood flow to the kidneys, resulting in a decrease in __(4)__. __(5)__ is the deposit of lipid in the tunica intima of arteries. These deposits become fibrotic and calcified, causing __(6)__, which interferes with normal blood flow and may lead to a __(7)__ (a clot or plaque formed inside a vessel). A piece of plaque, called an __(8)__, can break loose and lodge in smaller arteries, causing myocardial infarction or stroke. __(9)__, or cellular wear and tear, also contributes to aging. __(10)__ (responding to one's own antigens) or losing the ability to respond to a foreign antigen may be part of the aging process. Progeria may indicate there is a __(11)__ component to aging.

1. _____
2. _____
3. _____
4. _____
5. _____
6. _____
7. _____
8. _____
9. _____
10. _____
11. _____

QUICK RECALL

1. List the following structures in the order in which they form during development: blastocyst, embryonic disk, mesoderm, morula, primitive streak, zygote.

2. Name the structures derived from the inner cell mass and the trophoblast.

3. Name an organ system that develops entirely from ectoderm.

4. Name an organ system that develops entirely from mesoderm.

5. List the three kidneys that form during the development of the urinary system.

6. List the positive-feedback mechanism that occurs during parturition.

7. List the effects of estrogen, progesterone, prolactin, and oxytocin on the female breast.

8. List five circulatory changes that occur at birth.

9. List the eight life stages.

10. List four factors that may influence aging.

ANSWERS TO CHAPTER 29

CONTENT LEARNING ACTIVITY

Prenatal Development
1. Prenatal period; 2. Germinal period; 3. Embryonic period; 4. Fetal period; 5. Clinical age; 6. Postovulatory age

Fertilization
A. 1. Second meiotic division; 2. Female pronucleus; 3. Male pronucleus; 4. Zygote

B. 1. Corona radiata; 2. Zona pellucida; 3. Acrosomal reaction; 4. Fast block to polyspermy; 5. Slow block to polyspermy; 6. Perivitelline space

Early Cell Division: Morula and Blastocyst
1. Totipotent; 2. Pluripotent; 3. Morula; 4. Blastocyst; 5. Blastocele; 6. Trophoblast; 7. Inner cell mass

Implantation of the Blastocyst and Development of the Placenta
1. Implantation; 2. Placenta; 3. Syncytiotrophoblast; 4. Lacunae; 5. Cytotrophoblast

Formation of Germ Layers
1. Amniotic cavity; 2. Embryonic disk; 3. Ectoderm; 4. Yolk sac; 5. Gastrulation; 6. Primitive streak; 7. Mesoderm

Neural Tube and Neural Crest Formation
1. Neural plate; 2. Neural folds; 3. Neural tube; 4. Neural crest cells; 5. Mesenchyme

Formation of the Somites, the Gut, and Body Cavities
1. Somites; 2. Somitomeres; 3. Cloacal membrane; 4. Evaginations; 5. Branchial arches; 6. Pharyngeal pouches; 7. Celom

Limb Bud and Face Development
1. Apical ectodermal ridge; 2. Frontonasal process; 3. Maxillary processes; 4. Nasal placodes; 5. Primary palate; 6. Secondary palate

Development of the Organ Systems
A. 1. Mesoderm; 2. Neural crest cells; 3. Mesoderm; 4. Neural tube cavity
B. 1. Myoblasts; 2. Olfactory bulb and nerve; 3. Optic stalk and optic vesicle; 4. Posterior pituitary
C. 1. Thyroid gland; 2. Adrenal medulla; 3. Adrenal cortex; 4. Pancreas
D. 1. Blood islands; 2. Sinus venosus; 3. Bulbus cordis; 4. Foramen ovale
E. 1. Pronephros; 2. Mesonephros; 3. Cloaca; 4. Allantois; 5. Metanephros
F. 1. Primordial germ cells; 2. Paramesonephric ducts; 3. Mesonephric ducts
G. 1. Genital tubercle and urogenital folds; 2. Labioscrotal swellings; 3. Genital tubercle; 4. Urogenital folds

Growth of the Fetus
1. Embryonic; 2. Growing; 3. Lanugo; 4. Vernix caseosa; 5. Subcutaneous fat

Parturition
A. 1. First stage of labor; 2. Second stage of labor; 3. Third stage of labor
B. 1. Increases; 2. Increases; 3. Decrease; 4. Increase; 5. Increase; 6. Increase; 7. Increases; 8. Increases; 9. Increases

The Newborn
A. 1. Foramen ovale; 2. Fossa ovalis; 3. Ductus arteriosus; 4. Ligamentum arteriosum; 5. Umbilical vein; 6. Ligamentum teres; 7. Ductus venosus; 8. Ligamentum venosum; 9. Medial umbilical ligaments
B. 1. Meconium; 2. Stomach pH; 3. Lactose; 4. Amylase; 5. Bilirubin
C. 1. Congenital disorders; 2. Teratogens

Lactation
1. Estrogen; 2. Progesterone; 3. Prolactin; 4. Prolactin-releasing factor (PRF); 5. Colostrum; 6. Oxytocin; 7. Oxytocin

Life Stages
1. Germinal; 2. Embryo; 3. Fetus; 4. Neonate; 5. Infant; 6. Child; 7. Adolescent

Aging
1. Collagen; 2. Decrease; 3. Heart; 4. Filtration; 5. Atherosclerosis; 6. Arteriosclerosis; 7. Thrombus; 8. Embolus; 9. Cytologic aging; 10. Autoimmunity; 11. Genetic

Quick Recall

1. Zygote, morula, blastocyst, embryonic disk, primitive streak, mesoderm
2. Inner cell mass: embryo; trophoblast: placenta
3. Nervous system
4. Circulatory system. The muscular and skeletal systems are almost all mesodermal in origin, except for some muscles and bones of the head, which are derived from ectoderm (neural crest cells).
5. Pronephros, mesonephros, and metanephros
6. Stretch of the uterus stimulates oxytocin secretion, which stimulates uterine contraction, which increases uterine stretch.
7. Estrogen: development of duct system and fat deposition; progesterone: development of secretory alveoli; prolactin: milk production; oxytocin: milk letdown
8. Foramen ovale closes, ductus arteriosus closes, ductus venosus closes, and umbilical veins and arteries close.
9. Germinal, embryo, fetus, neonate, infant, child, adolescent, adult
10. Cross-linking of collagen, loss of functional cells, atherosclerosis and arteriosclerosis, cytologic aging, immune changes, and genetic components

Notes

Notes

Notes

Notes

Notes